Studies in Computational Intelligence

Volume 1226

The series "Studies in Computational Intelligence" (SCI) publishes new developments and advances in the various areas of computational intelligence—quickly and with a high quality. The intent is to cover the theory, applications, and design methods of computational intelligence, as embedded in the fields of engineering, computer science, physics and life sciences, as well as the methodologies behind them. The series contains monographs, lecture notes and edited volumes in computational intelligence spanning the areas of neural networks, connectionist systems, genetic algorithms, evolutionary computation, artificial intelligence, cellular automata, self-organizing systems, soft computing, fuzzy systems, and hybrid intelligent systems. Of particular value to both the contributors and the readership are the short publication timeframe and the world-wide distribution, which enable both wide and rapid dissemination of research output.

Indexed by SCOPUS, DBLP, WTI AG (Switzerland), zbMATH, SCImago.

All books published in the series are submitted for consideration in Web of Science.

Samarjit Roy · Palash Bairagi ·
Siddhartha Bhattacharyya · Debashis De
Editors

Humanized Intelligence in Sustainable Finance and Economic Behaviours

Concepts, Frameworks, and Applications

Springer

Editors
Samarjit Roy
Department of Software Engineering
School of Computing and Information Technology
Eastern International University
Becamex IDC Corp
Thu Dau Mot City, Binh Duong, Vietnam

Siddhartha Bhattacharyya
Faculty of Electrical Engineering and Computer Science
VSB Technical University of Ostrava
Ostrava, Czech Republic

Algebra Bernays University
Zagreb, Croatia

Palash Bairagi
Department of Business Management
Central University of Odisha
Koraput, Odisha, India

Debashis De
Department of Computer Science and Engineering
Maulana Abul Kalam Azad University of Technology, West Bengal
Kolkata, India

ISSN 1860-949X ISSN 1860-9503 (electronic)
Studies in Computational Intelligence
ISBN 978-981-95-0598-2 ISBN 978-981-95-0599-9 (eBook)
https://doi.org/10.1007/978-981-95-0599-9

This Springer imprint is published by the registered company Springer Nature Singapore Pte Ltd.
The registered company address is: 152 Beach Road, #21-01/04 Gateway East, Singapore 189721, Singapore

To the pioneers of Ethical Intelligence, whose wisdom bridges information with sustainable compassion....

Preface

In the recent years, the intersection of artificial intelligence (AI), finance, and sustainability has emerged as a focal point for researchers, practitioners, and policymakers alike. As we navigate the complex landscape of global finance and economic behaviours, it becomes increasingly evident that traditional approaches are no longer adequate to address the multifaceted challenges we face. The urgency to reconcile economic growth with environmental preservation, social equity, and ethical considerations has never been greater.

This edited volume, titled *Humanised Intelligent Analytics in Sustainable Finance and Economic Behaviours: Concepts, Frameworks, and Applications*, represents a collaborative effort to explore the potential of human-centred AI approaches in shaping the future of finance and economics. With a focus on sustainability, ethics, and human values, this book brings together diverse perspectives and cutting-edge research from leading experts in the fields of AI, finance, economics, and sustainability.

The chapters in this volume delve into various aspects of humanised intelligent analytics, offering conceptual insights, methodological frameworks, and practical applications. From incorporating human preferences and biases into algorithmic decision-making to leveraging AI for sustainable investment strategies, the contributions in this book reflect the breadth and depth of inquiry in this rapidly evolving field.

We initialise the groundwork by examining the theoretical foundations of humanised intelligent analytics and its implications for sustainable finance and economic behaviours. It contains a set of significant depiction as the systematic framework on the Humanised Intelligent Analytics in Sustainable Finance and Economic Behaviours, AI and Human-centric approaches towards the Financial Decision-Making and Sustainable Finance Governance.

This edited volume explores the critical frameworks and practical implications of Intelligent financial decision-making, risk management, and market dynamics, while also addressing the ethical challenges and societal implications of AI adoption in finance. It additionally illustrates AI-induced financial propositions, Digital Transformation, BPM Tools, and Techniques in the Industry 4.0.

Eventually, the contributions delve into the real-time case studies and real-world examples that showcase the transformative potential of human-centred AI approaches in driving sustainable development and fostering economic resilience. It significantly highlights the real-time system scenarios of consumer behaviour of online buying, empowering India's MSME Sector with the potential of green finance, sustainable economics for consumer electronics in agriculture, and sustainability disclosure based on GRI-G4 framework using MCDM mechanism, digital wallet, FinTech, and energy economics. We conclude this edited volume with the concluding remarks and the scope for future researches on humanised intelligence in sustainable finance and economic behaviours.

We hope that this volume serves as a valuable resource for researchers, practitioners, policymakers, and students who are interested in the intersection of AI, finance, and sustainability. By advancing our understanding of humanised intelligent analytics and its role in shaping the future of finance and economics, we aim to catalyse meaningful progress towards a more equitable, inclusive, and sustainable global economy. We extend our sincere gratitude to all the contributors who have generously shared their expertise and insights, as well as to the reviewers whose feedbacks have helped the contributing authors to enhance the quality of this volume. Special thanks also go to the editorial team at Springer for their support and guidance throughout the publication process.

Thu Dau Mot City, Vietnam — Samarjit Roy
Koraput, India — Palash Bairagi
Ostrava, Czech Republic / Zagreb Croatia — Siddhartha Bhattacharyya
Kolkata, India — Debashis De

Contents

About the Editors

Dr. Samarjit Roy is currently working as a faculty member in the Department of Software Engineering, School of Computing and Information Technology, Eastern International University, Vietnam. He has been serving as a visiting scientist at Multimedia University and Lincoln University, Malaysia. He served as an Associate Professor in the Department of CSE, JIS Group, Kolkata, India, and an Assistant Professor in the School of Computer Science, Engineering, and Applications, Department of Computer Science and Engineering at the D. Y. Patil International University, Pune, Maharashtra, India. His research areas include musical pattern recognition, information retrieval, context-aware recommender systems, speech, audio and signal processing, cloud computing, soft computing, machine and deep learning, internet of music things, and crowd-sensing. He has published 75 research articles, books, book chapters, etc., in numerous reputed National and International journals, conferences, and other publication societies.

Dr. Palash Bairagi serves as an Assistant Professor of Finance at the Department of Business management, SCMS, Central University of Odisha, Koraput, Odisha. Dr. Bairagi has a strong academic background, having earned his Ph.D. in Management from the Institute of Management Studies at Banaras Hindu University. His doctoral research probed into the nuanced dynamics of risk tolerance and risk perception in investment decisions, analysed through the lens of behavioural finance. With over six years of extensive teaching experience, he has previously contributed his expertise to institutions such as Symbiosis School of Banking and Finance, Symbiosis International University, School of Business, Dr. Vishwanath Karad MIT World Peace University, D. Y. Patil International University and Rama University in India. His scholarly contributions are significant, with numerous publications in peer-reviewed journals (Scopus and SCI) and chapters in academic books with Springers. His research interests span across Behavioural Finance, Computational Finance, Sustainable Finance, Corporate Social Responsibility, and Fintech. He is actively engaged in various professional organisations such as Indian Accounting Association (IAA), Global Association for Humanities and Social Science Research (GAHSSR) and Editorial Member Edwin Group of Journal of continues to influence the academic community with his innovative research and dedication to education.

Dr. Siddhartha Bhattacharyya [FRSA (UK), FIET (UK), FIEI (I), FIETE, LFOSI, FSCRS, SMIEEE, SMACM, SMAAIA, SMIETI] earned his Bachelor in Physics, Bachelor in Optics and Optoelectronics, and Master in Optics and Optoelectronics from the University of Calcutta, India, in 1995, 1998, and 2000, respectively. He completed his Ph.D. in Computer Science and Engineering from Jadavpur University, India, in 2008. He also received the Ph.D. (Habilitation) degree from VSB Technical University of Ostrava, Czech Republic, in 2023. He is the recipient of the University Gold Medal from the University of Calcutta for his Masters. He is the recipient of several coveted awards, including the Distinguished HoD Award and Distinguished Professor Award conferred by the Computer Society of India, Mumbai Chapter, India, in 2017, the Honorary Doctorate Award (D. Litt.) from The University of South America, and the South East Asian Regional Computing Confederation (SEARCC) International Digital Award ICT Educator of the Year in 2017. He has been appointed as the ACM Distinguished Speaker for 2018–2020. He was inducted into the People of ACM Hall of Fame by ACM, the USA, in 2020. He has been appointed as the IEEE Computer Society Distinguished Visitor for 2021–2024. He has been elected as a full foreign member of the Russian Academy of Natural Sciences (RANS) and the Russian Academy of Engineering (REA). He has been elected a full fellow of The Royal Society for Arts, Manufacturers and Commerce (RSA), London, UK. He is currently serving as a Senior Researcher in the faculty of electrical engineering and computer science at VSB Technical University of Ostrava, Czech Republic. He is also serving as the Scientific Advisor of Algebra Bernays University, Zagreb, Croatia. Before this, he served as the Principal of Rajnagar Mahavidyalaya, Rajnagar, Birbhum. He served as a Professor in the Department of Computer Science and Engineering of Christ University, Bangalore. He was the Principal of RCC Institute of Information Technology, Kolkata, India, from 2017 to 2019. He has also served as a Senior Research Scientist in the Faculty of Electrical Engineering and Computer Science of VSB Technical University of Ostrava, Czech Republic, from 2018 to 2019. Before this, he was a Professor of Information Technology at the RCC Institute of Information Technology, Kolkata, India. He served as the Head of the Department from March 2014 to December 2016. Before this, he was an Associate Professor of Information Technology at RCC Institute of Information Technology, Kolkata, India, from 2011 to 2014. Before that, he served as an Assistant Professor in Computer Science and Information Technology at the University Institute of Technology, The University of Burdwan, India, from 2005 to 2011. He was a Lecturer in Information Technology at Kalyani Government Engineering College, India, from 2001 to 2005. He is a co-author of 6 books and the co-editor of 109 books and has more than 400 research publications in international journals and conference proceedings to his credit. He has two PCTs and 20 patents to his credit.

He has been a member of several national and international conferences' organising and technical programme committees. He is the founding Chair of several international conferences like International Conference on Research in Computational Intelligence and Communication Networks (ICRCICN), International Symposium on Signal and Image Processing (ISSIP), Doctoral Symposium on Intelligence Enabled Research (DoSIER), the International Conference on Human-Centric Smart Computing (ICHCSC), International Conference on Computers, Information Processing and Advanced Education (CIPAE), International Conference on Intelligent Systems and Human Machine Collaboration ICISHMC), and Doctoral Symposium on Human-Centred Computing (HUMAN). He was the General Chair of several international conferences like WCNSSP 2016 (Chiang Mai, Thailand), ICACCP (2017, 2019) (Sikkim, India), ICICC 2018 (New Delhi, India), and ICICC 2019 (Ostrava, Czech Republic).

Debashis De is a Professor in the Department of Computer Science and Engineering at the Maulana Abul Kalam Azad University of Technology, West Bengal, India. He is a Senior Member-IEEE, Fellow IETE, and Life member of CSI. He was awarded the prestigious Boyscast Fellowship by the Department of Science and Technology, Government of India, to work at the Heriot-Watt University, Scotland, UK. He received the Endeavour Fellowship Award from 2008–2009 by DEST Australia to work at the University of Western Australia. He received the Young Scientist award in 2005 at New Delhi and in 2011 in Istanbul, Turkey, from the International Union of Radio Science, Belgium. In 2016, he received the JC Bose Research Award from IETE, New Delhi. In 2019, he received the Shiksha-Ratna Award from the Govt of West Bengal. He published in 400+ journals and 200+ conference papers, twenty books, and was granted four patents. He has projects sponsored by AICTE, UGC, WBDST, DST, WBDST, World Bank, AWS, and MeitY. He served as chairman of CSI and IETE Kolkata, and EC member of IEEE COMSOC Kolkata. His h-index is 51, his citation is 12,500. He is listed in the top 2% scientists of the world and ranked 122nd among the best scientists of India for 2025 by Research.com, USA. His research interests are Edge intelligence, IoT, and Quantum Computing.

Humanized Intelligent Analytics in Sustainable Finance and Economic Behaviours: A Systematized Outline

Samarjit Roy, Palash Bairagi, Shuddhasatta Chowdhury, Ambarish Deb, Siddhartha Bhattacharyya, Debashis De, Takaaki Goto, and Debadrita Panda

Abstract This chapter investigates the convergence of human-centric aspects and artificial intelligence within the sustainable finance and economic behaviours, organizing an outline for the humanized intelligent analytics. As the financial paradigms grasp with associating the environmental, social, and governance factors, the requirements for cutting-edge analytical schemas that encompasses both quantitative information and qualitative human discernments has recently become paramount. We address that fundamental needs in this chapter by specifying the architecture of the human-in-the-loop AI frameworks for enhancing decision-making processes,

S. Roy (✉)
Department of Software Engineering, School of Computing and Information Technology, Eastern International University, Becamex IDC Corp, Binh Duong, Vietnam
e-mail: samarjit.roy@ieee.org

Lincoln University College, Selangor, Malaysia

P. Bairagi
Department of Business Management, Central University of Odisha, Koraput, Odisha, India

S. Chowdhury · A. Deb
Symbiosis Centre for Information Technology, Symbiosis International (Deemed University), Pune, India

S. Bhattacharyya
Faculty of Electrical Engineering and Computer Science, VSB Technical University of Ostrava, Ostrava, Czech Republic

Algebra Bernays University, Zagreb, Croatia

D. De
Department of Computer Science and Engineering, Maulana Abul Kalam Azad University of Technology, Kolkata, West Bengal, India

T. Goto
Department of Information Sciences and Arts, Toyo University, Tokyo, Japan

D. Panda
Business Administration and Industrial Engineering, Department of Social Sciences, Technology and Arts, Luleå University of Technology, Luleå, Sweden

S. Roy et al. (eds.), *Humanized Intelligence in Sustainable Finance and Economic Behaviours*, Studies in Computational Intelligence 1226,
https://doi.org/10.1007/978-981-95-0599-9_1

advance risk assessment, and publicize ethical behaviours in the context of sustainable finance. We additionally incorporate several real-world case studies and illustrative examples to demonstrate how humanized intelligent analytics can be harnessed to various domains. We represent the data visualization aids, including the system architecture, flow of activities, and performance dashboards to elucidate multifaceted concepts and empirical findings. Eventually, we provide a comprehensive AI-human convergence outline in sustainable finance that offers a roadmap for future researchers to elaborate effective and ethically sound intelligent systems for advancement of sustainable finance and promoting an equitable and sustainable economic future.

Keywords Humanized intelligent analytics · Sustainable finance · Economic behaviours · Green finance · Human-in-the-loop AI · Risk assessment · Data-driven policy · Ethical decision-making

1 Introduction

The evolving panorama of global economics and finance, the sustainability is not a tangential task, rather this is a systematic progression. As societies and advanced technologies deal with climate variation, social disparity, and unpredictable markets, the necessity for finance frameworks should be both intelligent and intense ethical background. This chapter proposes an emerging conceptual demonstration: Humanized Intelligent Analytics (HIA)—an integration that fetches together intelligence, behavioural science, and economic sustainability to manage financial decision-making and economic innovations.

The convergence of human-centric computational paradigms with intelligent analytics is momentous in traversing the significant convolutions of the sustainable finance and perceptive economic behaviours, characterizing a paradigm shift from conventional to virtuously quantitative approaches [1]. This convergence enables a more nuanced and context-aware analysis, fostering financial strategies that are not only economically sound but also environmentally and socially responsible [2]. The democratization of financial and economic planning, emphasized by the imperative to abridge sophisticated financial decision-making systems, requires the assimilation of reachable and efficient demonstrations adapted to the diverse requirements of society 5.0 [3]. Within the generous realm of economic behaviour and sustainable finance, the usage of humanized intelligent analytics recommends an emerging lens through which to analyse predictable assumptions and designs, thereby enhancing decision-making schemas, considering behavioural refinements and ethical implications [4]. Proposed interdisciplinary and/or multidisciplinary approach recognizes the integral limitations of rational activity models, promoting for a more all-inclusive considerations of how human beings and organizations interact with typical financial systems and sustainable development experiments [5]. The enduring digital revolution, illustrated by the propagation of data-driven analytics and recent advancements in artificial intelligence, enhances the extraordinary prospects for refining the financial modelling and investment stratagems, commanding an in-depth assessment of the ethical considerations and societal impacts of emerging technologies. Moreover, the growth and deployment of AI-induced schemas in finance require precise attention to issues such as the bias, transparency, and accountability for ensuring fairness and associativity in financial markets [6].

Conventional demonstrations of economic behaviours and finance analysis often avoid the refined, human-centric components which can drive towards the lifelong sustainability. Typical models are dependent heavily on the historical insights and mechanised algorithms, effective for prediction, fell short in apprehending the ethical, sensitive, and behavioural aspects of real-world decision-making. In contrast, the projected HIA seeks to incorporate human values, responsible and explainable AI, and socio-environmental stewardships into the core of economic analytics and financial policy designs.

Our proposed systematized outline demonstrates the process of incorporating multi-source information—including the financial, economic, social, environmental, and behavioural inputs which can be fused using the intelligent algorithms and human-centred assumptions. We highlight the transformative principles of human behaviour analysis, sustainable finance modelling, and adaptive AI for determining investment policies which are not only economically reliable but also environmentally impactful.

Furthermore, we contextualize several case studies in this endeavour within real-time applications, such as the green finance, responsible investing, and the advancing economic policy. The advanced analytics with human-centric computational principles and planetary boundaries, proposed approaches offer a corridor to fostering sustainable financial and economic behaviours—a keystone of resilience in the all-encompassing progression of the twenty-first century.

In order to represent the abovementioned ones, we explore the system architecture, extensive methodologies, and real-time implications of the framework in the forms of case studies, to provide scholars, policymakers, and financial practitioners with a holistic lens to perceive sustainability-driven innovations in financial artefacts.

1.1 Motivation and Contributions

Our motivation behind the explorations of HIA reclines in the rising recognition which economic and financial schemes are highly knitted with human behaviour, social values, and environmental restraints. Proposed recognition difficulties include the conventional emphasis on extending shareholder values at the expense of wide societal well-being, promoting a more composed and sustainable approach to economic implementation. The limitations of conservative financial models, which habitually flop to account for outwardness, behavioural biases, and lifelong sustainability, have enhanced progressively apparent for the global challenges such as climate and weather change, social inequality, and financial instability. We claim that by indulging humanized insights and ethical considerations into statistical and analytical frameworks, it can be promising for developing more robust and resilient emerging financial frameworks that boost in sustainable development goals and stimulate comprehensive economic growth. Our proposed HIA is not a surrogate paradigm of conventional financial analytics and economic behaviours, rather the incorporation of Humanized intelligences can often enhance the analytical schema towards the sustainable financial models and economic behaviours.

The key contributions of our study are as follow: (a) a structured schema, demonstrated in Fig. 1, for incorporating human-centred design methodologies with advanced analytical strategies in the context of sustainable finance and economic behaviours, that offers a comprehensive backbone to develop AI paradigms that are impartial, transparent, and allied with societal disciplines, (b) this endeavour aims to tie together the disparity between technological capabilities and ethical considerations, providing application-induced guidance for financial organizations, policymakers, and researchers, who seek for leveraging AI in societal developments, (c) it presents a comprehensive exploration of intelligent systems that can augment personalized financial services and can lead to improve customer and/or end-users satisfaction and community engagement [7]. (d) this discussion is not only enhances the theoretical concepts of humanized intelligent analytics but also offers practical tools and methodologies for implementing these concepts in real-world scenarios, demonstrating their potential to transform the financial industry and contribute to a more sustainable and equitable future.

1.2 Chapter Organization

We prepare this chapter as follow: Sect. 2 presents the related works along with comparative state-of-the-art existing literatures. We also classify the interdisciplinary literatures into seven contextual subsections. Section 3 presents the proposed six-layered system architecture. We present the statistical background of the proposed approach in Sect. 4. The performance metrics shown in Sect. 5. The Sect. 6 demonstrates a case study on HIA-driven Cryptocurrency Transaction and Blockchain Transparency. Another systematised case study on Humanized Intelligence in Credit Card Fraud Detection is presented in Sect. 7. Eventually, this chapter concludes with the discussion, conclusion, and several scopes for future research in Sect. 8.

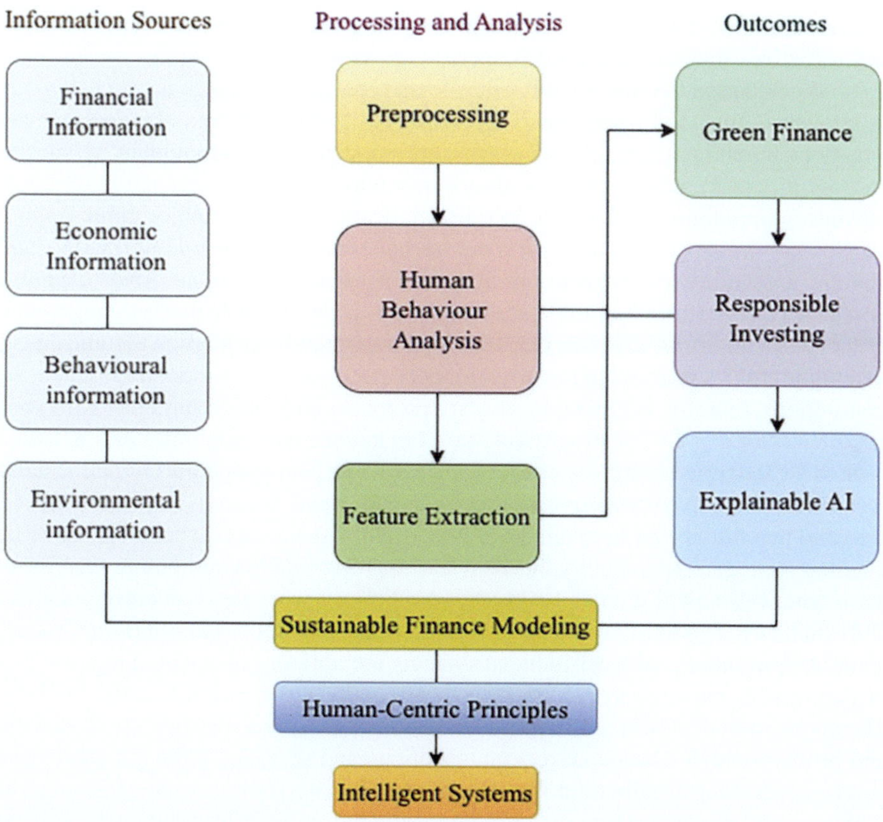

Fig. 1 Structured schema of HIA-driven sustainable finance and economic behaviours

2 Related Works

2.1 Integrating Intelligence and Human Insights in Financial Analytics

The confluence of AI and Machine Learning (ML) within the financial sector have catalysed a transformative shift, fundamentally altering conventional paradigms of risk management, fraud detection, and the provision of personalized financial services [8]. The convergence of AI with information analytics has enabled the automation of routine schedules, explored the precision boundaries of predictive data analytics, and highlighted a more sophisticated comprehension of intricating financial datasets and databases [9]. The recent scholarly contributions leverage the mounting incorporation of AI algorithms in diverse financial demonstrations and applications, that range from the algorithmic trading to customer credit scoring and service automation [10]. The enlarging benefits of AI in financial sectors have spurred significant academic and research interest, as the reflection in the rising number of research publications and initiatives that focused on this specific domain [11]. The studies use to highlight AI's potential to improve efficiency, lessen costs, and enhance decision-making in numerous financial operations [12]. However, the researchers also identify the inherent risks and potential challenges that are associated with adoption of AI, such as data security and privacy concerns, algorithmic biasness, and the necessity for robust regulatory paradigms [13]. The real-time applications of AI in algorithmic trading exemplifies the capacity to anatomize market patterns and historical information, and thereby, expediting trade executions at speeds is surpassing human-driven capabilities [14]. Yet, they introduce the space complexities, pertaining to market stability and the latent for algorithmic biases [15, 16]. Furthermore, AI's utilization in fraud detection highlights its prowess in the discernment of anomalous activities and fraudulent transactions, thereby, augmenting security protocols and mitigating financial losses are often sensed [17]. The progression of AI in finance and economics can be traced back to the latter part of the twentieth century, with initial applications fundamentally focused on rule-driven systems for automated trading and risk analysis [18]. However, the valid potential of AI in finance was unleashed with the advancement of machine learning strategies, specifically the deep learning that enabled computing systems to learn from billions of data and construct predictions with unprecedented accuracy and efficiency. The utility of AI in financial risk assessment has also admired significant advancements, with AI-induced systems are able of processing large number of datasets in real time to recognize potential risks and fraudulent activities [19]. The application of AI extends to develop customer experience through personalized and professionalized financial recommendations and services, although challenges persist concerning information for privacy and security [20]. In spite of the significant benefits, the integration of AI in finance is not without its encounters. Ethical considerations encompassing AI-driven decision-making, including fairness issues, data transparency, and service accountability, have become progressively prominent. There are typical concerns about the potential for algorithmic bias to disseminate existing inequalities in financial services, predominantly in the areas, such as, the lending and credit scoring [21]. The prerequisite for explainable AI in finance is increasingly emphasized to ensure transparency and facilitate human comprehension of AI-driven assessments [22]. Explainable AI aims to deliver insights into how AI models arrive at precise conclusions, thereby fostering trust and accountability. The requirement to address biases in AI algorithms and safeguard equitable outcomes is utmost [23]. The elaboration of fair and ethical AI resolutions in financial services is an industrial objective, and several ethical principles, parameters, and methodologies have been projected in recent years [24]. Regulatory paradigms must adapt to address the exclusive challenges posed by AI in finance, encompassing data privacy, algorithmic transparency, and consumer safeguard. International organizations, viz., the OECD and G20 have proposed principles for financial AI, highlighting the demand for responsible and effective implementation [12]. There are growing recognition of the importance of international cooperation in setting up standards and optimum practices for AI in finance to inhibit regulatory arbitrage and ensure reliable consumer protection across jurisdictions.

2.2 *Navigating Ethical Dimensions in AI-Driven Financial Frameworks*

The incorporation of AI in financial services depicts significant ethical considerations which should be vigilantly addressed to ensure fairness, transparency, and accountability [19]. AI-driven frameworks have the capability to involuntarily discriminate against specific groups, raising apprehensions about fairness in lending and risk assessment [19]. Algorithmic bias, occurring from prejudiced training data or flawed algorithms, can propagate and amplify prevailing societal inequalities [25]. To ensure the transparency in AI systems is pivotal to foster trust and enable stakeholders to comprehend how decisions are prepared. Explainable AI strategies can enrich transparency by affording insights into the decision-making progressions of AI models. Explainability and interpretability are perceived as mechanisms to progress algorithmic fairness, transparency, and accountability [26]. Setting up of transparent lines of accountability for AI systems is decisive, especially when errors or biases are led to adverse outcomes. Organizations should ensure that those are responsible in developing and deploying AI algorithms are accountable for the decisions and concurrent actions. This includes the addressing issues, such as, algorithmic accountability, responsibility for errors or biases, and liability in case of adverse outcomes [27]. To address the critical ethical issues in AI necessitates interdisciplinary collaboration, comprising experts from technology, ethics, law, and finance. Such alliance can promote responsible AI development and deployment by combining diverse perspectives and expertise. Ethical considerations should be cohesive throughout the AI development lifecycle, from data sensing and model design to deploy and visualize. Moreover, robust governance schemas and ethical guidelines are indispensable to harness the whole potential of AI while safeguarding against inadvertent consequences [22]. These frameworks should comprehend principles such as fairness, transparency, accountability, and respect for human rights. Enduring the monitoring and evaluation of AI-integrated systems are essential to recognize and mitigate significant biases and unintended consequences. Moreover, the end-users of AI tools necessitate to be vigilant to maintain the accuracy and integrity of the data generated and disseminated [28]. Incessant monitoring assists to ensure that AI system functionalities as intended and do not generate discriminatory or unfair results.

2.3 *The Intelligent Analytics Using HIA*

HIA demonstrates a sophisticated paradigm shift in how we approach for information analysis, moving beyond conventional statistical mechanisms to incorporate human-alike reasoning and understanding. Our methodology amalgamated the potentials of both humans and computing machines, leveraging AI's computing efficiency and information retrieval competences while preserving human judgment, clairvoyance, and contextual awareness. By supplementing human capabilities with intelligent systems, HIA enriches decision-making progressions and escorts to more perceptive and actionable outcomes. The key principle behind HIA is to create AI systems that can understand and respond to human benefits and preferences. This involves in developing AI models that can interpret complex information, recognize nuanced patterns, and provide explanations that are easily understood by humans during information retrieval. This is effective in domains, such as, sustainable finance and economic behaviour, where decisions often require integrating diverse information sources and considering multiple stakeholders. The prominence is on explainable AI, that aims to design AI decision-making processes transparent and interpretable, enabling humans to understand how AI models arrive at specific conclusions [29].

2.4 Human-Centred Design Principles in AI Analytics

Human-centric design principles are gradually recognized as necessity for generating effective and ethical AI analytics frames [30]. The doctrines prioritize the requirements, preferences, and values of human end-users, which certifies AI systems are manifested to be comprehensible, robust, and aligned with sustainable human-centric goals. Harnessing human-centric design to AI analytics encompasses understanding how human beings interact with AI peripherals, cherishing potential biases and ethical apprehensions, and emergent strategies to diminish these risks. One significant aspect of human-centric prototype is to focus on user experience. AI analytics schemas should be framed to be accessible to end-users with erratic levels of technical expertise, stipulating vibrant and concise explanations of outcomes and recommendations. An additional significant principle is to make certain that AI systems are transparent and explainable. End-users of AI tools entail to be attentive in upholding the efficiency, accuracy, sensitiveness, and reliability of the information generated and disseminated. By providing comprehensions into the decision-making procedures of AI models, users can understand better how AI-driven stratagem arrives at specific conclusions and construct apprised judgments on the reliability of the outcomes. Moreover, an approach on human-centred design emphasizes the consequence of involving diverse stakeholders in the expansion and evaluation of AI systems. This contains incorporating feedback from domain experts, end-users, and those who may be exaggerated by AI-driven decisions. Encompassing ethical considerations into the design process is also indispensable. Ensuring responsible AI involves not only the trustworthiness of the outcomes but also the design process, the reasons behind the design, and the individuals intricate in the design [31]. By addressing these issues proactively, the developers can safeguard that AI systems are aligned with human values and promote fairness, equity, and social moral [32].

2.5 Technical Solutions for Human-Centric AI

Numerous technical solutions have emerged to adopt the challenges of producing human-centric AI systems, outstandingly in the areas of privacy, transparency, and fairness [33]. These solutions range from algorithmic improvements to newly designed software tools and development processes, entirely aimed at constructing AI more accountable and aligned with human values. Explainable AI is one of the prominent areas of innovation, focused on developing AI models which can elucidate decisions and predictions. This is important in the domains of finance and economics as well, where understanding the reasoning behind AI-driven recommendations is imperative for building trust and certifying responsible decision-making [32]. Fairness-aware AI is an additional critical area, that aims to mitigate biases in AI systems and ensure that they do not discriminate against precise groups or individuals. This involves in the development algorithms and techniques for identifying and removing biases in data, models, and decision-making procedures. Privacy-preserving AI precisely focuses to protect sensitive information while still allowing AI systems to learn from training data. Moreover, the human-in-the-loop AI systems are newly designed to incorporate human inputs and oversight into the decision-making process. While a set of approaches are being projected to address the ethical challenges arising from AI in recognizing and implementing solutions remains complicated. A significant component is to foster an understanding among developers that their work is convincingly connected to ethical considerations and that play a pivotal role in shaping the technology's impact [34]. It consists of the tools that evaluate fairness, implement machine learning algorithms explainable, and make these toolkits available for third party to use [35].

2.6 Unpacking Recent Research on Humanized Intelligent Analytics

Scrutinizing the recent research landscape surrounding HIA provides a crucial foundation for understanding the current state of the art, categorizing key trends, and uncovering domains ripe for further exploration. A comprehensive literature review recommends a growing interest in AI's ethical implications and the prominence of incorporating human values into AI systems [36]. The recent work has also focused in developing methods and algorithms for explainable AI, fairness-aware AI, and privacy-preserving AI [37]. Agent-based AI evolution necessitates a balanced methodology between technological advancement and ethical governance, underscoring the importance of setting up robust frameworks for responsible deployment [38]. The surge in computational potential and data availability has been instrumental in forward-moving AI technologies, yet it is the careful integration of human insights and ethical considerations that sincerely unlocks its potential. Researchers are additionally exploring novel directions to design AI models more transparent and interpretable, permitting users to understand the reasoning behind their predictions and decisions. AI's increasing responsibility in decision-making imposes an understanding of its impacts on societal values, such as equity and autonomy [39]. Fairness-aware AI seeks in mitigating biases in sensed data and corresponding algorithms, to ensure that AI systems do not discriminate against certain individuals. Privacy-preserving AI directs in protecting sensitive data.

2.7 AI Governance Frameworks

The development of AI governance frameworks is also a key research domain. This framework provides guidelines and rules for the development and deployment of AI systems, ensuring the alignment with ethical principles and societal values [40]. Additional recent research articles have also demonstrated significant contributions to the contexts of HIA. For instance [41], performs a meta-analysis schema of 200 governance policies and provides ethical guidelines for AI usage, identifying at least 17 resonating principles prevalent in the policies and guidelines.

3 Proposed System Outline

We propose a detailed system architecture of the proposed Humanized Intelligent Analytics in Sustainable Finance and Economic Behaviours, that contains five independent layers, such as the (a) Data Acquisition Layer, (b) Data Pre-processing and Feature Engineering Layer, (c) AI Modelling Layer, (d) Human-in-the-Loop Integration Layer, (e) Visualization and Reporting Layer, and (f) Ethical and Governance Layer. A bottom-up demonstration of the proposed architecture is presented in Fig. 2. We additionally demonstrate the activity and sequence diagrams in Figs. 3 and 4, respectively, of the proposed system for elaborating the entire system behaviours and interactions among the significant system components. Associated six distinct layers of the proposed system are described as follows.

1. Data Acquisition Layer: The Layer 1 constructs the foundation of HIA system, specially within the domains of sustainable finance and economic behaviours. Layer 1 is responsible for collecting, integrating, and preparing the diverse information streams necessary for analysis and decision-making. Provided the multifaceted nature of sustainability and economic behaviour, the data acquisition layer is able of handling a wide array of data types, sources, and formats. Key Components of Layer 1 are the (a) information sources, such as financial markets data, Environment-Social-Governance (ESG) data, news and sentiment data, Economic Indicators, alternative data, internal information; (b) information collection methods, viz., APIs, web scraping, data aggregators, and information

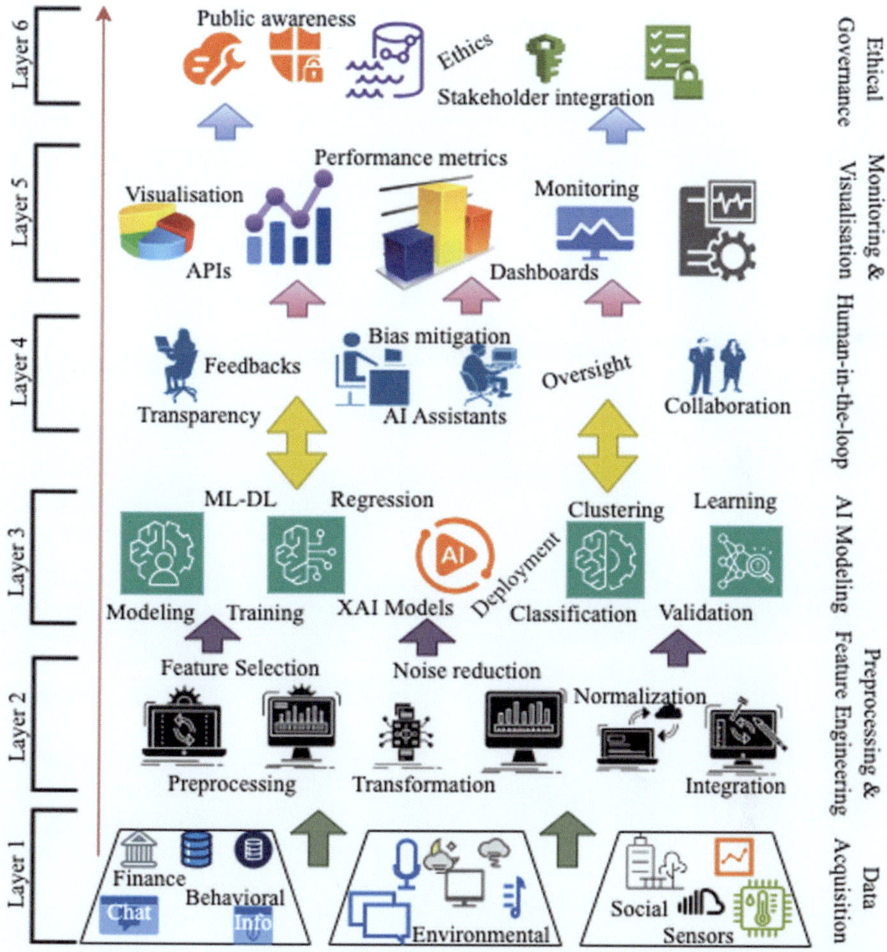

Fig. 2 Representation of proposed system architecture, consists of six distinct and interactive layers

sensing; (c) information fusion, such as the data integration, data standardization, cleansing, and transformation; (d) information pre-processing consists of feature engineering, normalization, and dimensionality reduction; (e) real-time considerations, viz., low latency and data streaming; (f) big data analytics, that consists of data quality, volume, variety, velocity, veracity. Instantaneously, the Data Acquisition Layer operates as the core consensus for accumulating and practicing the raw materials required for HIA in sustainable finance and economic behaviours. A well-equipped information acquisition layer plays a crucial role for endorsing that the Layer 2 and Layer 3 have entree to premium, relevant, and sensible information.

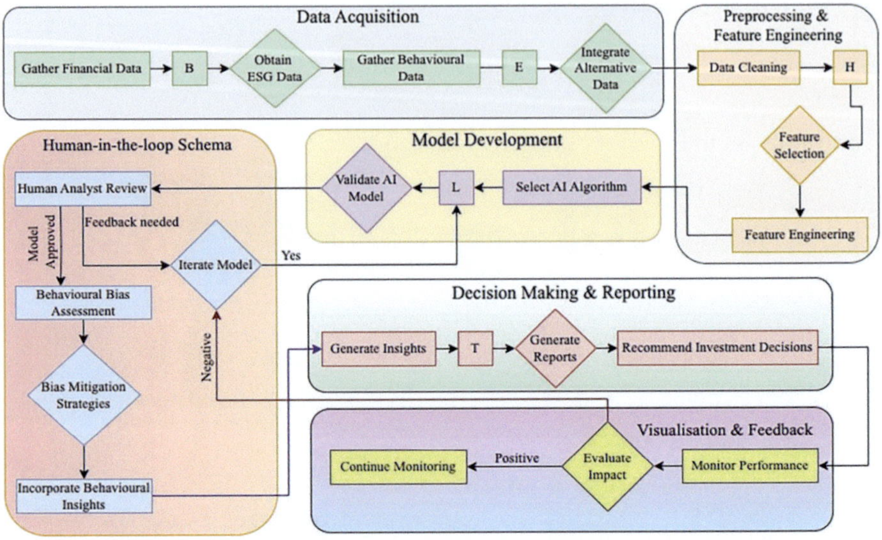

Fig. 3 Depiction of activity diagram to illustrate flow of activities involved in applying humanized intelligent analytics to sustainable finance and economic behaviours, incorporating insights from behavioural finance

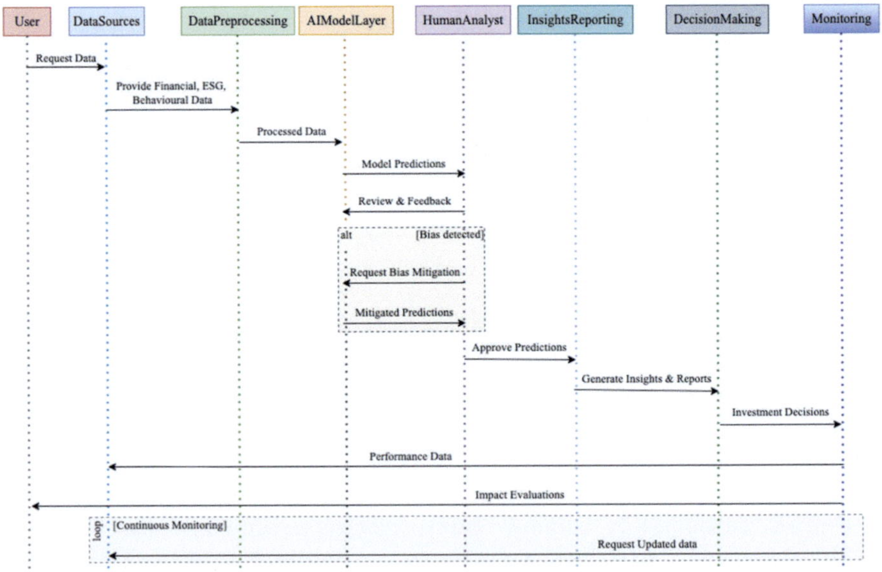

Fig. 4 Representation of the sequence diagram to display the sequence of interactions among diverge components of the proposed system

2. Information Pre-Processing and Feature Engineering Layer: Subsequent tier after the Layer 1, the Information Pre-processing and Feature Engineering Layer suggest for transforming raw information into a fitting format for analysis and data modelling. This layer involves in cleaning, data transforming, and enriching information to extract meaningful insights and improve the performance of AI models in Layer 3. Key components and considerations of Layer 2 consist of (a) Handling missing values, outlier detection and removal, noise reduction, information type conversion; (b) transformation, such as normalization, scaling, aggregation, discretization; (c) Feature Engineering for creating new features, like financial ratios, technical indicators, interaction terms, sentiment scores, and for selecting new features with univariate selection methods and recursive feature eliminations; (d) Time Series Specific Pre-processing, such as decomposition, stationary transformation; (e) Considerations for Sustainable Finance information, viz., ESG data integration, carbon footprint calculation, green bond analysis. The essential tools and techniques are considered for analysing this layer are the programming perspectives and data visualization. Layer 2 sets the stage for effective analysis and system modelling in the subsequent layers of the HIA schema.
3. AI Modelling Layer: The AI Modelling Layer in Layer 3 is where the pre-processed information are used to train, validate, and deploy AI models. It addresses specific problems in sustainable finance and economic behaviours. Layer 3 transforms information into actionable insights and predictions, providing a basis for informed decision-making mechanisms. The models within this layer are aligned with ethical principles and societal values. Essential components of this layers are: (a) Model Selection–regression, classification, clustering, time series, and deep learning models; (b) Model Training and Validation, such as data splitting, hyperparameter tuning, cross validation and regularization; (c) Explainable AI (XAI) consists of feature importance analysis, model visualization, XAI modelling; (d) Ethical Considerations, like Bias Detection-Mitigation, Transparency-Accountability, privacy protection, (e) Deployment, monitoring, retraining. The applications in Sustainable Finance can be the ESG Risk Assessment, Green Investment Opportunity modelling, and sustainable Portfolio Management. The AI Modelling Layer serves as the engine that drives HIA in sustainable finance and economic behaviours by addressing the components [42]. This layer transforms information into actionable insights and predictions, providing a foundation for informed decision-making, while the governance frameworks make sure responsible and ethical AI usage.
4. Human-in-the-Loop Integration Layer: The Human-in-the-Loop Integration Layer (Layer 4) is where the human expertise and oversight are seamlessly integrated into the AI system. This layer ensures that AI systems of Layer 3 are aligned with ethical principles, societal values, and human requirements. It highlights on combining the strengths of both humans and AI to achieve optimum outcomes compared to either working alone. The significant ingredients of this layer are: (a) human oversight, AI Assistance, and collaboration; (b) integration points, such as data input, model training, decision-making, monitoring and

evaluation; (c) Explainable AI, like transparency, interpretability, trust of transparency, and visualizations; (d) feedback mechanisms, such as human feedback loops and real-time adjustments, (e) Ethical Considerations, i.e. bias mitigation, fairness, and accountability, (f) interface design for user-friendly interfaces and contextual information. The tools and technologies are depended on platforms for incorporating human feedback into AI systems and essentially the XAI tools for explaining and visualizing AI decisions. With the design and implementation of Human-in-the-Loop Integration Layer, finance and economic organizations can ensure that AI systems are used in a responsible, ethical, and effective principles. The decision support systems in Layer 4 develop alerting systems and recommendation engines to provide personalized investment advice.

5. Visualization and Reporting Layer: The Visualization and Reporting Layer (Layer 5) transforms the perceptions, produced by the AI Modelling Layer in Layer 3 into easily understandable formats for numerous stakeholders. This layer is imperative for communicating complex information and model outputs in a way that enables informed decision-making and endorses transparency. Using effectual visualizations and reporting strategies, stakeholders can expand a vibrant understanding of the major findings, trends, and predictions derived from the AI models. Significant ingredients of Layer 5 are: (a) data visualization techniques, such as charts, graphs; (b) geospatial visualization, like maps, heatmaps, interactive dashboards, real-time monitoring; (c) reporting—automated report generation, customizable reports, and narrative reporting, (d) user interface design-intuitive interfaces, accessibility, and mobile Compatibility; (e) seamless and API integration. The specific applications in Sustainable Finance of this layer are such as ESG dashboards, climate risk reporting, and Sustainable Development Goal tracking. The Layer 5 makes certain that the insights generated by AI models are effectually communicated to stakeholders, stimulating informed decision-making and transparency. The incorporation of AI into information visualization can also largely improve the procedure and the quality of information visualization, making it capable to manage big, complicated, multi-dimensional datasets more efficiently.
6. Ethical and Governance Layer: The Ethical and Governance Layer, Layer 6, provides the overarching schema for ensuring that AI systems are aligned with ethical principles, legal requirements, and societal needs. This layer converges on setting up guidelines, policies, and methodologies for responsible AI development and deployment. It addresses concerns, such as the bias, fairness, transparency, accountability, and data privacy. The objective is to foster trust in AI models and warrant they are used in a pathway that benefits society entirely. The significant considerations in this layers consist of (a) Ethical Principles—fairness, transparency, accountability, privacy, and beneficence, (b) Governance frameworks—AI ethical guidelines, regulatory compliance, risk management, auditing and monitoring; (c) Stakeholder engagement, such as collaboration and public inputs; (d) Ethical AI design—value alignment, bias detection, mitigation, and XAI; (e) education, ethical training, and public awareness. Layer 6 offers an elastic and scalable paradigm to ensure the AI systems that are developed

and deployed in a responsible, ethical, and beneficial manner. This incorporates embedding robust ethical methods into their architecture and decision-making processes, a method that is in line with ethical governance. AI Governance Frameworks adheres to with ethical principles and societal importance. This comprises having a multifaceted methodology that considers technical, ethical, and social factors.

4 Statistical Background of HIA Schema

HIA in sustainable finance integrates three core domains, such as the statistical inference for behavioural and economic modelling, mathematical optimization for finance and policy design, and machine learning and explainable AI for predictive, human-aware systems.

Behavioural-Economic Regression Model used to model the relationship between behavioural traits (e.g. risk aversion, time preference) and sustainable financial choices.

$$\mathrm{SFS}_i = \alpha_0 + \alpha_1 \mathrm{RS}_i + \alpha_2 \mathrm{TP}_i + \alpha_3 \mathrm{IN}_i + \delta_i \tag{1}$$

where SFS_i denotes the sustainable financial investment score of individual i, RS_i is the risk aversion score, TP_i signifies the essential time preferences, IN_i represents as the income level, and δ_i is the error term.

Multi-objective optimization for the ESG portfolio management depends of the platform where the investment decisions must balance return (ρ), risk (σ), and sustainability, ESG score γ.

$$\text{The objective function for the ESG portfolio } = \max_m(\lambda_1 \rho(m) - \lambda_2 \sigma(m) + \lambda_3 \gamma(m)) \tag{2}$$

where $\sum m_i = 1, \quad m_i \geq 0 \forall i$; m_i is the portfolio weights; and $\lambda_1, \lambda_2, \lambda_3$ are decision-making preferences.

In the contexts of Human Behaviours and AI-driven systems, the Bayesian updating assists AI system incorporates the behavioural uncertainty.

$$P(\emptyset|\psi) = \frac{P(\psi|\emptyset).P(\emptyset)}{P(\psi)} \tag{3}$$

where $\emptyset$ represents the behavioural trait, such as, the trust levels or cooperativeness and ψ is the observed actions, such as, the sustainable purchases. This modelling allows the adaptive learning how the individuals adjust behaviours in response to economic sustainability policies.

For Explainable AI (XAI) in the contexts of economic decisions, SHapley Additive exPlanations (SHAP) method can be effective for the financial organizations to explain why a model recommends some certain sustainable investment and economic decisions.

$$\text{The SHAP contribution model prediction}, f(x) = \xi_0 + \sum_{i=1}^{n} \xi_i \tag{4}$$

where ξ_i is the contribution of feature i.

5 System Performance Metrics

Figure 5 demonstrates a 3D surface plot for the sustainable Financial Investment versus Risk Aversion and Time Preference to visualize the influence of two key behavioural factors, risk aversion and time preference on an individual's propensity to recommend sustainable financial investments (SFI). The X-axis denotes as the Risk Aversion, Y-axis is the Time Preference, and Z-axis represents the Sustainable Financial Investment Score. The surface plot in Fig. 5 demonstrates that (a) Higher risk aversion increases SFI, (b) Lower time preference also increases SFI, and (c) the plot surface slopes upwards from low-risk-averse and impatient individuals to high-risk-averse and patient ones, indicating greater likelihood of sustainable investing.

Figure 6 demonstrates the efficient frontier with ESG Overlay to illustrate how the sustainability factors, the ESG scores, affect the conventional risk-return trade-off in portfolio optimization. In sustainable finance, the investors aim to optimize Return, Risk, and Sustainability. In X-axis, Portfolio Risk (Standard Deviation) is denoted. In Y-axis, the expected Return is presented. The Colour Coding defines the ESG Score, where the higher score is denoted as the greener colour. In Fig. 6, the investors can visually identify portfolios that balance all three objectives. Further Points along the ESG colour spectrum, green colour, offer insight into how much sustainability can be gained for small trade-offs in risk or return.

Figure 7 demonstrates the Bayesian Updating in the context of Eco-Conscious Behaviour Modelling to model uncertain human behaviour using Bayesian inference, specifically in the context of how individuals respond to sustainability incentives or campaigns. In the graph visualization, Line Plot with Three Curves are visible. These are (a) Prior, the initial belief about behaviour, (b) Likelihood, based on new data, and (c) Posterior, the updated belief. As new behavioural data is gathered (e.g. participation in green programmes), the model updates belief in the individual's eco-consciousness. This enables adaptive policy design, the incentives can be personalized as beliefs shift over time.

Figure 8 presents the SHAP-like Feature Impact Plot, that depicts as an Explainable AI in ESG Investment Decision-making modelling to show which factors most strongly influence an AI model's prediction of whether an individual will invest

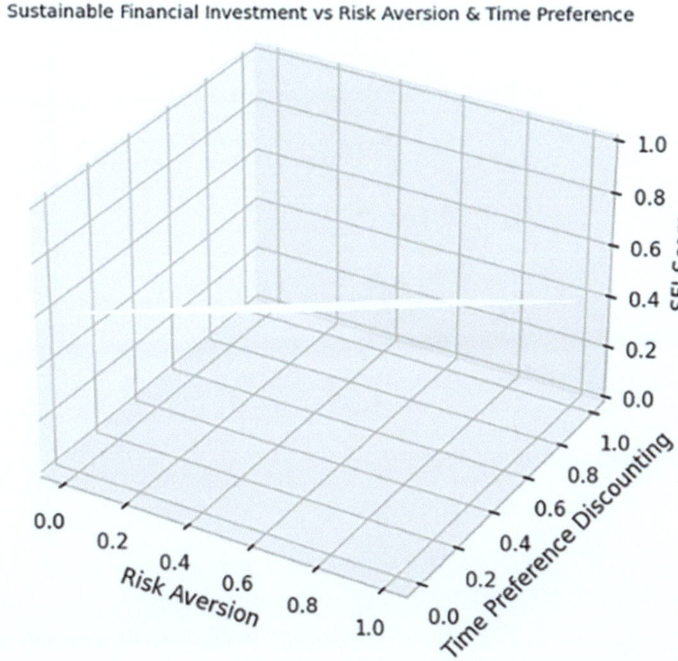

Fig. 5 Demonstration of the surface plot of for the sustainable financial investment versus risk aversion and time preference to visualize the influence of two key behavioural factors, risk aversion and time preference, on an individual's propensity to recommend sustainable financial investments

sustainably. SHAP values attribute the output of an AI model to its input features. Y-axis and X-axis demonstrate the Features, like ESG Awareness, Income, Risk Aversion and Average SHAP values, respectively. The colour indicates direction or relative impact. We observe from the plot interpretation: (a) features with higher SHAP values have greater influence on model predictions, (b) ESG awareness might be the most significant driver of ESG investment behaviour, (c) Time Preference might negatively impact likelihood where impatience reduces ESG investing, (d) Fig. 8 enables transparency, builds trust, and guides policy or educational intervention.

6 Case Study: HIA in Cryptocurrency Transaction and Blockchain Transparency

This case study tackles the complex challenge of reconciling user privacy with the transparency advantages offered by blockchain technology. Addressing issues such as scalability, legal compliance, governance, and user digitalization disparities, the study formulates research objectives centred on five distinct scenarios. Each scenario delineates problems, accompanied by corresponding research hypotheses,

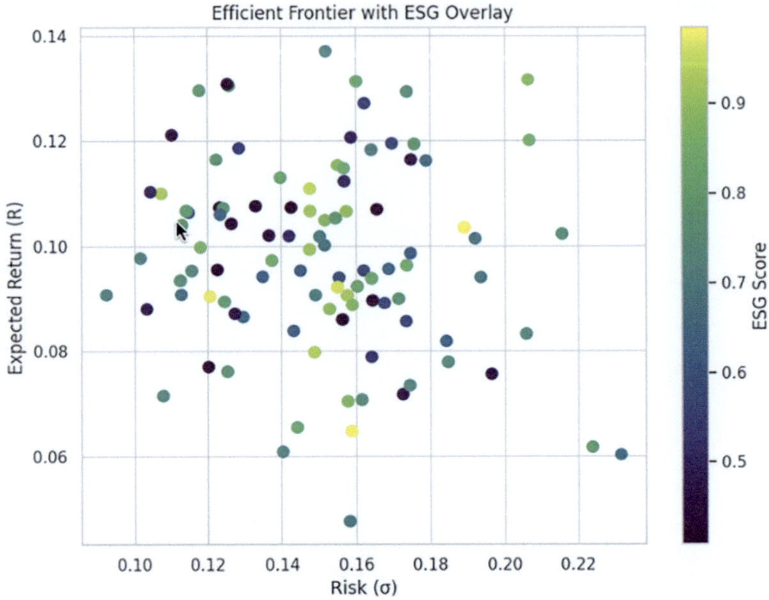

Fig. 6 ESG overlay demonstration to discuss sustainability factors and ESG scores affect traditional risk-return schema in portfolio optimization

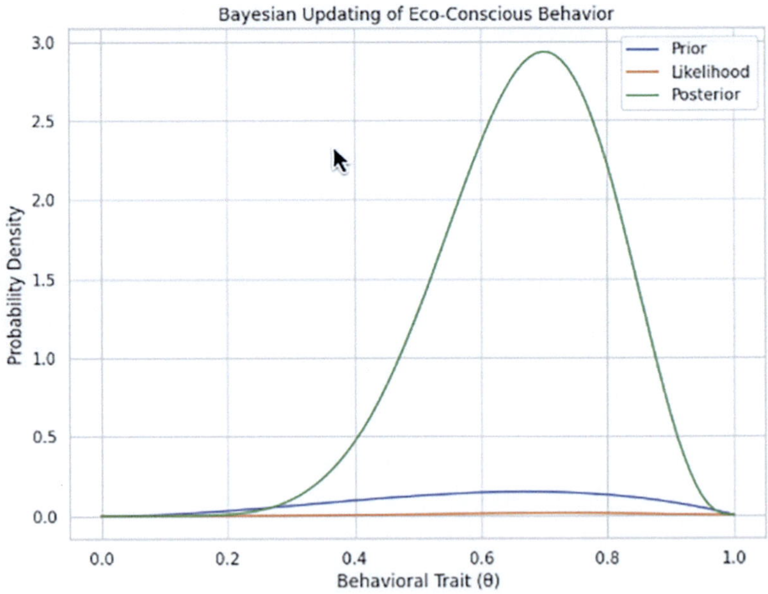

Fig. 7 Demonstration of the Bayesian updation in the context of eco-conscious behaviour modelling to model uncertain human behaviour using Bayesian inference

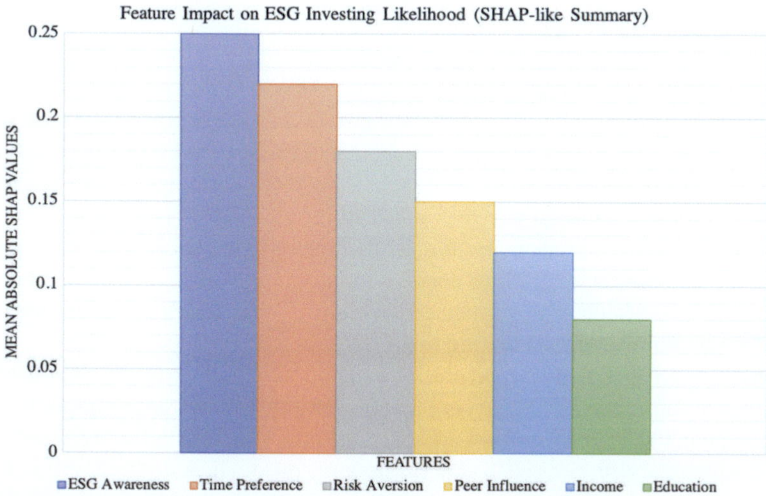

Fig. 8 Presentation of SHAP-like Feature Impact Plot, for depiction of XAI in ESG Investment Decision-making modelling to display which factors most strongly influence AI model's prediction of whether an individual will invest sustainably

and proposes a strategic pathway through a flow diagram, emphasizing the critical need to address identified challenges. Utilizing a data-driven method with a Naïve Bayes classifier, the proposed case study depicts the impact of blockchain transparency on cryptocurrency transaction amounts. Five solutions emerge from the classifier's analysis, focusing on privacy, scalability, legal compliance, governance, and user adoption challenges. The research concludes that these strategic solutions, validated through the classifier, play a crucial role in the success of blockchain platforms by effectively addressing multifaceted challenges. The rejection of null hypotheses provides insights into the delicate balance required between transparency and privacy, the necessity for scalable infrastructure, the importance of legal compliance and governance frameworks, and the influence of user demographics on blockchain adoption.

6.1 Case Study Scenarios

Scenario 1: A blockchain platform has emerged as a trusted intermediary where users willingly entrust sensitive financial and identity information. The platform aims to capitalize on the transparency benefits of blockchain while maintaining user privacy and ensuring that only authorized parties have access to the confidential data.

Scenario 2: A blockchain platform finds itself grappling with fluctuating transaction volumes. The challenge is to ensure the scalability of the blockchain solution to handle varying transaction loads while maintaining optimal performance levels.
Scenario 3: A crypto exchange platform operates in a landscape where legal frameworks surrounding cryptocurrencies and blockchain technology are continually evolving. The challenge is to ensure that the platform remains compliant with these changing legal requirements and institutional processes.
Scenario 4: In a dynamic ecosystem, a blockchain-based project involves a multitude of stakeholders, including users, regulators, and developers. The challenge is to establish effective governance mechanisms that foster collaboration, trust, and alignment of interests among these diverse stakeholders.
Scenario 5: A blockchain-based platform encounters a scenario where its user base consists of individuals with varying levels of digital literacy and engagement. The challenge is to ensure a uniform and user-friendly experience for both tech-savvy users and those with limited digital proficiency during the adoption of blockchain technology.

6.2 *Case Study Objectives*

Problem 1: The challenge lies in striking a delicate balance between leveraging the transparency advantages of blockchain technology and safeguarding sensitive information. Users demand a high level of privacy and security, fearing unauthorized access or malicious use of their financial and identity details.
Problem 2: The platform faces the critical issue of adapting to changing transaction volumes. As user engagement grows, the existing blockchain infrastructure struggles to scale seamlessly, leading to concerns about performance degradation. This poses a significant challenge as maintaining a high-performance level is crucial for user satisfaction and the overall success of the platform.
Problem 3: With the regulatory environment for cryptocurrencies constantly shifting, the crypto exchange platform faces the critical task of aligning its blockchain implementation with evolving legal frameworks. Failure to do so can result in regulatory scrutiny, legal complications, and reputational damage, making legal compliance a central concern for the platform's sustainability.
Problem 4: The project faces the complex task of managing and governing collaboration among stakeholders with varying interests and priorities. The absence of a robust governance framework can lead to conflicts, lack of transparency, and a breakdown in trust. It becomes crucial to address these challenges to ensure the sustained success and growth of the blockchain project.
Problem 5: The platform faces the complexity of catering to a diverse user demographic, ranging from tech-savvy individuals who readily embrace digital solutions to those with limited digital literacy. Ensuring a seamless and user-friendly experience for all users becomes imperative to drive widespread adoption of blockchain technology.

6.3 Research Hypotheses

i. H_0: Blockchain transparency does not compromise user privacy and financial identity security.
ii. H_0: Increasing transaction volumes do not adversely impact the performance of the blockchain infrastructure.
iii. H_0: Blockchain implementation alignment with legal frameworks does not significantly affect the regulatory compliance of the crypto exchange platform.
iv. H_0: The absence of a robust governance framework does not contribute to conflicts, transparency issues, and breakdowns in trust among stakeholders.
v. H_0: Diverse user demographics do not significantly impact the seamless and user-friendly adoption of blockchain technology on the platform.

6.4 Methodology

Objective: It is a data-driven approach where we are trying to find out the parameters that impact the most on the Transaction Amount of the Cryptocurrency Levels using Blockchain with five solutions provided below. It aims at gaining insights from the dataset so as to see how many cryptocurrency levels have responded positively to Blockchain's Transparency in securing and protecting the buyers and sellers.

Benefits: Efficient Resource Allocation, Improved Conversion Rates, Enhanced Customer Experience, Data-Driven Decision-Making.

Challenges: Data Quality, Privacy and Compliance, Model Interpretability and Continuous improvement.

The strategic solutions outlined across diverse facets of blockchain technology, when analysed through a Naïve Bayes classifier in Fig. 9, reveal a nuanced and effective approach to overcoming complex challenges. Each solution, meticulously tailored to specific scenarios, contributes significantly to the overall success of the blockchain platforms under consideration in Figs. 10, 11 and 12. Figure 13. Represents the system architecture that is presented how transaction amount is transferred in the Cryptocurrency Levels using a Blockchain maintaining Transparency.

Naïve Bayes Equation:

$$
\begin{aligned}
\mathrm{P(Transparency)} = {} & \mathrm{P(Transaction\ Amount = A1|Transparency)} \\
& \cdot \mathrm{P(Transaction\ Amount = A2|Transparency)} \cdot \ldots \\
& \cdot \mathrm{P(CryptocurrencyLevels = A1|Transparency)} \\
& \cdot \mathrm{P(CryptocurrencyLevels = A2|Transparency)} \cdot \ldots \\
& \cdot \mathrm{P(DebtToIncome = A1|Transparency)} \\
& \cdot \mathrm{P(DebtToIncome = A2|Transparency)} \cdot \ldots
\end{aligned}
$$

```
# To find the dtypes in the DataFrame of each columns
df.dtypes
```

```
Age                         int64
Education                   int64
Employment                  int64
Cryptocurrency Levels       int64
Transaction Amount          int64
DebtToIncome              float64
CreditToDebt              float64
OtherDebt                 float64
Transparent                 int64
dtype: object
```

Fig. 9 Screenshot of implementation code for explaining the data type of the variables chosen for Naïve Bayes classification

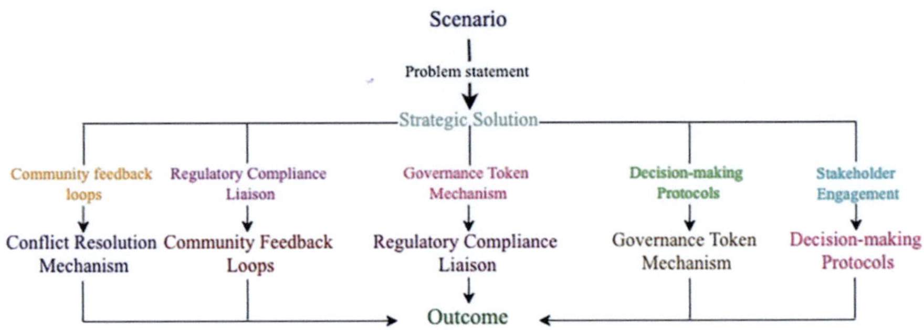

Fig. 10 Schema of sequential blockchain activities

6.5 *Scope for Future Research of the Proposed Case Study*

The future scope of this research is promising and diverse. Advanced deep learning techniques, including RNNs and attention mechanisms, warrant exploration to further enhance prediction accuracy. Incorporating sector-specific external factors and data sources, such as regulatory changes or industry-specific events, can offer a substantial boost to forecasting precision. Investigating ensemble models tailored to sector dynamics and hybrid approaches holds potential for achieving even more accurate predictions. As financial markets continually evolve, the ongoing refinement of forecasting models remains paramount. These models are instruments that empower decision-makers with timely and well-informed insights, ensuring a competitive edge in the ever-fluctuating world of finance.

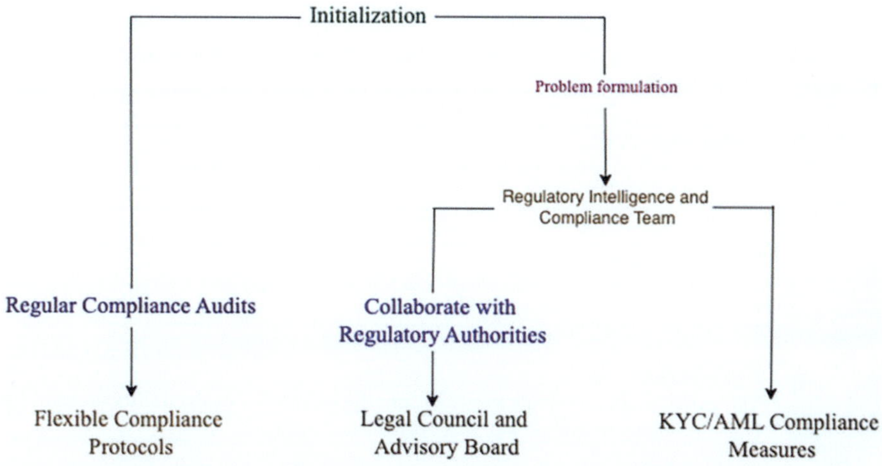

Fig. 11 Proposed schema representation in the blockchain-driven paradigm

6.6 The Key Five Solutions

Privacy and Security: The Naïve Bayes classifier reinforces the effectiveness of achieving a delicate balance between transparency and privacy. By leveraging sophisticated security measures, the platform not only safeguards user trust but also emerges as a leader in secure and transparent data management. The classifier attests to the robustness of the platform's privacy measures.

Scalability and Performance: Through the lens of the Naïve Bayes classifier, the comprehensive approach to scalability and performance emerges as highly reliable. The dynamic scaling mechanism, sharing, off-chain solutions, and optimized consensus algorithms collectively ensure sustained high performance, even during peak transaction volumes. The classifier validates the platform's scalability and performance capabilities.

Legal Compliance: Proactive legal compliance, as analysed by the Naïve Bayes classifier, establishes the crypto exchange platform as a responsible and compliant player. Collaboration with regulators, robust KYC/AML measures, and regular audits contribute to legitimacy and resilience in the face of evolving legal requirements. The classifier underscores the platform's commitment to navigating regulatory landscapes effectively.

Governance and Collaboration: Navigating collaboration complexities among diverse stakeholders, as evidenced by the Naïve Bayes classifier, ensures a harmonious environment. Stakeholder engagement, transparent decision-making, governance token mechanisms, and conflict resolution protocols foster understanding and trust. The classifier validates the effectiveness of these measures in promoting collaborative governance for sustainable success.

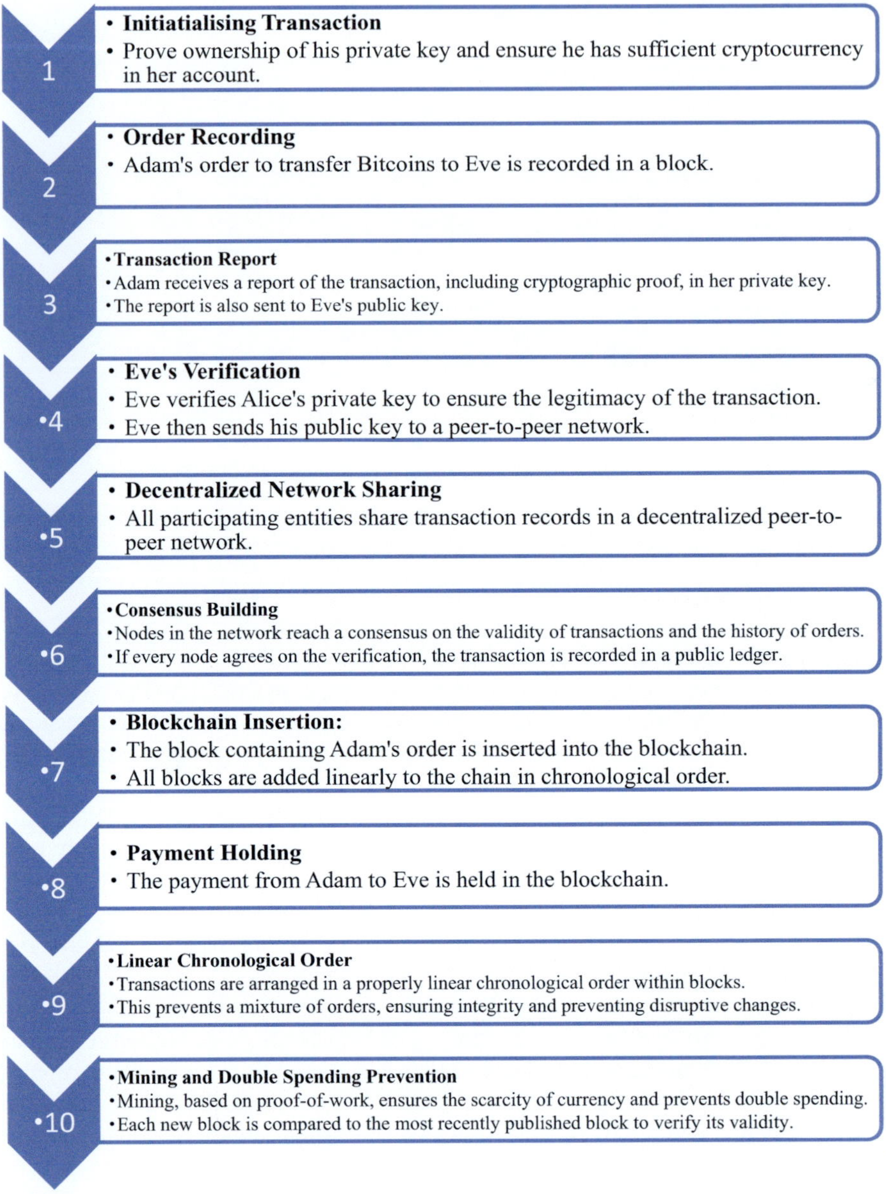

Fig. 12 Flow of activities of the proposed intelligent system-driven blockchain architecture

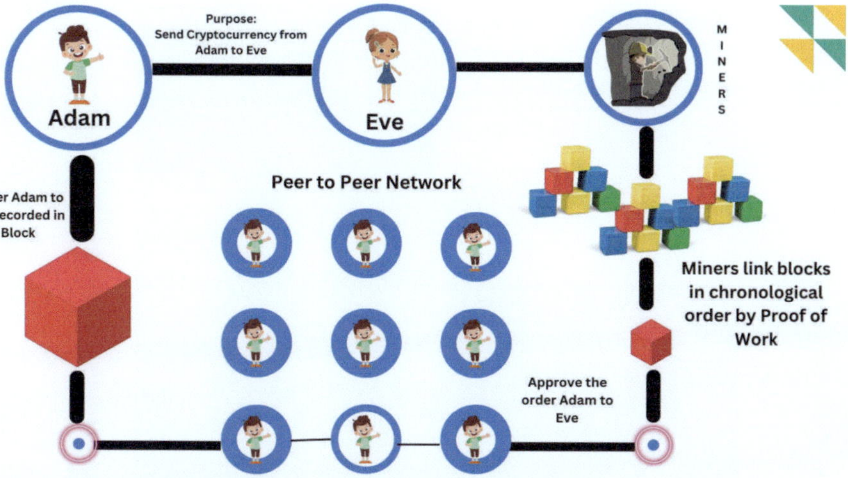

Fig. 13 Model representing how transaction amount is transferred in the cryptocurrency levels using a blockchain maintaining transparency

Digitalization Challenges: The Naïve Bayes classifier highlights the success in addressing heterogeneous levels of digitalization among users. User-centric design and educational outreach, including multilingual support, tiered assistance, progressive onboarding, and community-driven forums, contribute to a positive and collaborative user experience. The classifier affirms the platform's ability to bridge the digital divide and foster widespread adoption in Fig. 14.

Figure 15 presents the confusion matrix demonstration to emphasize the further effectiveness of the Naïve Bayes classifier in transparency prediction. The matrix, detailing true positives, true negatives, false positives, and false negatives, provides a comprehensive view of the model's performance. The highest estimates for the Transparent label affirm the classifier's proficiency in making accurate predictions (Table 1).

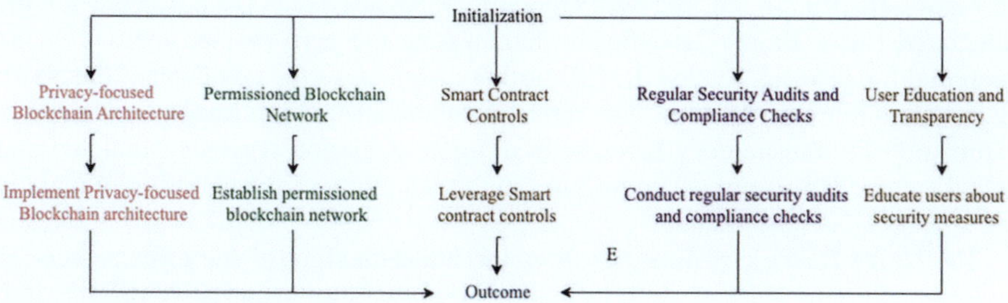

Fig. 14 Classification schema for predicting outcome in Blockchain-based system

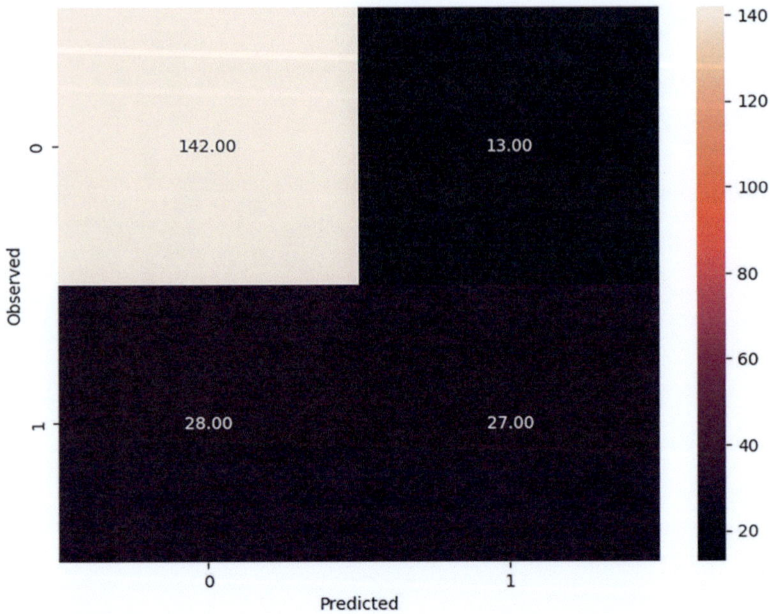

Fig. 15 The confusion matrix to derive classification effectiveness of the provided Naïve Bayes classifier in transparency prediction

Table 1 The performance metrics of Naïve Bayes classification

Accuracy: 0.7950				
	Precision	Recall	F_1-score	Support
Class 0	0.7241	0.9032	0.8038	93.000
Class 1	0.8928	0.7009	0.7853	107.000
Macro average	0.8084	0.8020	0.7945	200.000
Weighted average	0.8144	0.7950	0.7939	200.000

Figure 16 depicts the Receiver Operating Characteristics (ROC) curve for the integrated Naïve Bayes Classifier to demonstrate the performance metrics of the proposed framework. The low RMSE further underscores the precision of the Naive Bayes classifier's predictions. The model demonstrates a high level of accuracy in estimating the transparency label, with minimal deviation between predicted and actual values. This accuracy contributes to the reliability of the classifier in making transparent label predictions.

The Naive Bayes classifier enhances the understanding of the effectiveness of strategic solutions in blockchain technology by providing quantitative insights into the probability of success for each facet. This data-driven approach enriches the overall assessment of the blockchain platforms, positioning them as comprehensive, adaptable, and user-centric systems in the evolving landscape of blockchain

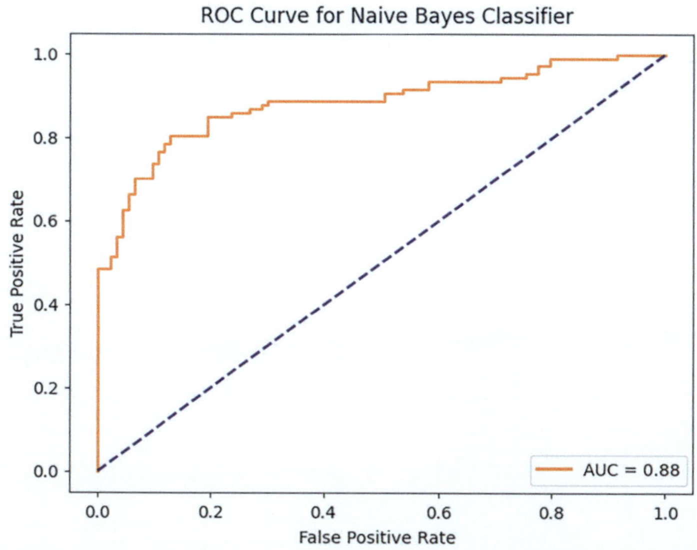

Fig. 16 The ROC curve for the integrated Naïve Bayes classifier mechanism

technology. Figure 17 refers the area under the ROC Curve (AUC-ROC) for the Transparent label indicates a strong discriminatory ability, reinforcing the classifier's efficacy in correctly identifying instances characterized by transparency.

6.7 Discussion

Rejecting the null hypotheses reveals crucial insights: Problem 1 underscores the delicate balance required between transparency and privacy in blockchain. Problem 2 emphasizes the imperative for scalable infrastructure to sustain growing transaction volumes, directly impacting performance. Problem 3 rejects the assumption that blockchain implementation remains unaffected by regulatory compliance, highlighting the need for alignment with evolving legal frameworks. Problem 4 challenges the notion that a governance framework's absence has no effect, emphasizing its pivotal role in fostering collaboration, transparency, and trust. Lastly, Problem 5 contradicts the idea that user demographics have no impact on blockchain adoption, stressing the necessity of tailored user experiences for widespread acceptance.

Implementing a privacy-focused blockchain architecture is crucial for addressing user concerns. By utilizing advanced cryptographic techniques like zero-knowledge proofs and ring signatures, the platform ensures transaction validation without compromising sensitive information, enhancing transparency in transaction history while upholding user data confidentiality. Similarly, establishing a permissioned blockchain network introduces a controlled access paradigm, restricting data access

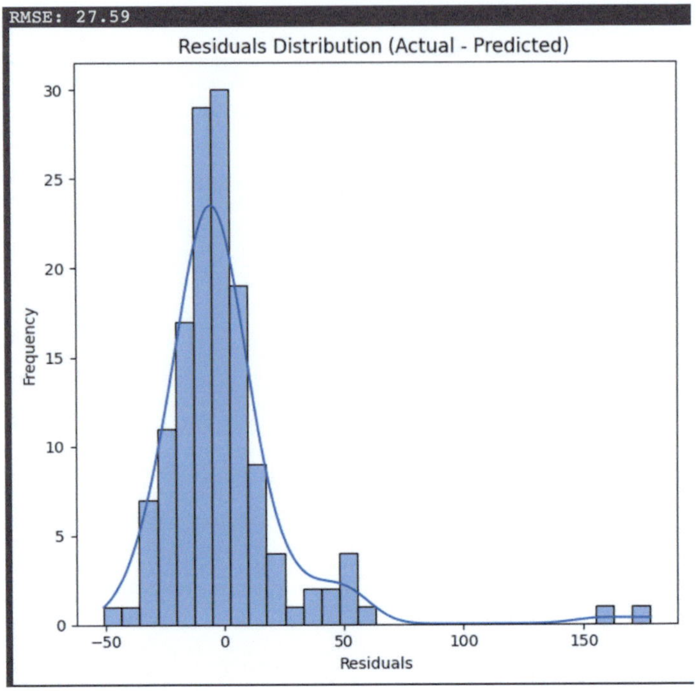

Fig. 17 The Residuals Distribution for the integrated Naïve Bayes classifier mechanism

to authorized entities. This enhances security, mitigates the risk of unauthorized access, and maintains a firm commitment to transparency and user data protection. Leveraging smart contracts for granular access controls defines rules that permit only predefined entities to access specific information, fortifying the overall security posture. Additionally, the incorporation of multi-factor authentication (MFA) significantly reduces the risk of unauthorized access, even with compromised credentials, strengthening overall security. Regular security audits and compliance checks, conducted by third-party experts, are integral to maintaining a trustworthy platform, identifying vulnerabilities, and ensuring adherence to data protection regulations, building user trust and positioning the platform as a leader in security and transparency.

7 Case Study: Humanized Intelligence in Credit Card Fraud Detection

Credit card fraud remains a widespread problem, bringing financial dangers to users and organizations. We offer a fraud detection system that combines powerful machine learning approaches, rule-based analysis, and transaction monitoring to address this ever-changing difficulty using Halving RandomizedSearchCV, a hyperparameter tuning method to improve the model's generalization ability. It is a powerful and efficient hyperparameter optimization method that optimizes speed and performance. It is a valuable tool for machine learning practitioners who want to optimize the performance of models trained on a large volume of data- as is the case with credit card transactions, within appropriate time and resource restrictions. A prototype of the proposed model was trained on 3.5 Million rows of data and returned promising results, having an accuracy of 99%, an F_1-Score of 99.61%, and a perfect ROC score of 1. Given the promising results, these methods were deemed feasible to be included in a Credit Card Fraud Detection process using Machine Learning techniques. However, further work is needed in a case-by-case performance comparison of traditional and experimental methods of model tuning in this use case, and the feasibility of these methods with Deep Learning techniques with the purpose of detecting and flagging fraud transactions.

A significant research gap is observed in finding efficient hyperparameter tuning methods of the machine learning models used for fraud detection among all the papers surveyed. Given that the sum total of credit card transactions in 2022 reached $9.56 Trillion in the United States of America alone [16], it would be reasonable to assume that the number of transactions needed per year to reach such a sum would number in the billions. To perform hyperparameter tuning on a machine learning model on millions of transactions using traditional methods such as GridSearchCV and RandomSearchCV would be highly time and resource-intensive, as a large number of combinations of hyperparameters would be tested on a model to find the best-performing model, each iteration involving millions of data points. Given the necessity of hyperparameter tuning in the machine learning process, we shall test the feasibility of newer and more experimental hyperparameter tuning methods on Fraud Prediction models, such as HalvingRandomizedSearchCV and HalvingGridSearchCV.

The dataset used in this study was obtained from an online open-source data repository, available at: https://www.kaggle.com/datasets/kartik2112/fraud-detection. The following section describes the columns of the dataset used in this research, specifically focusing on transaction and customer information.

7.1 Transaction-Related Components

- trans_date_trans_time: This identifier combines the date and time of each transaction for accurate chronological reference.
- cc_num: This column represents the anonymized credit card number used in the transaction, ensuring data privacy while retaining information relevant to fraud detection analysis.
- merchant: This column identifies the merchant where the transaction occurred, providing valuable context for spending patterns and potential risk factors.

- category: This column categorizes the transaction based on the type of goods or services purchased, offering insights into spending patterns and potentially fraudulent activities.
- amt: This column reflects the transaction amount in monetary units, serving as a crucial feature for anomaly detection and fraud identification.
- trans_num: This column serves as a transaction identifier.
- unix_time: This column provides the transaction's timestamp in Unix time format, enabling temporal analysis and identifying potential seasonal or time-based patterns in fraudulent activities.
- merch_lat: This column, if available, indicates the latitude coordinate of the merchant's location, aiding in geospatial analysis and potentially uncovering location-based patterns of fraudulent activity.
- merch_long: Similar to merch_lat, this column, if available, provides the longitude coordinate of the merchant's location for further geospatial analysis.

7.2 Customer-Related Components

- **first**: This column provides the first name of the customer.
- **last**: Similarly, this column provides the customer's last name.
- **gender**: This column provides the gender of the customer.
- **street**: This column contains the customer's street address.
- **city**: This column identifies the customer's city.
- **state**: Similarly, this column indicates the customer's state or province, enhancing location context and potentially uncovering regional trends in fraudulent activity.
- **zip**: This column provides the customer's ZIP code, offering additional location granularity for geospatial analysis and potentially identifying high-risk ZIP codes for focused fraud prevention efforts.
- **lat**: This column indicates the latitude coordinate of the customer's location, enabling geospatial analysis and potentially uncovering location-based patterns of fraudulent activity.
- **lng**: This column provides the longitude coordinate of the customer's location for further geospatial analysis.
- **city_pop**: This column reflects the population of the customer's city, potentially offering insights into the likelihood of fraud based on population density or other associated factors.
- **job**: This column, if available, provides the customer's job or occupation, potentially offering insights into income level and spending patterns related to fraud risk.
- **dob**: This column, if available, indicates the customer's date of birth, potentially enabling age-based analysis and identification of age-related patterns in fraudulent activity.

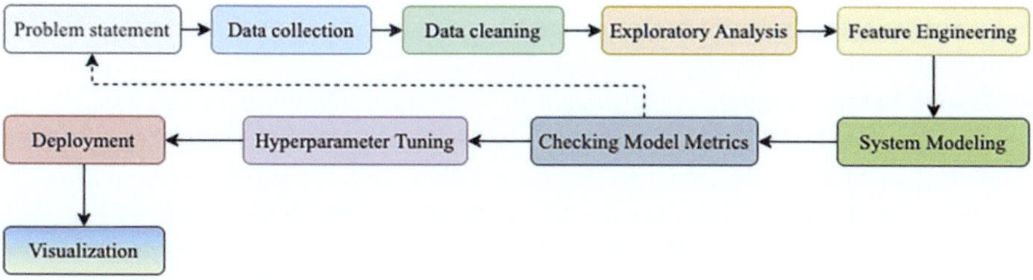

Fig. 18 Process flow diagram for the fraud detection model

7.3 *Proposed Methodology*

This process flow diagram in Fig. 18, is derived from the CRISP-DM framework of Data Mining [18], adapted for our use case of Fraud Detection in Credit Card Transactions. With the problem statement and Data Source finalized, we move on to the data cleaning process, including null value removal and sanity checks. After that, we visualize the data for question-answering via EDA. Then, we conduct feature engineering and select meaningful features for modelling. Given the imbalance in features, we apply random oversampling techniques to increase the representation of the minority class. After that, we trained Logistic Regression and Decision Tree models on the data and evaluated the performance of the models using metrics such as precision, recall, F_1-score, and the area under the receiver-operating-characteristic curve (AU-ROC). Moving forward, we select the Decision Tree models due to better performance.

To prevent overfitting, we apply hyperparameter tuning to ensure that the model can generalize to unseen data. Since the number of rows is around 3.5 Million, we use an experimental method of Hyperparameter tuning known as RandomizedsearchCV. It efficiently tunes hyperparameters by iteratively narrowing the search space: it starts with a wide range of values for each hyperparameter, evaluates a limited number of random combinations, discards poorly performing configurations, and then repeats this process with narrowed ranges for the remaining candidates, focusing computational resources on promising areas. This approach often converges to optimal hyperparameters more quickly than traditional RandomizedSearchCV, especially in large or complex hyperparameter spaces. Finally, the model can be saved and exported for the purpose of deployment.

7.4 *Observations and Insights*

- The distribution of transaction amounts showed an extreme right skewness, showing that a majority of the transactions were constrained to a small bracket on the left of the distribution. This means that transactions involving high amounts were much less frequent as compared to transactions of lower amounts, likely to be fraudulent transactions.

- Despite having a better performance in learning the patterns in the data, it was observed that the Decision Tree models had adapted to the noise in the data as well, leading to a perfect accuracy score on seen data. This is a problem, as adapting to noise leads to the model learning non-existent patterns, which ends up lowering the generalization ability of the model on unseen data. In a scenario where millions of credit card transactions happen on a daily basis, even a single misclassification can cost a company a lot, both in terms of money and client goodwill. Moreover, adapting to noise leads to increasing model complexity, resulting increased time costs.
- Hyperparameter tuning techniques were employed to resolve overfitting and complexity issues. However, given the training dataset having around 3.5 Million instances, traditional techniques of model tuning were not feasible due to the huge time and resource costs involved. Experimental forms of hyperparameter tuning returned a more generalized model, while retaining its predictive capability.

7.5 Performance Metrics

In Fig. 19, there were no training instances which were wrongly classified by the initial Decision Tree model, which is an indicator of overfitting. In Fig. 20, the tuned model generated 552,832 True Positive and 551,722 True Negative values. False Positive cases were constrained to 1092 while count of False Negatives were 0. Meaning, not a single unseen observation labelled as 'fraud' went unrecognized by the model. The classification report returned an overview of precision, recall and f1-score values for each class, 1 representing instances of fraud transactions and 0 representing instanced of non fraud transactions. All values returned as 1.

Fig. 19 Confusion matrix for the untuned model on visible training data

Predicted / Actual	Not Fraud	Fraud
Not Fraud	1289911	0
Fraud	0	1289929

Predicted / Actual	Not Fraud	Fraud
Not Fraud	552832	0
Fraud	1092	551722

Fig. 20 Confusion matrix of the final model on unseen data

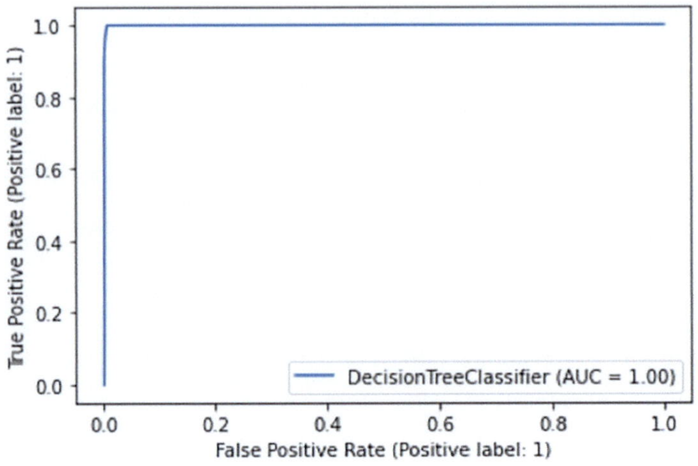

Fig. 21 The ROC curve for the final model

The ROC graph in Fig. 21 shows the performance of a classification model at all classification thresholds. This graph shows that at all decision thresholds, TPR remains constant at near 1, showing that the model is able to distinguish between positive and negative cases for any threshold. The tuned Decision Tree model achieved an accuracy of 99% on unseen testing data, indicating strong generalization ability and potential for real-world application in detecting credit card fraud. Furthermore, the F_1-score of 99.6% demonstrate the model's effectiveness in identifying both true positives and true negatives, highlighting its suitability for credit card fraud detection tasks.

7.6 Discussion

Detecting Fraud in Credit Card transactions is an important business use case in the Banking, Financial Services, and Insurance (BFSI) industry to preserve business value and retain client trust. It is ever-evolving, with newer techniques of committing fraud and even newer techniques of detecting and flagging them. Given the excellent performance of the Machine Learning model achieved with HalvingRandomizedSearchCV, these methods are indeed feasible to be included in a Credit Card Fraud Detection process using Machine Learning techniques. However, further work is needed in a case-by-case performance comparison of traditional and experimental methods of model tuning in this use case, and the feasibility of these methods with Deep Learning techniques with the purpose of detecting and flagging fraud transactions.

8 Conclusion and Scope for Future Research of the Proposed System

The assessment of HIA within the realms of sustainable finance and economic behaviours reveals a multifaceted landscape suitable with potential and fraught with challenges. A desperate observation is the relative negligence of theoretical underpinnings concerning investors' irrational behaviour within behavioural finance research and analytics, signalling a requirement for more robust theoretical paradigms. The framed domains of behavioural finance acknowledges the potential impacts of behavioural biases on investor activities and human decision-making, moving away from conventional economic theories that presume perfect rationality. These biases, stemming from sentiments and cognitive progressions, can guide to suboptimal investment adoptions and market inefficiencies. The incorporation of psychological insights into financial models represent pivotal steps towards understanding and predicting market behaviour with greater efficiency. Investors are not always logical, and their self-containment is constricted; moreover, they will always be swayed by their own biases. When formulating a monetary or financial decision, psychological factors and emotional influence human behaviour, and when investors are stressed, they misplace their capability to create a correct and rational decision, and thus they frequently make disfavoured decisions.

Furthermore, the confluence of behavioural finance into decision-making paradigms is becoming increasingly pivotal, as it challenges conventional finance assumptions by confessing the task of investor psychology in perceiving market anomalies. As markets progress and emerging financial instruments emerge, understanding behavioural underpinnings becomes even more critical for both individual investors and portfolio managers. The incorporation of behavioural finance principles into active portfolio management depicts a practical implecation of academic insights, with portfolio managers using behavioural annotations to navigate market anomalies and manage client funds more effectively. It has become transparent that understand behavioural finance for enabling finance managers to take better decisions. By comprehending the psychology of investors and markets, financial managers are capable to better management, mitigate risk, and eventually spawn higher profits. Future research may explore the effects of cultural and social differences on behavioural biases to enrich our compassion of investor behaviour globally, specially as financial markets turn out to be more interconnected.

'Behaviouralizing' finance involves substituting conventional neoclassical assumptions with more behaviourally realistic counterparts. This approach augments the ecological validity of financial artefacts, permitting them to better capture the complexities of real-world financial phenomena. A more comprehensive model of financial behaviour and investment decisions can be framed by obtaining into account the interactions between cognition and sentiment, as well as the biases and heuristics that affect investors' judgements. Future research should focus on generating models that can adapt to shift market dynamics and investor emotion to advance practical insights for financial decision-makers and policymakers. Moreover, empirical studies can validate and refine validated models, confirming their relevance and applicability in real-world settings.

In conclusion, the assessment of humanized intelligent analytics in sustainable finance and economic behaviours serves as a clarion call for interdisciplinary collaboration, urging researchers and practitioners to transcend habitual boundaries in pursuit of a more holistic and nuanced consideration of economic and financial decision-making. Proposed discussion not only enriches our theoretical backbones but also equips us with the tools to navigate the complexities of advanced financial markets, fosters greater resilience and make sustainable in the face of evolving economic landscapes. By leveraging the psychological factors that influence investor behaviour and analytics, financial professionals can generate more informed decisions, manage risk more efficiently, and eventually enhance enduring financial outcomes. The upsurge of behavioural finance demonstrates a return to a more comprehensive approach to financial modelling, incorporating insights from psychology to understand potentiality in financial phenomena.

References

1. Moustati I, Gherabi N (2025) Bridging behavioral insights and automated trading: an internet of behaviors approach for enhanced financial decision-making. Information 16:338. https://doi.org/10.3390/info16050338
2. Wong EY, Wines T, Li S (2017) Sustainable data analytics for environmental performance monitoring in dynamic supply chain infrastructure. Manag Stud 5. https://doi.org/10.17265/2328-2185/2017.06.001
3. de Zarzà I, de Curtò J, Roig G, Calafate CT (2023) Optimized financial planning: integrating individual and cooperative budgeting models with LLM recommendations. AI 5:91. https://doi.org/10.3390/ai5010006
4. Coşkun A, Karakoç M (2020) A conceptual framework for behavioral accounting. Emerald Publishing Limited eBooks, p 77. https://doi.org/10.1108/978-1-80043-095-220201007
5. Levine J, Chan KMA, Satterfield T (2015) From rational actor to efficient complexity manager: exorcising the ghost of Homo economicus with a unified synthesis of cognition research. Ecol Econ 114:22. https://doi.org/10.1016/j.ecolecon.2015.03.010
6. Wang Y (2024) Ethical considerations of AI in financial decision. Comput Artif Intell 2:1290. https://doi.org/10.59400/cai.v2i1.1290
7. Bhuiyan MS (2024) The role of AI-enhanced personalization in customer experiences. J Comput Sci Technol Stud 6:162. https://doi.org/10.32996/jcsts.2024.6.1.17
8. Rane N, Choudhary S, Rane J (2024) Artificial intelligence and machine learning in business intelligence, finance, and e-commerce: a review. SSRN Electron J. https://doi.org/10.2139/ssrn.4843988
9. Mhlanga D (2020) Industry 4.0 in finance: the impact of artificial intelligence (AI) on digital financial inclusion. Int J Finan Stud 8:45. https://doi.org/10.3390/ijfs8030045
10. Ridzuan NN, Masri M, Anshari M, Fitriyani NL, Syafrudin M (2024) AI in the financial sector: the line between innovation, regulation and ethical responsibility. Information 15:432. https://doi.org/10.3390/info15080432
11. Roy P, Ghose B, Singh PK, Tyagi PK, Vasudevan A (2025) Artificial intelligence and finance: a bibliometric review on the trends, influences, and research directions. F1000Res https://doi.org/10.12688/f1000research.160959.1

12. Byambaa O, Yondon C, Enkhbat R, Darkhijav B, Rahman M (2025) An empirical examination of the adoption of artificial intelligence in banking services: the case of Mongolia. Future Bus J 11. https://doi.org/10.1186/s43093-025-00504-y
13. Ostmann F, Dorobantu C (2021) AI in Financial Services. Alan Turing Inst. https://doi.org/10.5281/zenodo.4916041
14. Buchanan B (2019) Artificial intelligence in finance. Eur Organ Nucl Res. https://doi.org/10.5281/zenodo.2612537
15. Li Y, Wang S, Ding H, Chen H (2023) Large language models in finance: a survey. arXiv (Cornell University). https://doi.org/10.48550/arxiv.2311.10723
16. Gupta A, Puri M, Keshan M, Tiwari V (2025) AI in financial decision-making: revolutionizing investment strategies and risk management. https://doi.org/10.2139/ssrn.5085764
17. Patil D (2025) Artificial intelligence in financial services: advancements in fraud detection, risk management, and algorithmic trading optimization. https://doi.org/10.2139/ssrn.5057412
18. Quinn B (2023) Explaining AI in finance: past, present, prospects. arXiv (Cornell University). https://doi.org/10.48550/arxiv.2306.02773
19. Patil D (2025) Artificial intelligence in financial risk assessment and fraud detection: opportunities and ethical concerns. Available at SSRN 5057434. https://doi.org/10.2139/ssrn.5057434
20. Ebbage A (2018) Banking on artificial intelligence. Eng Technol 13:66. https://doi.org/10.1049/et.2018.1008
21. Lakshmi YS (2024) The impact of artificial intelligence on financial literacy. ShodhKosh J Vis Perform Arts 5. https://doi.org/10.29121/shodhkosh.v5.i5.2024.1606
22. Bajpai A (2024) Evaluating the impact of artificial intelligence on enhancing tax compliance and financial regulation. SSRN Electron J. https://doi.org/10.2139/ssrn.4922459
23. Kingsman N (2021) Debiasing credit scoring using evolutionary algorithms. arXiv (Cornell University). https://doi.org/10.48550/arxiv.2110.12838
24. Kurshan E, Chen J, Storchan V, Shen H (2021) On the current and emerging challenges of developing fair and ethical AI solutions in financial services, vol 1. https://doi.org/10.1145/3490354.3494408
25. Fritz-Morgenthal S, Hein B, Papenbrock J (2022) Financial risk management and explainable, trustworthy, responsible AI. Front Artif Intell 5. https://doi.org/10.3389/frai.2022.779799
26. Cath C (2018) Governing artificial intelligence: ethical, legal and technical opportunities and challenges. Philos Trans Roy Soc A Math Phys Eng Sci 376:20180080. https://doi.org/10.1098/rsta.2018.0080
27. Dhopte A, Bagde H (2023) Smart smile: revolutionizing dentistry with artificial intelligence. Cureus https://doi.org/10.7759/cureus.41227.
28. Williamson S, Prybutok VR (2024) The era of artificial intelligence deception: unraveling the complexities of false realities and emerging threats of misinformation. Information 15:299. https://doi.org/10.3390/info15060299
29. Schmidt A (2020) Interactive human centered. Artif Intell. https://doi.org/10.1145/3399715.3400873
30. Sheridan TB, Verplank WL (1978) Human and computer control of undersea teleoperators. https://doi.org/10.21236/ada057655
31. Dignum V (2022) Responsible artificial intelligence—from principles to practice. arXiv (Cornell University). https://doi.org/10.48550/arxiv.2205.10785
32. Hernández E (2024) Towards an ethical and inclusive implementation of artificial intelligence in organizations: a multidimensional framework. arXiv (Cornell University). https://doi.org/10.48550/arxiv.2405.01697
33. Lepri B, Oliver N, Pentland A (2021) Ethical machines: the human-centric use of artificial intelligence. IScience. https://doi.org/10.1016/j.isci.2021.102249
34. Borenstein J, Howard A (2020) Emerging challenges in AI and the need for AI ethics education. AI Ethics. https://doi.org/10.1007/s43681-020-00002-7

35. Ali SJ, Christin A, Smart A, Katila R (2023) Walking the walk of AI ethics: organizational challenges and the individualization of risk among ethics entrepreneurs. In: 2022 ACM conference on fairness, accountability, and transparency, p 217. https://doi.org/10.1145/3593013.3593990
36. Gao DK, Haverly A, Mittal S, Wu J, Chen J (2024) AI ethics: a bibliometric analysis, critical issues, and key gaps. arXiv. https://doi.org/10.48550/ARXIV.2403.14681
37. Alfredo R, Echeverría V, Jin Y, Yan L, Swiecki Z, Găsević D, Martínez-Maldonado R (2024) Human-centred learning analytics and AI in education: a systematic literature review. Comput Educ Artif Intell 6:100215. https://doi.org/10.1016/j.caeai.2024.100215
38. Viswanathan P, Ai A (2025) A comprehensive framework for autonomous decision-making systems in artificial intelligence. Int J Comput Eng Technol 16:862. https://doi.org/10.34218/ijcet_16_01_069
39. Chen Y, Ahn MJ, Wang Y (2023) Artificial Intelligence and public values: value impacts and governance in the public sector. Sustainability 15:4796. https://doi.org/10.3390/su15064796
40. Sigfrids A, Nieminen M, Leikas J, Pikkuaho P (2022) How should public administrations foster the ethical development and use of artificial intelligence? A review of proposals for developing governance of AI. Front Hum Dyn. https://doi.org/10.3389/fhumd.2022.858108
41. Corrêa NK, Galvão C, Santos JW, Pino CD, Pinto EP, Barbosa C, Massmann D, Mambrini R, Galvão L, Terem E, de Oliveira N (2023) Worldwide AI ethics: a review of 200 guidelines and recommendations for AI governance. Patterns. https://doi.org/10.1016/j.patter.2023.100857
42. Nishant R, Kennedy M, Corbett J (2020) Artificial intelligence for sustainability: challenges, opportunities, and a research agenda. Int J Inf Manag 53:102104. https://doi.org/10.1016/j.ijinfomgt.2020.102104

Sustainable Finance: Insights and Directions

Anindita Chakraborty

Abstract Sustainable finance (SF) has established itself as a distinct topic of research in finance. As the concern about environmental, social, and governance (ESG) issues is growing, investors are encouraging companies to act more sensibly, which has sparked a surge toward more research and publishing in this field. Although a variety of literature is available in this field, still there is a necessity for comprehensive work to understand the development of this field—where we have reached and what we need to do next. This paper tries to conduct a logical literature review with the help of SPAR-4-SLR to understand the development of sustainable finance and its sub-fields. The study deconstructs the most significant papers, top contributing journals, authors, and nations for conducting research in the sustainable finance field. The study comes out with several themes of sustainable finance outcome-based research. Thus, this study proposes several suggestions for the future development of this field.

Keywords Sustainable finance · Impact investing · ESG · Sustainability · Systematic literature review · SPAR-4-SLR

1 Introduction

1.1 Background

The world of finance is incredibly dynamic, and there is a constant evolution of financial tools and instruments in response to the demands placed on them by society and those in charge of policymaking. With all United Nations member states focus on Sustainable Development Goals (SDGs) being a part of the 2030 Agenda for Sustainable Development [1], sustainable finance has attracted huge attention. Since

A. Chakraborty (✉)
Institute of Management Studies, Banaras Hindu University, Uttar Pradesh, Varanasi, India
e-mail: aninditachakraborty19@gmail.com

S. Roy et al. (eds.), *Humanized Intelligence in Sustainable Finance and Economic Behaviours*, Studies in Computational Intelligence 1226,
https://doi.org/10.1007/978-981-95-0599-9_2

the last decade, the UN Global Compact has collaborated with businesses, investors, and UN agencies on sustainable finance. Sustainable finance has appeared to be an important notion between finance and the SDGs [2]. With more and more attention from investors on environmental, social, and governance (ESG) in recent years, many investment decisions are inclined toward green-energy projects and investing in those companies that determine social values such as social-inclusion or good-governance [3]. Sustainable finance is defined as "the process integrating in due account of environmental, social, and governance (ESG) factors […] into investment decision-making within the financial sector, which encourages a shift toward longer-term responsible investment. This approach supports, allocation of sustainable economic activities and capital projects that are environmentally sustainable and socially inclusive" [4]. Sustainable finance instruments such as green bonds, social bonds, sustainability bonds, sustainable development goal bonds, sustainability-linked bonds, or ESG-linked bonds can act as an intersection between finance and SDGs [5]. These types of initiatives taken by the corporations. It permits flow of capital to meet corporate practice ESG commitments, and it attract institutional investors (IIs) as they perceive financial guarantees to sustainability as a good indicator of a corporates sustainable progress. With the rapid increase in green finance, companies have the chance to positively influence sustainable development and are working to integrate their operations with several other SDGs, including climate action [5]. The field of sustainable finance recently has also seen many transformations, moving from the established area of ESG investing to the more results-driven strategy of impact investing. This transformation marks a turning point in the financial sector and can positively impact the environment and society.

1.2 Motivation

Various review-based papers have been conducted in this domain, including [2, 6–8] and tried to cover various aspects of the concept but were limited to the method of conducting the review. Research papers [2, 6] tried to provide a comprehensive outlook. However, this paper will try to add to the knowledge base by using the advanced systematic literature review method SPAR-4-SLR [6] along with bibliometric analysis by extending the period of the study. The chapter begins with a systematic review of literature which is the "*identification, selection and critical appraisal of research […] to answer the 'formulated questions'*". The systematic literature review should adhere to a well-defined protocol before the start of the review process—bibliometric analysis, progressing to a systematic literature review for a more in-depth examination. The second part of the study includes bibliometric analysis for identifying core research or authors, and their relationship, by covering all publications related to a given topic or field" [9]. This provides a scholarly network and research trajectory that will be of immense use for the fundamental knowledge of the subject [10]. The chapter begins with a systematic literature review which is the "identification, selection and critical appraisal of research in order to answer a formulated question" [11]. The systematic review should adhere to a well-defined protocol before the start of the review process. bibliometric analysis, progressing to a systematic literature review for a more in-depth examination. The second part of the study includes bibliometric analysis, which is "a scientific computer-assisted review methodology that can identify core research or authors, as well as their relationship, by covering all the publications related to a given topic or field". This provides a scholarly network and research trajectory that will be of immense use for the fundamental knowledge of the subject.

1.3 Chapter Contribution

Various review-based papers have been conducted earlier, but this study specifically limited its search process to "sustainable finance." In contrast, a recent review by extended its purview to "carbon credit", "carbon tax", "climate finance", "sustainable finance", etc. [2]. So, we tried to limit the scope of the study to "sustainable finance". This research will add to what is already known about sustainable finance and outline future studies in this area. Therefore, by conducting the bibliometric analysis along with SLR, we focus on the following research questions:

RQ1: What is the recent research trend in sustainable finance?
RQ2: Which authors, documents, journals, and countries contributed highly to this field?
RQ3: What is the conceptual structure of sustainable finance?
RQ4: What are the major themes covering in this field?
RQ5: What are the future directions of research of this field?

1.4 Chapter Organization

The chapter is organized as follows—methodologies are presented first, followed by findings and commentary. Further, it concludes with a discussion of where more study is needed and what suggestions can be implemented in the future research.

2 Methodology

The chapter is conducted in two steps. Firstly, a Systematic Literature Review (SLR) is adopted to collect the data, and for that, the SPAR-4-SLR protocol is used. The SPAR-4-SLR protocol is Scientific Procedures and Rationales for logical and

Systematic Review given by [12] for collecting relevant data from the SCOPUS database, which is considered the most recent protocol comprising assembling, arranging, and assessing. In step two, bibliometric analysis is conducted using Biblioshiny software. It provides a web interface for Bibliometrix using R Studio [13].

2.1 Research Protocol

The study employed citation analysis and thematic mapping to understand the research domain. Figure 1 represents the SPAR-4-SLR protocol adopted for the study.

1. **Assembling** The initial search was conducted on 27 September 2023, using the search string, includes "article title, abstract and keywords" with the keywords "Sustainable finance" OR "ESG" OR "Green bonds" OR "Green finance" OR "Climate finance" on the Scopus database. Scopus was chosen as a search database because it provides wide coverage of social science papers rather than the Web of Science and previous studies [2, 10, 12, 14] also suggested that. After searching, a total of 835 documents were extracted.

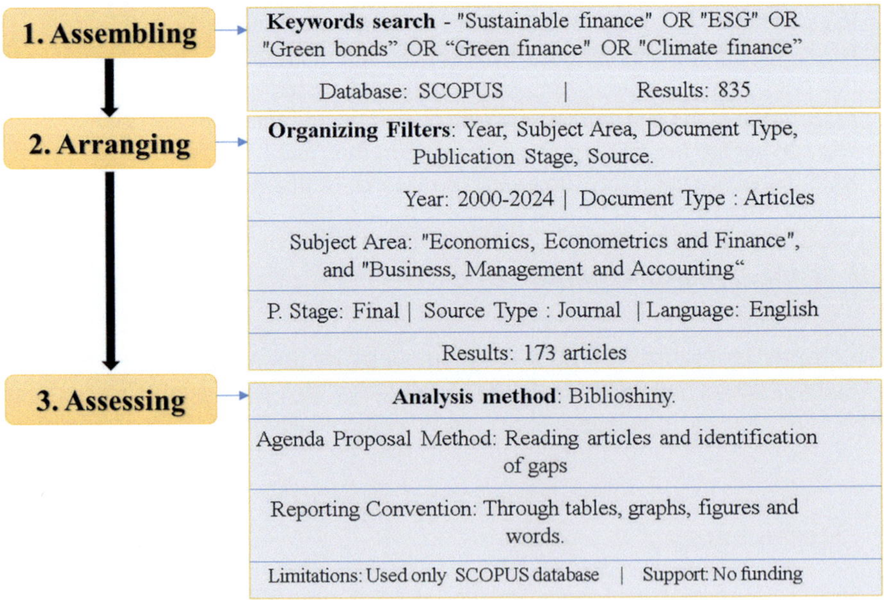

Fig. 1 SPAR-4-SLR Protocol. *Source* Author's compilation

2. **Arranging**: To arrange the 835 documents, we again returned to the assembling stage and tried to limit the search results based on (a) *year*, (b) *subject-area*, (c) *document-type*, (d) *publication-stage*, (e) *source-type, and* (f) *language.* The search results were filtered and limited to "2000–2024", domain of "economics, econometrics and finance," "business, management, and accounting," "article," "final," "journal," and "English" as recommended by [2, 15]. This reduced the number of documents; the final result was 174. The bibliometric analysis was run on 174 documents.
3. **Assessing**: The assessment of 174 documents was done through bibliometric analysis. Bibliometric analysis as a tool for systematic literature review is very common in finance [16]. For bibliometric analysis, the chapter used performance analysis techniques and science mapping [2]. Performance analysis was done by identifying publication trends, most cited articles, journals, authors, and countries. In comparison, science mapping was done through word cloud and network analysis.

3 Findings

3.1 Descriptive Features

Table 1 gives the descriptive characteristics of sustainable finance research. The period taken for the study was 2009–2024; 174 documents were there, with 566 keywords.

Table 1 Descriptive characteristics of Sustainable finance research

Timespan	2009:2024
Documents	174
Citations per doc (Average)	13.12
Keywords (ID)—Plus	267
Keywords (DE)-Author's	566
Authors	412
Authors of single authored (docs)	37
Authored docs (Single)	40
Co-Authors (per Doc.)	2.53
International co-authorships (in %)	30.64

4 Performance Analysis

4.1 Publication Development for Sustainable Finance

Figure 2 presents the year-wise publication trend. The period taken for this study was 2009–2024. The data shows that in 2009, there was only one publication, which was raised in the year 2017 when three documents were published, then it reached 8 in the year 2019, and 51 documents were reported in the year 2022. This rise in publications was seen from 2017, and after the Paris agreement related to climate change, and SDGs were signed by United Nations Member States (UNMS) in 2015. The results were in line with the agreement signed, which shows the interest of researchers regarding climate change issues and sustainable finance and how it affects society and corporations because the period between 2017 and 2022 shows an exponential rise in publications related to these domains. Figure 3 infers that a maximum number of articles were written on the topic of "sustainable finance," followed by "ESG" and "Green bonds," and most of them were written in the "Journal of Sustainable Finance and Investment."

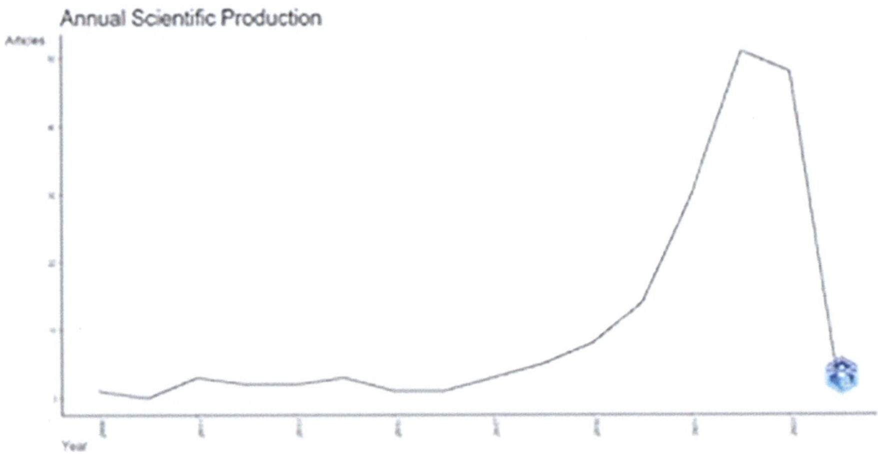

Fig. 2 Year-wise publication

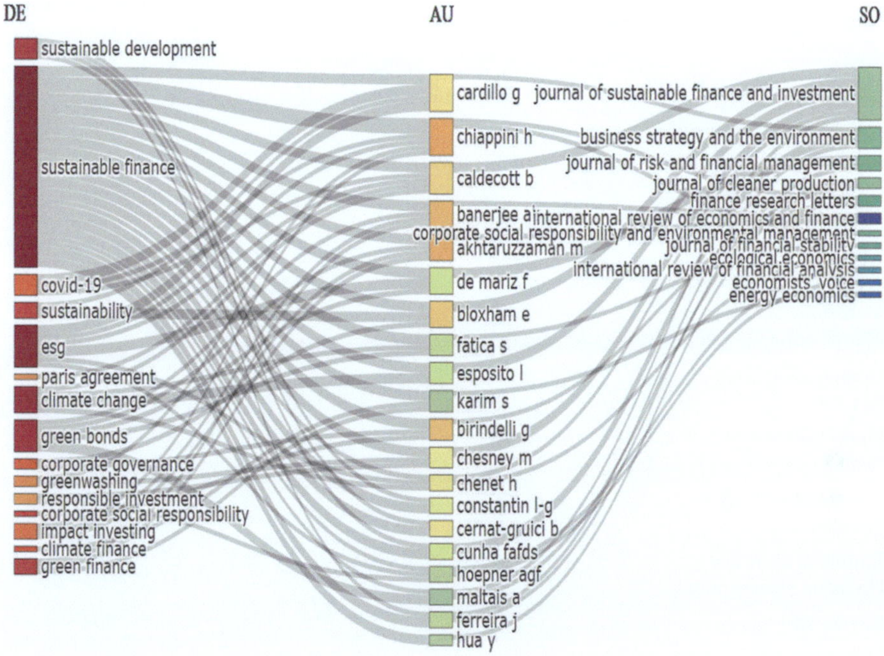

Fig. 3 Three-field plot (left: keywords, middle: authors, right: source)

4.2 *Most Contributing Journals in Sustainable Finance*

Table 2 provides the list of the top 10 journals in the field of sustainable finance, including the names, h-index, g-index, m-index, the total number of citations (TC), net production (NP), and starting year of the journal (PY-Start). One hundred seventy-three documents were published in 68 journals between 2009 and 2024. The table gives that the *Journal of Sustainable Finance and Investment (JSFI)*, *Journal of Cleaner Production (JCP)*, and *Business Strategy and the Environment (BSE)* have the highest citations, 417, 355, and 167, with net production of 29, 12, and 8, respectively.

4.3 *Most Cited Articles in Sustainable Finance*

Table 3 represents the most cited articles for sustainable finance research. Table 3 shows that "*Corporate Green Bonds*" by Flammer [17] is the most cited article from the *Journal of Finance Economics*. It has been cited 321 times since its publication, and the total number of citations (TC) per year is 107 times since 2021. The second article is "*Sustainable Business Model (SBM) Archetypes for the Banking Industry*"

Table 2 List of top ten journals

Element	h_index	g_index	m_index	TC	NP	PY_start
Journal of Sustainable Finance and Investment	11	20	0.846	417	29	2011
Journal of Cleaner Production	7	12	1.167	355	12	2018
Business Strategy and the Environment	5	8	0.556	167	8	2015
CSRand Environmental Management	3	4	1.5	24	5	2022
European Business Organization Law Review	3	4	0.6	29	4	2019
Finance Research Letters	3	5	0.75	120	5	2020
Journal of Financial Stability	3	5	1	112	5	2021
Journal of Risk and Financial Management	3	5	0.75	26	6	2020
Review of Finance	3	3	1.5	25	3	2022
Accounting, Economics, and Law: A Convivium	2	2	0.5	17	2	2020

by Angus W.H. Yip and Nancy M.P. Bocken [18], with 171 citations and 28.5 total citations (TC) per year since 2018. The results also revealed that the top three articles are on green bonds and sustainable development models. The top 25 highly cited articles consist of 1569 citations, showing the importance of sustainable finance in the research domain.

4.4 *Top Authors Works on Sustainable Finance*

Table 4 presents the top contributing authors in sustainable finance. The table gives that Chenet, H., Fatica, S., and Cunha, F.A.F.D.S. [27] are the top contributing authors with two, one, and one publication and 94, 41, and 39 citations, respectively. Collectively, there are 28 authors with 33 publications and 338 citations.

4.4.1 Influence of Author

Table 5 and Fig. 4 present "Lokta's law," which states that "for a given number of authors who have written one article, the number of writing multiple articles can be

Table 3 Most cited articles

Authors	Article works on	Journal name	Total citations	TC per year
Flammer [17]	Corporate green bonds	Journal of Finance Economics	321	107
Yip and Bocken [18]	Sustainable business model in banking Industry	Journal of Cleaner Production	171	28.5
Pham and Huynh [19]	Investor attention → green bond market	Finance Research Letters	96	24
Fatemi and Fooladi [20]	Sustainable finance	Global Finance Journal	86	7.818182
Azhgaliyeva et al. [21]	Green bonds → for financing renewable energy	Journal of Sustainable Finance and Investment	73	18.25
Maltais and Nykvist [22]	Green bonds in advancing sustainability	Journal of Sustainable Finance and Investment	73	24.33333
Lee [23]	Green Finance and Sustainable Development Goals	Journal of Asian Finance Economics and Business	65	16.25
Bento et al. [24]	Crowdfunding for sustainability ventures	Journal of Cleaner Production	64	12.8
Roncoroni et al. [25]	Climate risk and financial stability in banks and investment funds	Journal of Financial Stability	52	17.33333
Chenet et al. [26]	Finance, climate change and radical uncertainty	Ecological Economics	51	17

Table 4 Top authors

Author	TP	TC	TCBY
Chenet, H	2	94	27.75
Fatica, S	1	41	13.667
Cunha, F. A. F. D. S	1	39	13
Ferreira, J	2	37	12.333
Cardillo, G	1	21	21
Caldecott, B	2	16	8
Esposito, L	1	16	3.2
Chesney, M	1	14	4.667
Akhtaruzzaman, M	1	12	6
Banerjee, A. K	1	12	6

Table 5 Lokta's law

Documents written	N. of authors	Proportion of authors
3	1	0.002
2	28	0.068
1	383	0.93

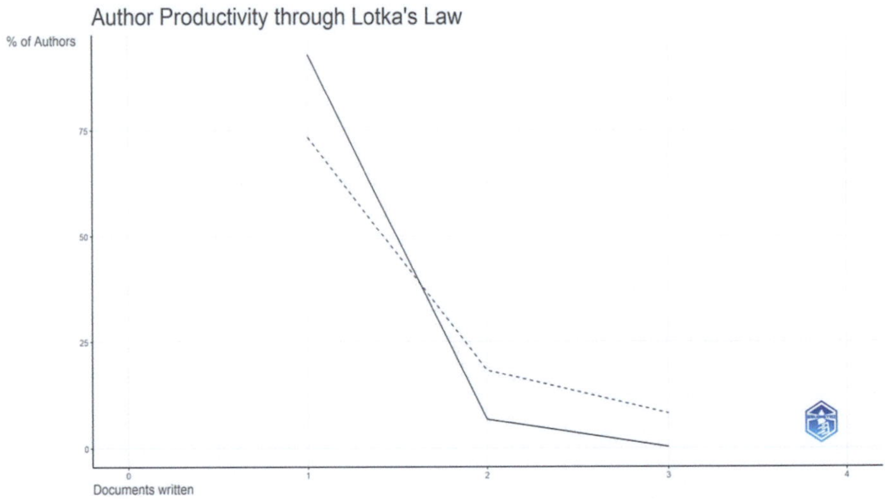

Fig. 4 Author's productivity through Lokta's law

predicted" [28]. The results show that only 1 author out of 411 authors has written 3 documents, while 383 authors have written a single document.

4.5 *Top Contributing Institutions in the Domain of Sustainable Finance*

Table 6 presents the top contributing institutions in this domain. Table 6 shows that the most dominant institution is the University of Bologna, Italy, with 8 articles published since 2009 in sustainable finance, followed by the Bocconi University, Italy, and Hunan University, China.

4.6 *Top Contributing Countries in the Domain of Sustainable Finance*

Table 7 presents the top contributing countries in this field. It is important to note that Italy has the highest number of publications during the study period, i.e., 2009–2024, with 72 publications with 203 citations, while China is in second position with 34 publications over the period with 316 citations. Similarly, the UK is third, with 20 publications in this domain with 123 citations.

However, Singapore and Portugal have the highest average article citations of 39.50, with 7 articles and 33.70 for 5 articles, respectively.

Table 6 Top institutions contributions

Affiliation	Country	Articles
University of Bologna	Italy	8
Bocconi University	Italy	6
Hunan University	China	6
American University of the Middle East	Kuwait	5
Stockholm Environment Institute	Sweden	5
University of Kassel	Germany	5
University of Zurich	Switzerland	5
Columbia University	USA	4
Copenhagen Business School	Denmark	4

Table 7 Top contributing countries

Country	TP	TC	Average article citations
Italy	72	203	8.10
China	34	316	24.30
United Kingdom	20	123	12.30
USA	18	168	18.70
Sweden	12	114	22.80
Portugal	5	101	33.70
Germany	34	89	5.90
Singapore	7	79	39.50
Switzerland	6	55	18.30
Brazil	6	42	21.00

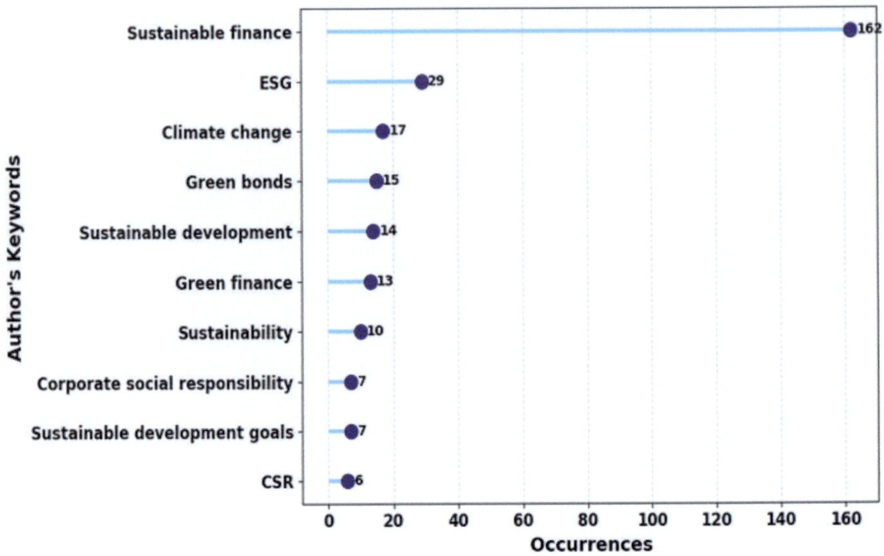

Fig. 5 Occurrences of author's keywords

4.7 *Science Mappinf*

Science mapping consists of a graphical representation of the existing knowledge and how they are interrelated to understand the domain under study better [29]. The science mapping of sustainable finance is done through coupling, word frequency, word cloud, network analysis, and collaboration network.

4.7.1 Most Frequent Word

Figure 5 and Table 8 show the frequency of the author's keywords. The data infers that "sustainable finance" has occurred 162 times in the article keywords; next is "ESG" followed by "climate change" with 29 and 17 times, respectively. Out of 571 authors' keywords, the other prominent keywords are "green bonds," "sustainable development," "green finance," "sustainability," "corporate social responsibility," and "sustainable development goals."

4.7.2 Words' Frequency Over the Time

Figure 6 also illustrates the word frequency throughout 2009–2024, and the figure depicts that the word "sustainable finance" has a speedy growth from 2015, the year the Paris Agreement was signed and reached its highest point in 2023. Similarly, the words "ESG," "green bonds," and "climate change" have taken a steep growth in

Table 8 Occurrences of author's keywords

Words	Occurrences
Sustainable Finance	162
ESG	29
Climate Change	17
Green Bonds	15
Sustainable Development	14
Green Finance	13
Sustainability	10
Corporate Social Responsibility	7
Sustainable Development Goals	7
CSR	6

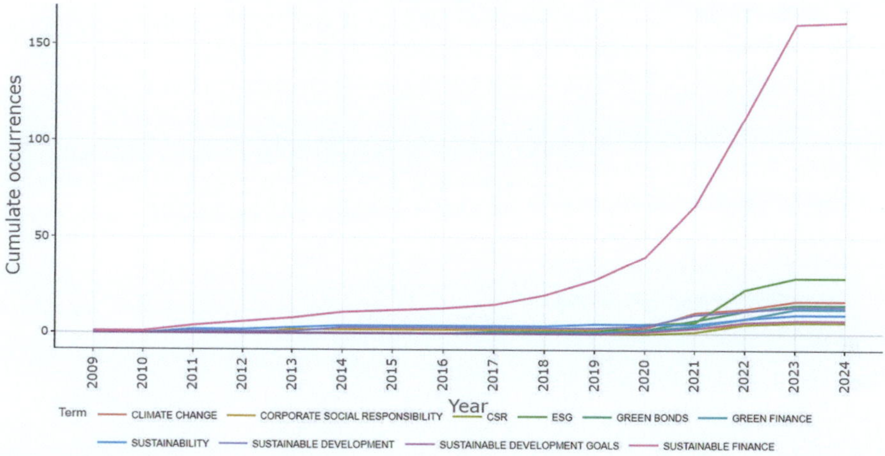

Fig. 6 Words' frequency over the time

literature since 2021. This also shows the importance of climate change, which is a multifaceted problem that the collective efforts of individuals, governments, and organizations can solve. For that, different financial instruments are introduced in the financial system to make investors more concerned about the environment.

4.7.3 Word Cloud

The word cloud of sustainable finance research using keywords plus. The figure shows some prominent words like "sustainable finance", "investments", "finance", "sustainability", "climate change", "green finance", "European Union", etc. We can analyze the word cloud by seeing the size of the words: "The bigger in size the word shows, the more it occurs in literature" [30] (Fig. 7).

Fig. 7 Word cloud of sustainable finance research

4.8 *Clustering by Coupling*

To identify the developing themes and burgeoning clusters in sustainable finance research, bibliographic coupling is conducted. The concept was first presented by [31]. Bibliometric coupling tries to identify an association between two articles that have a high proportion of similarity in keywords, description, and citation [32]. The unit of analysis in the case of bibliographic coupling can be either the author, document, or source [33]. For investigation, the study selected the "document" as the unit of analysis. Figure 8 and Table 9 reveal that there are nine clusters in sustainable finance research. These nine clusters are of different colors and sizes. The highest cluster is "Sustainable Finance, Sustainable Development, Corporate Social Responsibility", with 111 words frequency in 174 documents, and the figure also reveals that. These clusters also open the scope for future research.

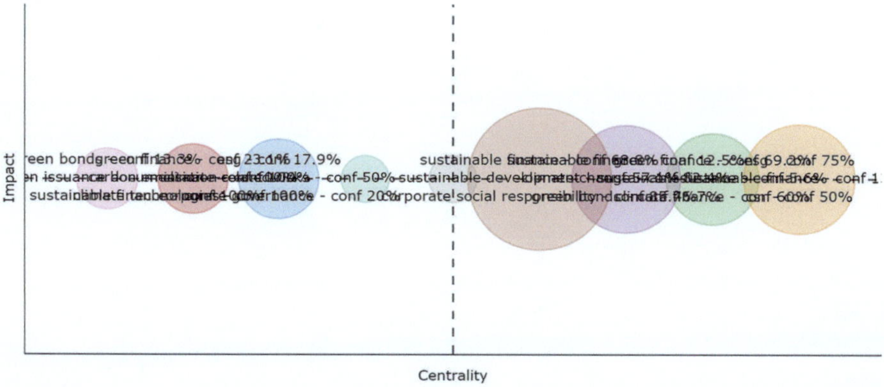

Fig. 8 Bibliographic coupling map

Table 9 Coupling map clusters

Label	Group	Freq	Centrality
Green finance, carbon emission, climate technologies	1	3	0.139147084
ESG, climate-related risks, corporate governance	2	5	0.187124757
Green finance, sustainable finance, climate finance	3	9	0.685638484
Sustainable finance, climate change, green bonds	4	20	0.630592247
ESG, sustainable finance, CSR	5	22	0.68744435
Sustainable finance, sustainable development, corporate social responsibility	6	111	0.57687853
Green bonds, disclosure of green issuance documentation, green and sustainable finance	7	2	0.134513501
CSR, economic history, economic thought	8	1	0.239495488
Carbon pricing, climate change, green finance	9	1	0.198528602

4.9 Network Analysis

The network analysis is conducted to test the intelligence in intellectual network structure of research. With the help of co-occurrence analysis, the study tries to unveil the evolution of the concept from 2009 to 2024. The major themes that emerged from

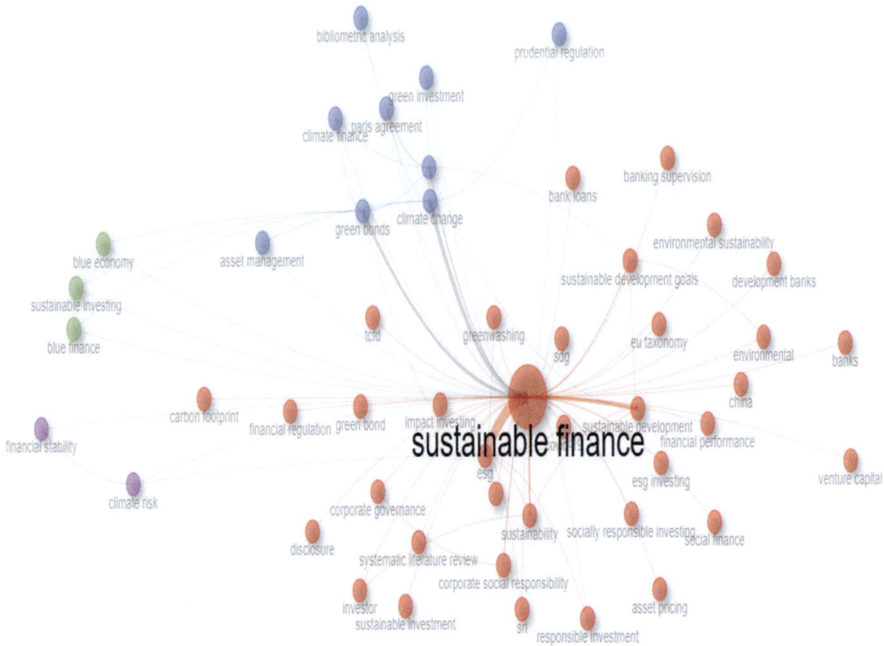

Fig. 9 Keyword network

the co-occurrence analysis of keywords are presented in Fig. 9. It is a graphical representation to visualize and draw inferences regarding relationships between entities in sustainable finance research. The figure reveals that there are four clusters, namely sustainable finance (first and red color), climate finance (second and blue color), sustainable investing (third and green color), and climate risk (fourth and purple color).

1. **Cluster 1 (Red)**: It indicates the largest cluster works on sustainable finance. The most popular keywords of this cluster are "ESG", "Sustainable Development (SD)", "Sustainability", "CSR", "SDGs", "Corporate Governance", "Impact Investing", "ESG Investing", "Green Bonds", "Greenwashing", "Socially Responsible Investing", "Social Finance", "EU Taxonomy", and "Financial Performance." This shows that under this cluster, researchers are more concerned about sustainability and SDGs. The research mainly revolves around the financial instruments of sustainability like sustainable finance, green bonds, impact investing, ESG investing, social finance, socially responsible investing, and their financial performance [15, 17, 34–38]. The training of the market participants for a better understanding of the sustainability issue and how it affects the valuation of the company is focused [39]. Similarly, corporate governance should be aligned with sustainability, and investors should cautiously assess the disclosures

to invest sustainably [40, 41]. Aria and Cuccurullo [13] concluded that standardization of the concept of green finance is very much necessary in building an analogous of capital markets for investment opportunities in sustainable infrastructure around the world to avoid "greenwashing". Kandrács [42] highlighted the issue of green transition in an economy. They stated that this type of transition required a huge amount of investment and financing from both the private sector and central banks, and the government can go for the PPP model [43].

2. **Cluster 2 (Blue)**: It is the second largest cluster about climate finance. The most popular keywords of this cluster are "green bonds", "climate change", "climate finance", "Paris Agreement", and "green investment". Under this cluster, researchers mainly focused on issues relating to climate change and the need for financial instruments like green bonds and green investments to mitigate the effects of greenhouse gases and the regulations and agreements about that [13, 44–47]. It has been stated by [13] that "generally accepted standards and definitions of 'green finance' are crucial to attracting investors in sustainable infrastructure". The consequences of green bond financing on the economy and environment, as well as talked about the legal and institutional issues [48]. Some research found that there is judgment effect between green bonds and investor attention. The factor that stimulates investors to invest in green bonds offers an additional channel for corporates to fulfill the financing requirements for transformation to a low-carbon economy [19]. The need for the introduction of social bonds after COVID-19 was also pointed out [49]. The study also reported that there was an unprecedented increase in the issuance and performance of the bonds over the conventional ones. They also highlighted the risk of "social washing" and the need to develop regulations to avoid it [49].
3. **Cluster 3 (Green)**: The third cluster pertains to sustainable investing. The most popular keywords of this cluster are "sustainable investing", "blue economy", and "blue finance". How investors' sustainable investing practices through environmental, social, and governance (ESG) and exclusionary screening integration affect the asset returns of the company and can bring the ecological transition was elaborated by [50]. Another important keyword that emerged from this cluster is the blue economy, which is one of the SDGs of the United Nations- "*life below water*". It is defined as "*an economy that comprises a range of economic sectors [...] and related policies that together determine whether the use of ocean resources is sustainable*". The research works connected to this show marine ecosystems' importance and the issuance of blue bonds to increase financing toward a sustainable economy [51, 52].
4. **Cluster 4 (Purple)**: The fourth cluster is the smallest in size, and it pertains to "climate risk" and "financial stability." This cluster is mainly connected to climate risk and its repercussions on the financial stability of an economy [53]. A framework for financial institutions and policymakers to deal with the stability and financial risks generated by climate change was suggested by [54, 26].

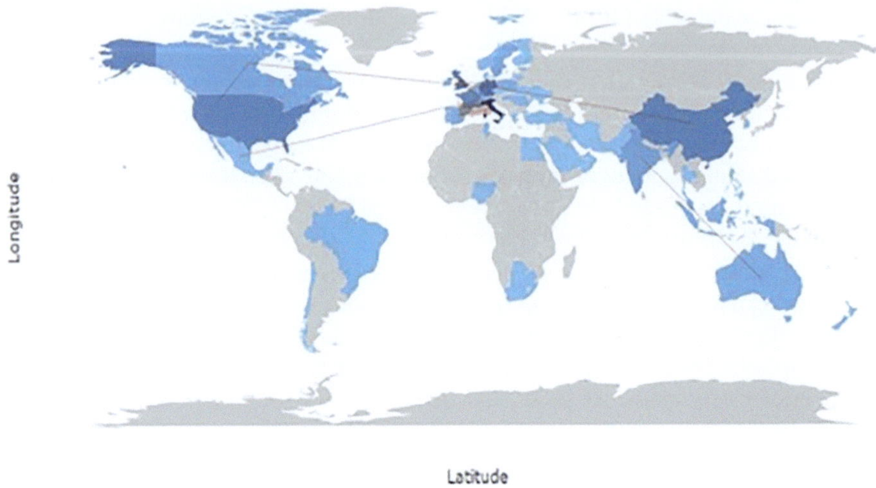

Fig. 10 Countries collaboration World Map

4.9.1 Collaboration Network

The collaboration network shows the association between the researchers from different countries. This collaboration also shows the networking among eminent-scholars researching the same topic in different countries and their collaborative work [55]. Figure 10 displays the pathway of global collaboration. The blue color on the map and the connecting lines shows the existence of global research networks. The figure illustrates that the USA has a collaboration with Canada, the UK with Belgium and China, Switzerland with Mexico, Italy with France, Germany with Canada, and India with Australia, which means that researchers from these countries collaborate to produce scientifically meaningful outcomes.

5 Conclusions of the Study

Plentiful papers have been published on sustainable finance since 1986 when the first paper by [6] was written to address the concern about social investing. The present paper mainly focused on the work between 2009 and 2024, and 174 papers were there. The study applied SLR, bibliometric, and content analysis to find the major contribution of this concept during this period. The results show that the evolution of sustainable finance goes beyond the traditional financial models and tries to integrate ESG considerations into investment and develop new instruments like green bonds, impact investing, SDG bonds, blue finance, etc., to reshape the investment landscape. The bibliographic coupling and cluster analysis show the influential works of the researchers. These clusters also revealed the upcoming future scope of sustainable finance works, which has huge potential and is multifaceted.

6 Limitations and Scope for Future Research

The study conducted the bibliometric analysis through the Scopus database for the period 2009–2024. Future studies can be conducted using databases like Web of Science (WoS), Google Scholar, EBSCO, JSTOR, etc., by expanding the time period. In this study, very few search words are used. Future researchers can include other keywords like "carbon credit," "green economy," "sustainable economy," etc., for conducting their study. Further, upcoming researchers need to focus on the exploration of innovative financial instruments, their performance, alignment with the profit motives of the corporates, and, at the same time, how these are fostering a greener economy at the global level. Extensive investigation is also required in the regulatory frameworks of sustainable finance instruments, sustainability reporting by corporates, and the role of technology like blockchain and artificial intelligence. Additionally, researchers can also try to identify the role of these instruments in the promotion of financial inclusion in emerging economies. Another important facet is to understand the psychology of green investors from the point of view of behavioral finance.

References

1. United Nation (2020) The sustainable development goals report 2020. https://unstats.un.org/sdgs/report/2020/The-Sustainable-Development-Goals-Report-2020.pdf
2. Kumar S, Sharma D, Rao S, Lim WM, Mangla SK (2022) Past, present, and future of sustainable finance: insights from big data analytics through machine learning of scholarly research. Ann Oper Res 1–44
3. World Economic Forum (2022) What is sustainable finance, and how is it changing the world? https://www.weforum.org/agenda/2022/01/what-is-sustainable-finance/
4. Nicholls A (2021) Sustainable finance: a primer and recent developments. Asian Dev Outlook 1–51
5. Bhatia M (n.d.) Sustainable development goals and green financing. https://www.thesustainability.io/sustainable-development-goals-green-financing
6. Ferris SP, Rykaczewski KP (1986) Social investing and portfolio management. Bus Soc 25(1):1–7
7. Kashi A, Shah ME (2023) Bibliometric review on sustainable finance. Sustainability 15:7119. https://doi.org/10.3390/su15097119
8. Rodriguez-Rojas MP, Clemente-Almendros JA, El Zein SA, Seguí-Amortegui L (2022) Taxonomy and tendencies in sustainable finance: A comprehensive literature analysis. Front Environ Sci 10:940526. https://doi.org/10.3389/fenvs.2022.940526
9. De Bellis N (2009) Bibliometrics and citation analysis: from the science citation index to cybermetrics. Scarecrow Press
10. Priyan PK, Nyabakora WI, Rwezimula G (2023) A bibliometric review of the knowledge base on capital structure decisions. Vision 27(2):155–166

11. Dewey A, Drahota A (2016) Introduction to systematic reviews: online learning module *Cochrane Training*. https://training.cochrane.org/interactivelearning/module-1-introduction-conducting-systematic-reviews
12. Paul J, Lim WM, O'Cass A, Hao AW, Bresciani S (2021) Scientific procedures and rationales for systematic literature reviews (SPAR-4-SLR). Int J Consum Stud 45(4):O1–O16
13. Aria M, Cuccurullo C (2017) Bibliometrix: an R-tool for comprehensive science mapping analysis. J Informet 11(4):959–975
14. Mongeon P, Paul-Hus A (2016) The journal coverage of Web of Science and Scopus: a comparative analysis. Scientometrics 106:213–228
15. Park J (2009) Sustainable consumption and the financial sector: analysing the markets for responsible investment in Hong Kong and Japan. Int J Consum Stud 33(2):206–214
16. Durisin B, Puzone F (2009) Maturation of corporate governance research, 1993–2007: an assessment. Corp Governance Int Rev 17(3):266–291
17. Flammer C (2021) Corporate green bonds. J Financ Econ 142(2):499–516
18. Yip AW, Bocken NM (2018) Sustainable business model archetypes for the banking industry. J Clean Prod 174:150–169
19. Pham L, Huynh TLD (2020) How does investor attention influence the green bond market? Financ Res Lett 35:101533. https://doi.org/10.1016/j.frl.2020.101533
20. Fatemi AM, Fooladi IJ (2013) Sustainable finance: a new paradigm. Glob Financ J 24(2):101–113
21. Azhgaliyeva D, Kapoor A, Liu Y (2020) Green bonds for financing renewable energy and energy efficiency in South-East Asia: a review of policies. J Sustain Finan Investment 10(2):113–140
22. Maltais A, Nykvist B (2020) Understanding the role of green bonds in advancing sustainability. J Sustain Finan Investment 1–20
23. Lee JW (2020) Green finance and sustainable development goals: the case of China. J Asian Finan Econ Bus 7(7):577–586
24. Bento N, Gianfrate G, Thoni MH (2019) Crowdfunding for sustainability ventures. J Clean Prod 237:117751
25. Roncoroni A, Battiston S, Escobar-Farfán LO, Martinez-Jaramillo S (2021) Climate risk and financial stability in the network of banks and investment funds. J Financ Stab 54:100870
26. Chenet H, Josh RC, Lerven FV (2021) Finance, climate-change, and radical uncertainty: towards a precautionary approach to financial policy. Ecol Econ 183:106957. https://doi.org/10.1016/j.ecolecon.2021.106957
27. Cunha FAFDS, Meira E, Orsato RJ (2021) Sustainable finance and investment: review and research agenda. Bus Strateg Environ 30(8):3821–3838
28. Schorr AE (1975) Lotka's law and map librarianship. J Am Soc Inf Sci 26(3):189–190
29. Donthu N, Kumar S, Mukherjee D, Pandey N, Lim WM (2021) How to conduct a bibliometric analysis: an overview and guidelines. J Bus Res 133:285–296
30. Shaukat K, Alam TM, Hameed IA, Luo S, Li J, Aujla GK, Iqbal F (2020) A comprehensive dataset for bibliometric analysis of SARS and coronavirus impact on social sciences. Data Brief 14(33):106520. https://doi.org/10.1016/j.dib.2020.106520
31. Kessler M (1963) Bibliographic coupling between scientific papers. Am Doc 14(1):10–25. https://doi.org/10.1002/asi.5090140103
32. Weinberg B (1974) Bibliographic coupling: a review. Inf Storage Retrieval 10(5–6):189–196. https://doi.org/10.1016/0020-0271(74)90058-8
33. Van Eck NJ, Waltman L (2019) VOSviwer manual version 1.6.10. In: CWTS meaningful metrics
34. Díaz-Caro C, Crespo-Cebada E, Goenechea BE, Mirón Sanguino ÁS (2023) Trinomial: return-risk and sustainability: is sustainability valued by investors? A choice experiment for spanish investors applied to SDG 12. Risks 11(8):149. MDPI AG. Retrieved from https://doi.org/10.3390/risks11080149
35. Mazzuca M, Panzera E, Ruberto S (2023) The investor's participation in social impact bonds. Int Rev Econ Financ 86:349–363. https://doi.org/10.1016/j.iref.2023.03.017

36. Meira E, Cunha FAFDS, Orsato RJ, Miralles-Quirós MM, Miralles-Quirós JL (2023) The added value and differentiation among ESG investment strategies in stock markets. Bus Strateg Environ 32(4):1816–1834
37. Naeem MA, Karim S, Rabbani MR, Bashar A, Kumar S (2023) Current state and future directions of green and sustainable finance: a bibliometric analysis. Qual Res Financ Markets 15(4):608–629. https://doi.org/10.1108/QRFM-10-2021-0174
38. Nahar HS, Mohamad M, Abd Rahman NR (2023) Sustainable finance: exploring the frontiers of current research. J Account Finan Emerg Econ 9(2):1–22. https://doi.org/10.26710/jafee.v9i2.1985
39. Waygood S (2011) How do the capital markets undermine sustainable development? What can be done to correct this? J Sustain Finan Investment 1(1):81–87. https://doi.org/10.3763/jsfi.2010.0008
40. Bloxham E (2011) Corporate governance and sustainability: new and old models of thinking. J Sustain Finan Investment 1(1):77–80. https://doi.org/10.3763/jsfi.2010.0003
41. Bloxham E (2011) The knowledge gap between investors and companies. J Sustain Finan Investment 1(2):156–158. https://doi.org/10.1080/20430795.2011.582326
42. Kandrács C (2023) Financing a sustainable economy in Hungary, opportunities and challenges: decarbonisation, green transition, sustainable finance, central bank. Public Finance Quarterly = Pénzügyi Szemle 69(1):29–45. https://doi.org/10.35551/PFQ_2023_1_2
43. Bak C, Bhattacharya A, Edenhofer O, Knopf B (2017) Towards a comprehensive approach to climate policy, sustainable infrastructure, and finance. Economics 11(1):20170033
44. Fuest CS, Meier V (2023) Sustainable finance and climate change: wasteful but a political commitment device? J Environ Econ Manag 118:102795. https://doi.org/10.1016/j.jeem.2023.102795
45. Liyanage SIH, Netswera FG, Motsumi A (2021) Insights from EU policy framework in aligning sustainable finance for sustainable development in Africa and Asia. Int J Energy Econ Policy 11(1):459–470. https://doi.org/10.32479/ijeep.9865
46. Ozili PK (2022) Green finance research around the world: a review of literature. Int J Green Econ 16(1):56–75
47. Scott Cato M, Fletcher C (2020) Introducing sell-by dates for stranded assets: ensuring an orderly transition to a sustainable economy. J Sustain Finan Investment 10(4):335–348
48. Cheong C, Choi J (2020) Green bonds: a survey. J Deriv Quant Stud 선물연구 28(4):175–189. https://doi.org/10.1108/JDQS-09-2020-0024
49. Torricelli C, Pellati E (2023) Social bonds and the "social premium." J Econ Finan 47:600–619. https://doi.org/10.1007/s12197-023-09620-3
50. Zerbib OD (2022) A sustainable capital asset pricing model (S-CAPM): evidence from environmental integration and sin stock exclusion. In: Proceedings of Paris December 2020 finance meeting EUROFIDAI—ESSEC
51. Bosmans P, de Mariz F (2023) The blue bond market: a catalyst for ocean and water financing. J Risk Financ Manag 16(3):184. MDPI AG. Retrieved from https://doi.org/10.3390/jrfm16030184
52. Thompson BS (2022) Blue bonds for marine conservation and a sustainable ocean economy: status, trends, and insights from green bonds. Mar Policy 144:105219. https://doi.org/10.1016/j.marpol.2022.105219
53. Wiklund S (2023) Evaluating physical climate risk for equity funds with quantitative modelling—how exposed are sustainable funds? J Sustain Finan Investment 13(2):893–918. https://doi.org/10.1080/20430795.2021.1894901
54. Caldecott B (2022) Climate risk management (CRM) and how it relates to achieving alignment with climate outcomes (ACO). J Sustain Finan Investment 12(4):1167–1170. https://doi.org/10.1080/20430795.2020.1848142
55. Abbas AF, Jusoh A, Mas'od A, Alsharif AH, Ali J (2022) Bibliometric analysis of information sharing in social media. Cogent Bus Manag 9(1):2016556. https://doi.org/10.1080/23311975.2021.2016556

A Bibliometric Analysis on Green Finance for Sustainability

Girija Nandini, Patjoshi Pramod Kumar, Apeksha Sahay, Mishra Padma Charan, and Alka Samantaray

Abstract Regulators and the government are paying more attention to climate change and global warming problems and putting pressure on companies to adopt green innovation (GI) and green finance (GF). The importance of GI increased as traditional methods and technologies was not able to achieve sustainable goals. GI promotes sustainability through the adoption of new technology. GF allocates funds for green projects and for goods and services that are environment-friendly. Sustainable development can be achieved if more financing is done for sustainable projects and the development of new and advanced technology like artificial intelligence and blockchain. GI and GF can support the organizations to design products that will reduce wastage and environmental impact while preserving economic feasibility. This paper thoroughly explains how GI and GF are important for sustainable development. Data are taken from an academic database, i.e., SCOPUS, and VOSviewer software is used for network analysis. We have taken the leading, impactful authors and the articles that are cited most in the area of GI and GF. Also, leading, and impactful Co-Authorship analysis, major Co-citations of the author's analysis, and important Co-occurrence of authors' keywords analysis have been taken for the network analysis. The result of this study will be helpful for the government, researchers, entrepreneurs, and investors to understand how the use of GI and GF will be helpful for sustainable development. A collaboration between academia, industries, and government is required to achieve sustainable development. The research will influence budding green research to generate value for corporations and the sustainable development of society at large.

G. Nandini · P. Pramod Kumar · A. Sahay
School of Management, Centurion University of Technology and Management, R.Sitapur, Odisha, India

M. Padma Charan (✉)
School of Commerce and Management Studies, Central University of Odisha, Koraput, Odisha, India
e-mail: pcmishra@cuo.ac.in

A. Samantaray
Institute of Business and Computer Studies, Siksha O Anusandhan (Deemed to Be University), Bhubaneswar, India

S. Roy et al. (eds.), *Humanized Intelligence in Sustainable Finance and Economic Behaviours*, Studies in Computational Intelligence 1226,
https://doi.org/10.1007/978-981-95-0599-9_3

Keywords Green finance · Green innovation · Sustainable development · VOSviewer · Green investment · Bibliometric analysis · Network analysis

1 Introduction

Creating a low-carbon economy is a major concern for countries worldwide. Regulators and the government are paying more attention to reducing the problem. Some of the countries accelerated their efforts for the growth of green industries and GI to reduce the global temperature [1]. The increasing pollution in the world requires GI and GF for sustainable development. The major concern for the industries to adopt GI and GF is the limitations of finance. Problems in getting financial resources disrupt the investment in the GI [2]. The investment required is high, the return is low and the uncertainty is more in the projects related to the GI of industries. The regulations related to GF and loans that can support investment in green innovation are important for the economy. The main objective of the regulating authority is to create an environment, which can support green growth [3]. Sustainability is a regulating concept that directs social connections with the environment and highlights responsibility for preserving it for the generations coming in the future [4]. The search for sustainable development is vital for achieving future outcomes [5]. Additionally, the limitations of finance are a major constraint, even though GF policies are there to encourage GI. GF generates prospects to permit diverse agencies to adopt environmental responsibility. For adopting environmental responsibility, the industries require sufficient financial support which can produce more renewable energy to achieve sustainable development goals. Due to globalization and fast change, companies must optimize the use of resources, and minimize waste to sustain in the competitive world [6]. For achieving sustainability in the organization, a circular economy business model is highly essential [7]. The major barriers to adopting a circular economy are the shortage of the latest technology and sufficient funds [8]. The company that wants to adopt circular practices has to adopt GI and GF [9, 10]. As the global population increases, the environment and climate are degrading, and GF and GI getting more attention for sustainable development [11]. Green finance is highly required for the existence of a low-carbon economy [12]. Many of the sustainable projects have high risk, low return, and mostly untested records. Support of GF can reduce the risk of the companies, that are trying to adopt green innovation [13]. Green finance can support the development of products that use new technology to achieve sustainable goals [14]. Green product innovation is getting more attention from regulators, organizations, and also by society at large. Due to the need, the research has been increased in this area in recent years. Many of the factors like top management assurance, increasing knowledge of green products, increased sales, high profit, increased exports, and cost savings are important for the successful development of green product innovations [15]. Companies are facing problems in getting funds from commercial banks for low-carbon projects. Additional policies should be implemented by the regulators specifically for green projects to enhance investment in such projects [16]. Companies should adopt green innovation practices for sustainable development [17]. Advanced digitalization in the country can develop sustainability in the urban areas with the support of green innovation [18]. The result of this study will be helpful for the government, researchers, entrepreneurs, and investors to understand how the use of GI and GF will be helpful for sustainable development.

2 Facets of GI and GF

Green innovations promote sustainability through the adoption of new technology [19]. With the use of green technologies and green innovations, energy efficiency can be maximized and greenhouse gas releases can be minimized [20]. The importance of GI increased as traditional methods and technologies were not able to achieve sustainable goals [21]. Ecological pollutions give importance to the requirement for GI in the economy. GF allocates funds to the research and development of green projects; it harmonizes with the GI for ecological safety [22]. The status of GI increased due to the need for sustainability and the reduction of human actions on the atmosphere [23]. When the financing is done by the investors for sustainable projects, it is known as green finance. The sustainable projects may be for renewable energy, sustainable transportation, or energy efficiency [24]. By developing new products and processes, GI will support sustainable development [25]. GI can create new jobs and new markets by developing new business strategies and promoting sustainability [26]. GI and GF are highly essential for accomplishing sustainable growth and reducing the problems of climate change. Sustainable development can be achieved if more financing is done for sustainable projects, and the development of new and advanced technology like artificial intelligence, blockchain, and green products. GI and GF can support organizations to design products that will reduce wastage and environmental impact while preserving economic feasibility [27]. GI and GF play an important role in getting a sustainable future. When the financial resources are allotted for sustainable, environmentally friendly projects, it is known as green finance. These sustainable projects may be related to sustainable transportation, renewable energy, or energy efficiency [28]. Environmentally sustainable projects are promoted when the capital flows to green projects [29]. Traditional finance is not sufficient for achieving sustainable goals, it requires green finance to achieve the goal [16]. Green finance plays a major role in achieving sustainable development by combining financial investments with the benefits of the environment [30]. The importance of green finance tools increased in past years to achieve sustainable goals. Most of the countries are in support of green finance for achieving sustainability [31]. Green finance is mainly focused on the investment of funds in green projects for achieving sustainable goals [13, 32]. It comprises different financial instruments for sustainable projects, which include green bonds and green insurance, and the use of blockchain technology for effective monitoring and control [33], as per the Sustainable Development Goals [34, 35]. As the ecological decline challenges and the decline of natural resources the need for green finance and green innovation highly increased. Green finance increases the production of green products leading to sustainable development. The establishment of green finance rules by the regulators will increase green finance and green projects [36].

3 Methodology

Bibliometric analysis is the statistical analysis of articles, books, book chapters, and other publications which supports researchers to explore more regarding the trends, status, and growth of the research in a particular field. Bibliometric analysis is done for this study. Data are taken from the SCOPUS database, and VOS viewer software is used for network analysis. We have taken the leading, impactful authors and the articles that are cited most in the area of GI and GF. Also, leading, and impactful Co-Authorship analysis, major Co-citations of the authors' analysis, and also important Co-occurrence of authors' keywords analysis has been taken for the network analysis. This study is mainly focused on 5 research questions. Data was collected to answer the research questions developed for the study.

4 Analysis and Discussion

The Scopus database is taken for the analysis of the research. 2341 samples have been taken for the research on the area of GI and GF. Table 1 explains the process of article selection and search criteria for the research. By using the keywords GI and GF, 2341 publications were selected from the Scopus database. By choosing the subject area, 450 articles were rejected and 1891 were accepted. After that, in the next step by taking only articles, and conference papers, 345 were rejected and 1546 were accepted. After language and content screening, 568 articles were taken for the analysis.

The research questions for the bibliometric analysis are:

1. What is the publication trend in the area of GI and GF?
2. Who are the leading and influential authors in the area of GI and GF?
3. Which are the important, and dynamic Co-Authorship?
4. What are the major Co-citations of the authors' analysis?
5. Which is the important Co-occurrence of authors' keywords?

4.1 Publication Trend Analysis

Figure 1 shows the existing movements of publication in the area of GI and GF. It is found from Table 2 and Fig. 2 that the publication in the area of GI and GF increased

Table 1 Article collection and exploration

Cleaning conditions	Reject	Accept
Cleaning conditions		
Cleaning database—Scopus		
Exploration date: 20 Dec 2021		
Search term: ("Green finance" OR "Green innovation")		2341
Subject area: "Business, management and accounting", "Social sciences"	450	1891
File nature: "Articles", "Conference papers"	345	1546
Publication selection		
Language selection: English only	220	1326
Year (2018–2024)	758	568
Content screening	0	568

from the year 2018 to 2024. The publication has increased from 28 in the year 2018 to 140 in the year of 2024.

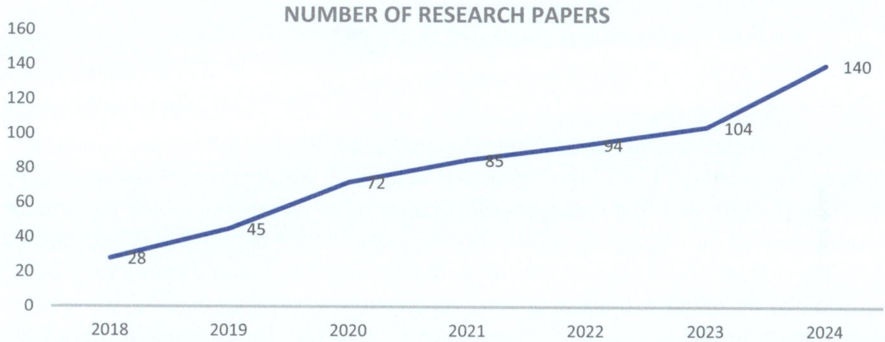

Fig. 1 Publication trend

Table 2 Existing movements of publication in the Area of GI and GF

Year of publication	No. of research papers
2018	28
2019	45
2020	72
2021	85
2022	94
2023	104
2024	140

Source Authors' calculation with the use of VOSviewer

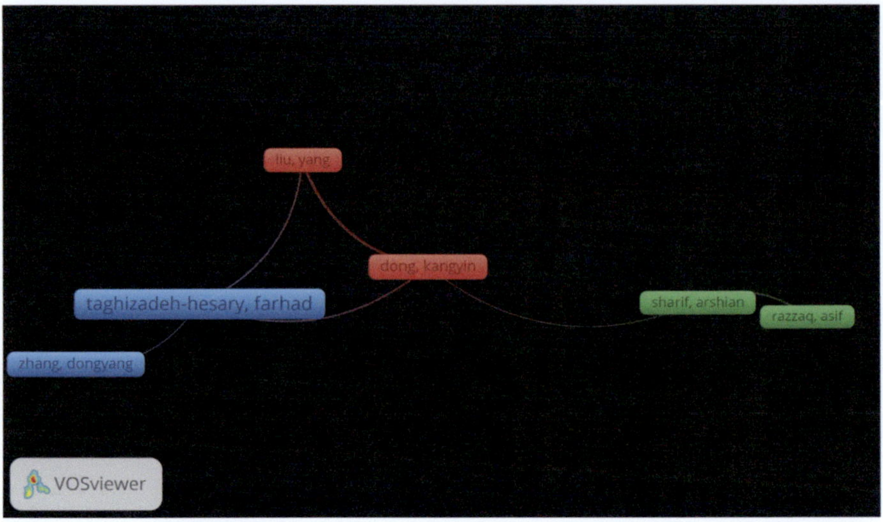

Fig. 2 Co-authorship network

4.2 Leading, Influential Authors Analysis

Table 3 discloses the top 5 Articles in the area of GI and GF. The top article discovered that organizations are facing problems regarding environmental issues. Companies realized that GI and GF are highly required to achieve sustainable development. The researchers suggested that managers must create an environment for the sustainable development of the organization [19]. Private participation is required for GF and GI [37]. The researcher developed a theoretical framework to understand green product innovations by organizations to achieve sustainable development [38]. Knowledge management in the area of green innovation is important for all sizes of firms [39]. The interaction amid external legitimacy pressure and internal profit of the organization positively affects green innovation leading to more innovative and sustainable development [40].

4.3 Co-authorship Network Analysis

Collaborative research forms between countries, authors, and institutions are known as co-authorship analysis in the bibliometric analysis. It supports identifying the researchers, collaboration patterns, and research networks in a particular area. Co-authorship of the authors explains about the authors who are strong and also who are weak in collaboration with other authors in a field of research.

As per our analysis, 3 clusters were formed as given in Table 4. In cluster 1 (Red), authors Dong and Kangyin did the highest work together with a weight document

Table 3 Leading, influential authors in the area of GI and GF

#	Authors	Title	Year	Cited by
1	Song W.; Yu H	Green Innovation Strategy and Green Innovation: The Roles of Green Creativity and Green Organizational Identity	2018	508
2	Taghizadeh-Hesary F.; Yoshino N	The way to induce private participation in green finance and investment	2019	504
3	Dangelico R.M.; Pujari D.; Pontrandolfo P	Green Product Innovation in Manufacturing Firms: A Sustainability-Oriented Dynamic Capability Perspective	2017	502
4	Abbas J.; Sağsan M	Impact of knowledge management practices on green innovation and corporate sustainable development: A structural analysis	2019	487
5	Li D.; Zheng M.; Cao C.; Chen X.; Ren S.; Huang M	The impact of legitimacy pressure and corporate profitability on green innovation: Evidence from China top 100	2017	453

Source Authors' Calculation from Scopus Database

of 7 and a weight total link strength of 6. In cluster 2 (green), authors Sharif and Arshian did the highest work together with a weight document of 6 and a weight total link strength of 3. In cluster 3 (blue), authors Taghizadeh-Hesary and Farhad did the highest work together with a weight document of 13 and a weight total link strength of 5.

Table 4 Co-authorship analysis

Label	Cluster	Weight < links >	Weight < total link strength >	weight < documents >
Dong, Kangyin	1	3	6	7
Liu, Yang	1	2	5	6
Razzaq, Asif	2	1	2	5
Sharif, Arshian	2	2	3	6
Taghizadeh-Hesary, Farhad	3	3	5	13
Zhang, Dongyang	3	1	1	7

Source Authors' calculation from Scopus database

Table 5 Co-citations of authors' analysis

Label	Cluster	Weight < links >	Weight < total Link strength >	weight < citations >
Wang y	1	999	119,460	1179
Zhang y	1	999	85,207	847
Li y	1	999	79,277	770
Liu y	1	999	69,614	664
Umar m	2	995	67,334	577
Wang x	1	999	67,112	660
Zhang j	1	999	65,208	611
Wang j	1	999	60,682	573
Taghizadeh-Hesary f	2	990	59,382	650
Li j	1	999	59,194	556

Source Authors' calculation from Scopus Database

4.4 Co-citations of Authors' Networks Analysis

If the authors of two publications obtain a reference from a similar author in a third article, it is identified as a co-citation network of authors [41]. Co-citations mutually control the number of papers quoted [42]. We have taken 25 as the minimum no. of citations of an author for this analysis.

Table 5 contains the supporting numerical information for this analysis. We have displayed only the top 10 co-citations based on the weight and total link strength. Based on our experience, four major clusters were identified, as shown in Fig. 3. Cluster 1 (red) consisted of 420 cited authors, cluster 2 (green) consisted of 314 cited authors, cluster 3 (Blue) and cluster 4 (yellow) consisted of 236 cited authors and 33 cited authors, respectively. Cluster 1 (red) has the highest co-citations and the topmost wang y. is having the highest weight citations of 1179, and the weight total link strength of 119,460 followed by Zhang y. 85,207. Cluster 2 (green) Umar m. has the highest weight total link strength of 67,334 followed by Taghizadeh-Hesary f. with a weight total link strength of 59,382. Cluster 3 (Blue) Abbas j. has the highest weight total link strength of 33,688 followed by Hair j. f. with the weight total link strength of 16,963. Cluster 4 (yellow) Hao y. has the highest weight total link strength of 19,723 followed by Tang k. with the weight total link strength of 10,198.

4.5 Analysis Based on Keywords

The co-occurrence of keywords is a type of analysis used for the bibliometric analysis. The network analysis for the study is done with the help of the VOSviewer.

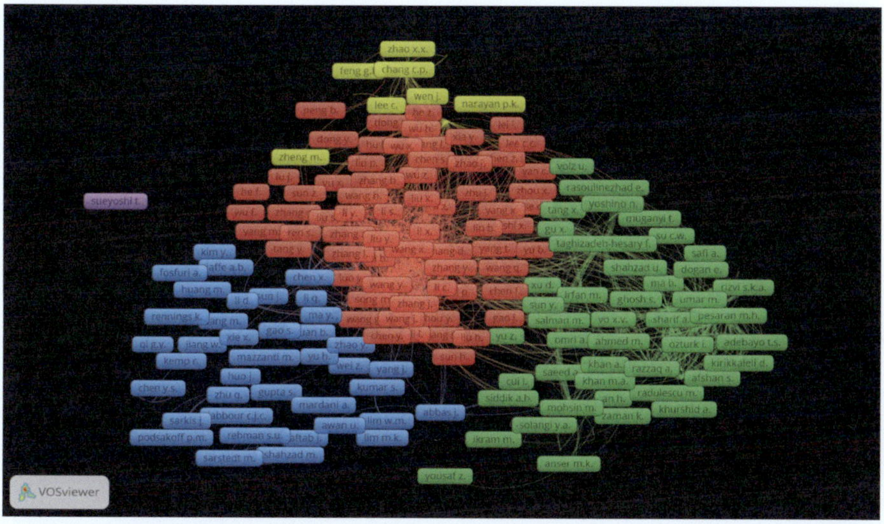

Fig. 3 Co-citations of authors' networks. *Source* Authors' Calculation from Scopus Database

Table 6 gives the 10 most used keywords for the study in GI and GF. The top three keywords used for the study are sustainable development, innovation, and green economy. Figure 4 shows the network analysis of the most used keywords. It is found that in cluster 1 (Red), "sustainable development", "innovation", "green economy" "green innovation", and "environmental economics" are having highest occurrences. Cluster 2 (Green) "Investments" and "Economic growth" are having highest occurrences. Cluster 2 (Blue) "green finance", "Sustainability", and "Finance" are having highest occurrences.

5 Conclusions

Economic sustainability can be possible through the development of the green industry. This article mainly focused on the importance of GI and GF for sustainable development and how the research has been growing in this area in recent years. Data are taken from an academic database, i.e., SCOPUS, and VOSviewer software is used for network analysis. We have taken the leading, impactful authors and the articles that are cited most in the area of GI and GF. Also, leading, and impactful Co-authorship analysis, major Co-citations of the author's analysis, and important Co-occurrence of authors' keywords analysis have been taken for the network analysis. This study is mainly focused on 5 research questions. Data was collected to answer the research questions developed for the study. The publication trend in green finance and green innovation has increased for the research period 2018 to 2024. Additionally, the result of this study will be helpful for the government, researchers, entrepreneurs,

Table 6 Co-occurrence of anuthors' keywords

Label	Cluster	Weight < links >	Weight < total link strength >	Weight < occurrences >
Sustainable development	1	95	5683	973
Innovation	1	95	2825	427
Green innovation	1	95	2066	415
Green finance	3	94	1843	328
Green economy	1	95	2281	311
Sustainability	3	94	2151	296
Environmental economics	1	94	1679	197
Finance	3	94	1402	168
Investments	2	90	1291	147
Economic growth	2	90	1118	123

Source Authors' calculation from Scopus Database

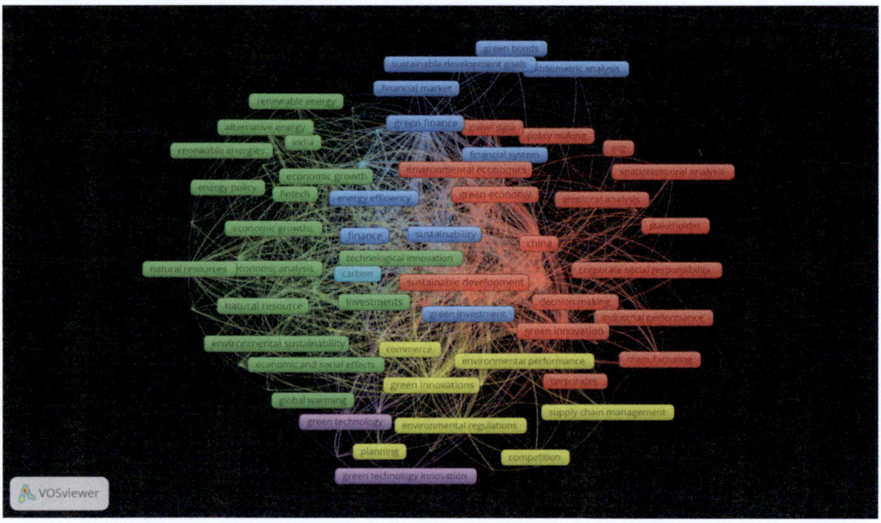

Fig. 4 Co-occurrence of anuthor's keywords

and investors to understand how the use of GF and GI will be helpful for sustainable development. There are also a few limitations of the research. Such as only the publications are taken from the Scopus database. Scopus data do not cover all the pertinent literature; thus, the information taken for the research is not inclusive. Other databases like Google Scholar and WOS can be taken for inclusive research. Other methodologies like SLR, and other network analyses like bibliometric coupling, citations have not been taken for the research.

5.1 *Implications*

The government should invest more in the research and development of the green sector, more capital expenditure in green projects, and encourage foreign direct investment in green innovation. A collaboration between academia, industries, and government is required to achieve sustainable development. The research will influence budding green research to generate value for corporations and the sustainable development of society at large. The results deliver some real-world suggestions for companies, regulators, bankers, investors, government officials, and investors. The government should do a literacy program and provide information to the stakeholders on green finance and green innovation. Financial institutions and banks should give loans for green projects to companies having high creditworthiness to reduce the credit risk associated with green finance. The bankers, financiers, and investors should take appropriate training on the procedure of green technology. The investors mainly make decisions based on risk and return, they have to take in their investment.

5.2 *Scope for Future Research*

Further research can be done on the perspective of risk and return of the investors while investing in green projects. The analysis of investors' awareness about green projects in different countries can be an interesting research work. Also, more future studies can be done on the different policies adopted by different countries and states for GI and GF. Moreover, the research may discover the growth of GF and GI with the use of artificial intelligence and blockchain analysis.

References

1. Yang YC, Nie PY, Huang JB (2020) The optimal strategies for clean technology to advance green transition. Sci Total Environ 716:134439. https://doi.org/10.1016/J.SCITOTENV.2019.134439
2. Andersen DC (2017) Do credit constraints favor dirty production? Theory and plant-level evidence. J Environ Econ Manag 84:189–208. https://doi.org/10.1016/J.JEEM.2017.04.002
3. Wang Y, Zhi Q (2016) The role of green finance in environmental protection: two aspects of market mechanism and policies. Energy Procedia 104:311–316. https://doi.org/10.1016/J.EGYPRO.2016.12.053
4. Bekuma T, Mamo G, Regassa A (2023) Indigenous and improved adaptation technologies in response to climate change adaptation and barriers among smallholder farmers in the East

Wollega Zone of Oromia, Ethiopia. Res Globalization 6:100110. https://doi.org/10.1016/j.resglo.2022.100110
5. Rahman MH, Rahman J, Tanchangya T, Esquivias MA (2023) Green banking initiatives and sustainability: a comparative analysis between Bangladesh and India. Res Globalization 7:100184. https://doi.org/10.1016/j.resglo.2023.100184
6. Moore SB, Manring SL (2009) Strategy development in small and medium sized enterprises for sustainability and increased value creation. J Clean Prod 17(2):276–282. https://doi.org/10.1016/j.jclepro.2008.06.004
7. Pieroni MPP, McAloone TC, Pigosso DCA (2019) Business model innovation for circular economy and sustainability: a review of approaches. J Clean Prod 215:198–216. https://doi.org/10.1016/j.jclepro.2019.01.036
8. Gedam VV, Raut RD, de Sousa L, Jabbour AB, Tanksale AN, Narkhede BE (2021) Circular economy practices in a developing economy: Barriers to be defeated. J Clean Prod 311:127670. https://doi.org/10.1016/j.jclepro.2021.127670
9. Triguero Á, Cuerva MC, Sáez-Martínez FJ (2022) Closing the loop through eco-innovation by European firms: circular economy for sustainable development. Bus Strateg Environ 31(5):2337–2350. https://doi.org/10.1002/bse.3024
10. Acquah ISK, Baah C, Agyabeng-Mensah Y, Afum E (2023) Green procurement and green innovation for green organizational legitimacy and access to green finance: the mediating role of total quality management. Glob Bus Organ Excell 42(3):24–41. https://doi.org/10.1002/joe.22183
11. Vaka M, Walvekar R, Rasheed AK, Khalid M (2020) A review on Malaysia's solar energy pathway towards carbon-neutral Malaysia beyond Covid'19 pandemic. J Cleaner Prod 273:122834. https://doi.org/10.1016/j.jclepro.2020.122834
12. Dorfleitner G, Braun D (2019) Fintech, digitalization and blockchain: possible applications for green finance. In: Migliorelli M, Dessertine P (eds) The rise of green finance in Europe. Springer Nature, pp 207–237. https://doi.org/10.1007/978-3-030-22510-0_9
13. Bhatnagar M, Taneja S, Özen E (2022) A wave of green start-ups in India—the study of green finance as a support system for sustainable entrepreneurship. Green Finan 4(2):253–273. https://doi.org/10.3934/GF.2022012
14. Sadiq M, Amayri MA, Paramaiah C, Mai NH, Ngo TQ, Phan TTH (2022) Retracted article: how green finance and financial development promote green economic growth: deployment of clean energy sources in South Asia. Environ Sci Pollut Res 29(43):65521–65534. https://doi.org/10.1007/s11356-022-19947-9
15. Dangelico RM (2016) Green product innovation: where we are and where we are going. Bus Strateg Environ 25(8):560–576. https://doi.org/10.1002/bse.1886
16. Campiglio E (2016) Beyond carbon pricing: the role of banking and monetary policy in financing the transition to a low-carbon economy. Ecol Econ 121:220–230. https://doi.org/10.1016/j.ecolecon.2015.03.020
17. Weng HH, Chen JS, Chen PC (2015) Effects of green innovation on environmental and corporate performance: a stakeholder perspective. Sustainability 7(5):4997–5026. https://doi.org/10.3390/su7054997
18. Luo S, Yimamu N, Li Y, Wu H, Irfan M, Hao Y (2023) Digitalization and sustainable development: how could digital economy development improve green innovation in China? Bus Strateg Environ 32(4):1847–1871. https://doi.org/10.1002/bse.3223
19. Song W, Yu H (2018) Green innovation strategy and green innovation: the roles of green creativity and green organizational identity. Corp Soc Responsib Environ Manag 25(2):135–150. https://doi.org/10.1002/csr.1445
20. Gao D, Li Y, Li G (2022) Boosting the green total factor energy efficiency in urban China: does low-carbon city policy matter? Environ Sci Pollut Res 29(37):56341–56356. https://doi.org/10.1007/s11356-022-19553-9
21. Rennings K (2000) Redefining innovation—eco-innovation research and the contribution from ecological economics. Ecol Econ 32(2):319–332. https://doi.org/10.1016/S0921-8009(99)00112-3

22. Agrawal R, Agrawal S, Samadhiya A, Kumar A, Luthra S, Jain V (2024) Adoption of green finance and green innovation for achieving circularity: An exploratory review and future directions. Geosci Front 15(4):101669. https://doi.org/10.1016/j.gsf.2023.101669
23. Ibrahim RL, Huang Y, Mohammed A, Adebayo TS (2023) Natural resources-sustainable environment conflicts amidst COP26 resolutions: investigating the role of renewable energy, technology innovations, green finance, and structural change. Int J Sust Dev World 30(4):445–457. https://doi.org/10.1080/13504509.2022.2162147
24. Soundarrajan P, Vivek N (2016) Green finance for sustainable green economic growth in India. Agric Econ (Zemeˇdeˇlská Ekonomika) 62(1):35–44. https://doi.org/10.17221/174/2014-AGRICECON
25. Jesic J, Okanovic A, Panic AA (2021) Net zero 2050 as an EU priroty: modeling a system for efficient investments in eco innovation for climate change mitigation. Energy Sustain Soc 11:1–16. https://doi.org/10.1186/s13705-021-00326-0
26. Hsu CC, Tan KC, Mohamad Zailani SH (2016) Strategic orientations, sustainable supply chain initiatives, and reverse logistics: empirical evidence from an emerging market. Int J Oper Prod Manag 36(1):86–110. https://doi.org/10.1108/IJOPM-06-2014-0252
27. Geissdoerfer M, Morioka SN, de Carvalho MM, Evans S (2018) Business models and supply chains for the circular economy. J Clean Prod 190:712–721. https://doi.org/10.1016/j.jclepro.2018.04.159
28. Liu L, Peng Q (2022) Evolutionary game analysis of enterprise green innovation and green financing in platform supply chain. Sustainability 14(13):7807. https://doi.org/10.3390/su14137807
29. Richardson BJ (2009) Climate finance and its governance: moving to a low carbon economy through socially responsible financing? Int Comp Law Q 58(3):597–626. https://doi.org/10.1017/S0020589309001213
30. Yadav N, Pathania A, Ramesh KM. Green finance and its role in achieving sustainable development goals: a bibliometric analysis. FOCUS WTO 26(2):64–85. http://publication.iift.ac.in/focuswto_archive.asp
31. Maria MR, Ballini R, Souza RF (2023) Evolution of green finance: a bibliometric analysis through complex networks and machine learning. Sustainability 15(2):967. https://doi.org/10.3390/su15020967
32. Hafner S, Jones A, Anger-Kraavi A, Pohl J (2020) Closing the green finance gap—a systems perspective. Environ Innov Soc Trans 34:26–60. https://doi.org/10.1016/j.eist.2019.11.007
33. Kaur G, Aadheesh (2023) Application of blockchain for sustaining green finance. In: Trivedi S, Aggarwal R, Singh G (eds) Perspectives on blockchain technology and responsible investing. IGI Global, pp 226–235. https://doi.org/10.4018/978-1-6684-8361-9.ch011
34. Verma D, Kalra R, Baheti SS (2023) Examining the domain of green finance through bibliometric research analysis of 22 years (2000–2022): an analytical retrospective. Vis J Bus Perspect. https://doi.org/10.1177/09722629231157470
35. Sengupta R, Patil AA (2023) Green finance products and investments in the changing business world. In: Goel R, Baral SK (eds) Advances in business strategy and competitive advantage. IGI Global, pp 344–357. https://doi.org/10.4018/978-1-6684-8969-7.ch020
36. Jiakui C, Abbas J, Najam H, Liu J, Abbas J (2023) Green technological innovation, green finance, and financial development and their role in green total factor productivity: empirical insights from China. J Clean Prod 382:135131. https://doi.org/10.1016/j.jclepro.2022.135131
37. Taghizadeh-Hesary F, Yoshino N (2019) The way to induce private participation in green finance and investment. Financ Res Lett 31:98–103. https://doi.org/10.1016/j.frl.2019.04.016
38. Dangelico RM, Pujari D, Pontrandolfo P (2017) Green product innovation in manufacturing firms: a sustainability-oriented dynamic capability perspective. Bus Strateg Environ 26(4):490–506. https://doi.org/10.1002/bse.1932
39. Abbas J, Sağsan M (2019) Impact of knowledge management practices on green innovation and corporate sustainable development: a structural analysis. J Clean Prod 229:611–620. https://doi.org/10.1016/j.jclepro.2019.05.024

40. Li D, Zheng M, Cao C, Chen X, Ren S, Huang M (2017) The impact of legitimacy pressure and corporate profitability on green innovation: evidence from China top 100. J Clean Prod 141:41–49. https://doi.org/10.1016/j.jclepro.2016.08.123
41. Baber H, Fanea-Ivanovici M (2024) Fifteen years of crowdfunding–a bibliometric analysis. Tech Anal Strateg Manag 36(6):1248–1262. https://doi.org/10.1080/09537325.2022.2089548
42. Galletta S, Mazzù S, Naciti V (2022) A bibliometric analysis of ESG performance in the banking industry: From the current status to future directions. Res Int Bus Financ 62:101684. https://doi.org/10.1016/j.ribaf.2022.101684

Empowering MSME Sector: Unleashing Green Finance

Swati Srivastava, Mayank Kumar, Pallavi Thacker, Pranay Raj Rathore, Palash Bairagi, and Samarjit Roy

Abstract Green finance plays a pivotal role in empowering India's micro, small, and medium enterprises by supporting their transition toward sustainability, reducing their ecological impact and enabling them to tap into expanding markets for eco-friendly products and services. With the facility of accessing the funds, high-tech know-how, and innovative marketplaces for green offerings, green finance fosters innovation and enables sustainable development. The benefits that green finance offers MSME are manifold, including enhanced access to capital for sustainable business ventures, cost savings through efficient resource, energy utilization, entry into emerging markets, improved corporate reputation, compliance with environmental standards, and opportunities for knowledge enhancement. With the support of green finance, MSME not only gain the capability to drive growth through sustainable practices and boost their profitability, but also contribute to a greener, more sustainable Indian economy. The collaboration between MSME and green finance is paving the way for the future sustainability of India's economy, demonstrating the potential for a prosperous, environmentally conscious business landscape.

S. Srivastava (✉)
School of Management, Babu Banarasi Das University, Lucknow, Uttar Pradesh, India
e-mail: swati.srivastava24@gmail.com

M. Kumar
Shri Ramswaroop Memorial College of Engineering and Management, Uttar Pradesh, Lucknow, India

P. Thacker
IMS University of Lucknow, Lucknow, Uttar Pradesh, India

P. R. Rathore
Indian Institute of Management, Nagpur (IIM-Nagpur), MIHAN, Nagpur, Maharashtra, India

P. Bairagi
Department of Business Management, Central University of Odisha, Koraput, Odisha, India
e-mail: pbairagi@cuo.ac.in

S. Roy
Department of Software Engineering, School of Computing and Information Technology, Eastern International University, Becamex IDC Corp, Thu Dau Mot, Binh Duong Province, Vietnam

S. Roy et al. (eds.), *Humanized Intelligence in Sustainable Finance and Economic Behaviours*, Studies in Computational Intelligence 1226,
https://doi.org/10.1007/978-981-95-0599-9_4

Keywords Green finance · MSME · Sustainable growth · Economy · Green initiatives

1 Introduction

Green finance serves as a vital key for driving the development of micro, small, and medium enterprises in India through their sustainable transition and decreased environmental impact and market access to eco-friendly products and services. The MSME sector receives financial backing alongside technical support to enter green markets which results in innovation and helps India achieve its sustainability goals.

Through green finance adoption, MSME create sustainable growth patterns which lead to improved financial results and enable them to actively participate in building an environmentally conscious and resilient Indian economy. The sustainability of the country and its future depend on the fundamental of green finance and MSMEs' performance. The financial system of every country including India must adopt green practices to meet the zero emission targets. Within this process, organizations have duty to include environmental elements with the financial choices to support long-term growth. The combination of green bonds with carbon pricing and sustainable investment strategies is leading India toward a greener economy to achieve economic and ecological equilibrium [1]. The development of green finance in the Indian economy is critical to achieving net-zero emissions. This can be achieved only through green financial resources in order to support environmental projects and initiatives. MSMEs take advantages from green finance by acquisition of better financial resources form sustainable projects and practices, along with resource and energy efficiency, cost reductions, market expansion, and brand enhancement through environmental agreements and knowledge-based development opportunities [2].

These initiatives focus on the upgradation of energy efficiency [3], renewable energy projects [4], sustainable development in infrastructure sectors [5], and other initiatives focused on both the mitigation of climate change and its adaptation [6].

Further, a platform of green finance in India helps MSME access through development banks like Small Industries Development Bank of India (SIDBI), or through capital appreciation practices through green bonds. According to Institute for Energy Economics and Financial Analysis (IEEFA) reports, the India has the second highest number in the world after China, which has 63 million of MSMEs, which is about 30% of India's GDP. Similarly, 40% of its exports come from MSME, which provides jobs for approximately 110 million people [7].

Further, in 2023, during World MSME discussions, the theme "Empowering MSME financial innovation through sustainable development" was adopted. Underlining this key role of financial innovation that supports the sustainable growth of MSMEs brings green finance into the global spotlight, addressing toward environmental challenges and transformative potential of small enterprises [8]. Further, green finance helps MSMEs to analyze their carbon emission practices, use resources efficiently, and become more economical and resilient. These missions support the broader international sustainability in business operations, particularly for smaller firms that struggle to access traditional financial market. Hence, it can be said that "The world market of MSMEs proves to be an expression of support for the development of sustainable and environment-friendly [8]."

2 Motivation

The chapter focuses on the key role that green finance plays in promoting sustainability and environmental stewardship in India's micro, small, and medium enterprises (MSME). It acknowledges international drive toward sustainability and India's commitment to combating climate change and shifting to renewable energy. The chapter highlights the significance of enabling MSME through finance, technical know-how, and green finance advantages and opportunities. It seeks to present a wisdom of understanding on:

> How green finance facilitates MSME to adopt sustainability, mitigate environmental footprint, and access the expanding a market for greener products and services.

A discussion of principal advantages guides policy recommendations and promotes sustainability with sustainable development goals. The chapter also demonstrates green finance toward sustainable growth, economic performance, and resilient to Indian economy growth. Finally, a larger dialogues on the practical implementation and challenges provide a roadmap for business and stakeholders to maximize the transformative potential of green finance in MSME sectors.

3 Green Finance Contributions

Green finance provisions the MSME sectors by enhancing access to the funding, reducing operational costs, unlocking new market opportunities, and encouraging business credibility [24]. It additionally emphasizes the process of green finance to align with shifting in consumer expectations. It evolves policies that support sustainability. Significant initiatives, such as the green bonds, sustainable financial paradigms, and government-led programs, are depicted for demonstrating country's ongoing commitment toward sustainable development and climate changes. Furthermore, the chapter incorporates green finance to the societal Sustainable Development Goals (SDGs), majorly in areas, viz., clean energy production, sustainable infrastructure, and responsible power consumption. In addition, it depicts the broader economic impacts, like the reduction of carbon emissions, boosting energy security, creating employment opportunities, and enhancing environmental health along with the country's renewable energy schemas and sustainability goals. The proposed chapter terminates with underscoring the pivotal role of capacity building initiatives. These initiatives direct in raising awareness, construct technical and financial expertise, foster innovation, ensure regulatory compliance, and inspire collaboration empowering MSME to tap into the benefits of green finance for lifelong sustainable growth.

3.1 Empowering Green Finance in the MSME Sectors

In India, green finance emerges as a pivotal factor in bolstering the micro, small, and medium enterprises (MSME) sector, which plays an indispensable role in the nation's economy. MSME are recognized for their substantial contributions to employment generation, innovation, and overall economic growth [9]. Therefore, green finance initiatives not only facilitate sustainable practices but also hold the potential to empower and fortify the MSME sector, thus serving as a catalyst for India's economic development and resilience. This can be depicted in Fig. 1.

Green finance has a central role to play in enabling the MSME sector to adopt sustainability, minimize their environmental footprint, and capitalize on the opportunities arising from the increasing demand for green products and services [4]. The green finance empowers the MSME sector as:

1. **Access to Capital**—Green finance projects give MSME better access to capital specifically for funding green and sustainable projects. This promotes MSME investments in clean energy, energy efficiency, waste management, and other sustainable projects, in effect enabling them to overcome financial barriers and promoting the implementation of sustainable practices [4].

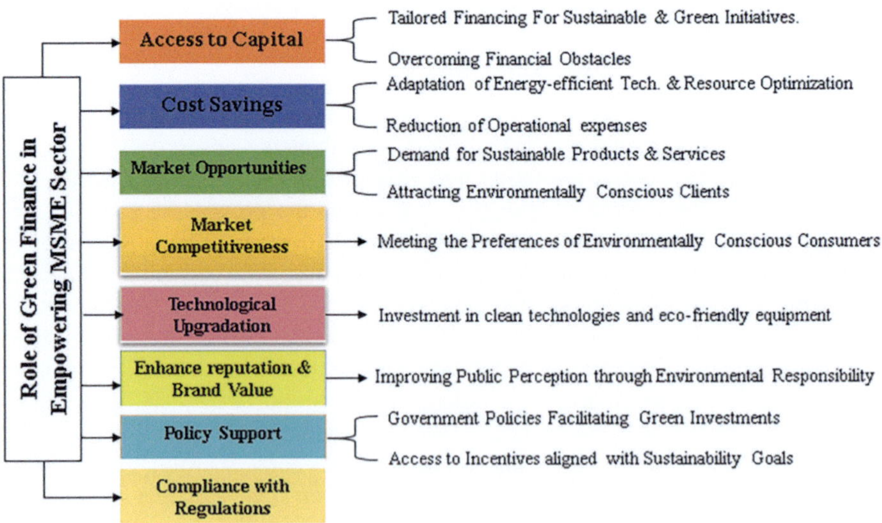

Fig. 1 Green finance and its role in empowering MSME sectors

2. **Cost Savings**—Green finance promotes the use of energy saving know-how, renewable energy systems, and efficient processes. This enables MSME to reduce energy consumption, increase waste minimization, and lower operating expenses. As a result, the company savings obtained from resource optimization and energy efficiency can be reinvested in expanding the business [3].
3. **Market Opportunities**—The growing need for green sustainable products and services offers a growing market segment. MSME that adapt green methods can serve the rising base of consumers, gain environmentally responsible clients, and differentiate distinguish themselves from their competitors. Green finance can support MSME investments in green production techniques, eco-certifications, and green supply chains, hence allowing them to leverage market opportunities and become more competitive [5]
4. **Market Competitiveness**—Consumers and businesses alike increasingly favorable sustainable and eco-friendly products. By adopting green practices, MSME are able to enhance their competitiveness and respond to the demands of this growing market segment. Green finance can assist MSME in adopting cleaner production practices, eco-certifications, and sustainable supply chain strategies, hence better positioning them to catch up with the changing tastes of environmentally friendly consumers.
5. **Technological Upgradation**—Green finance projects generally allocate financing for technological modernization and innovations for sustainability. MSME can avail themselves of such funds for investing in clean technology, environmentally friendly equipment, and process optimization systems. Such technological innovations can improve productivity, decrease environmental impacts, and improve business efficiency as a whole [27].
6. **Improved Reputation and Brand Value**—Elaborating green finance practices through the assistance of green finance can enrich MSME repute. Enthusiastic environmental concern and sustainability can boost the public image, get socially responsible investors, and construct stronger stakeholder relations. An upright brand image on sustainability can stimulate improved customer loyalty and trust, ultimately enhancing the growth and success of MSME.
7. **Policy Support**—The government organization has vigorously buoyed green finance and sustainability by way of several policy initiatives. The government has started social programs, such as the Green Climate Fund, Sustainable Finance paradigms, and Green Bonds to provision green investments. All policies prepare conveniently for MSME in gaining the access toward green finance, attract stewardships, and dovetail sustainable development intentions [9].
8. **Regulatory Compliance**—Green finance initiatives directly increase the compatibility with conventional environmental policies and sustainable behaviors. By accessing the green finance, MSME can be capable in ensuring regulatory compliance with environmental norms and laws, thereby avoiding associated risks which are converged with fines, law suits, and brand reputation loss owing to noncompliance [23]. Furthermore, the proactive implementation of sustainable green protocols assists the MSME standards themselves significantly for further regulatory adjustments under environmental sustainability aspects [33].

9. **Capacity Building**—Green financial products involve capacity building, including the capabilities and training in sustainable practices, green technology, and finance management. Training and workshop programs will facilitate how to effectively and efficiently use the green finance products and schemes [6].

3.2 Key Developments and Initiatives for Encouraging Green Finance

India is shining a spotlight as a nation that is committed in the direction of sustainable development and environmental consciousness through significant efforts taken toward green finance. The Securities and Exchange Board of India has played a pivotal role by introducing detailed regulations governing the issuance and listing of green bonds as well as establishing a structured framework to support green projects. Additionally, SEBI's involvement in the National Clean Energy Fund (NCEF) has been instrumental in advancing renewable energy initiatives across the nation. Also, the Central Electricity Regulatory Commission (CERC) runs Renewable Energy Certificates (RECs) creating a market-driven mechanism to monetize investments in renewable energy sources.

Financial organizations are united with the United Nations Sustainable Development Goals (SDGs) into their operations to abet the global sustainability. Through the incorporation, the India-Denmark Green Strategic Partnership has been emphasized by the India's green finance initiatives. The involvement of statutory organizations, viz., the NITI Aayog, in experimenting with carbon pricing demonstrates a strong dedication to promoting green finance and steering India toward a more environment-friendly and sustainable economic path [32].

A key milestone in advancing green finance in India is the development of green bonds. Green bonds are financial tools designed to raise funds specifically for eco-friendly projects [28].

The issuance of green bonds by Indian entities such as the National Thermal Power Corporation (NTPC) and the State Bank of India (SBI) serves as a notable demonstration of their commitment to embracing the principles of a green economy. The bonds finance projects for promoting renewable energy, energy efficiency, and sustainable infrastructure, demonstrating a clear linkage between financial investment and eco-friendly initiatives [10].

Besides green bonds, India has also witnessed the expansion of sustainable finance frameworks that align financial institutions with responsible investment and lending strategies. The Reserve Bank of India (RBI) launched the Reserve Bank of India, Sustainable Finance Framework in 2021, with a focus on integrating sustainability factors into banking business. A revolutionary shift toward sustainability can be achieved through this framework that is designed to compel banks to integrate environmental and social risk analysis in lending [11].

India has indeed made significant strides toward promoting green finance and sustainable development in the past decades. A chain of initiatives indicates that

India is committed toward climate change mitigation as well as transitioning to renewable power. The most crucial advancements and initiatives in promoting green finance play a pivotal role in increasing sustainability, growth and resilience.

They enable businesses, including MSME, to adopt environmentally friendly practices while levering emerging market opportunities and adhering to changing regulations. Principal initiatives in fostering green finance in India are detailed in Table 1.

These initiatives reflect India's acknowledgment of the immediate imperative to combat climate change and ensure sustainable development. India's proactive initiatives can address its commitment toward mitigating climate change effects and fostering sustainable development [25]. By prioritizing green finance, clean energy, and climate change mitigation strategies, India aimed to reduce greenhouse gas emissions, bolster energy security, and drive global efforts against climate change. These initiatives also reflect a growing emphasis on green finance and sustainability within the country, as stakeholders increasingly recognize the importance of providing targeted financial support to enable MSME to undertake environmentally friendly projects [26].

Committed efforts are required to unlock the entire potential of MSME in India and enhance its tasks in recruitment opportunities and economic expansion. By imposing the qualitative support, the MSME sectors can significantly heighten productivity and competitiveness. The incorporation of green finance advances India's Sustainable Development Goals by sponsoring the clean energy, sustainable infrastructure, climate action, and responsible power consumption. Through these approaches, India can accomplish sustainable and resilient Indian economy, via diminution in carbon emissions, strengthen secured energy, foster job generation, innovation, and improve environment quality. By embracing green finance, MSME can strategically improve their competitiveness, adapt to evolving market dynamics, and drive economic growth and job creation in the respective regions.

3.3 Configuration of Green Finance with India's Sustainable Development Goals

Green finance is a keystone for India's accomplishing Sustainable Development Goals (SDGs) in diverse sectors, demonstrating a mutually supportive incorporation among economic development and environmental preservation. In the realm of clean energy, green finance vehicles, including green bonds and sustainable finance infrastructures, facilitate investments in the renewable energy projects, fitting impeccably into SDG 7 (Affordable and Clean Energy). Likewise, in the domain of sustainable infrastructure, the innovative projects, viz., the Green Energy Corridor and AMRUT, supported by green finance, formulate substantial contributions to SDG 9 (Industry, Innovation and Infrastructure) and SDG 11 (Sustainable Cities and Communities) [19]. Furthermore, green finance assists in responsible consumption patterns by

Table 1 Key initiatives in promoting green finance in India

Initiative	Description	Reference
Carbon pricing initiatives	National and regional efforts in carbon pricing	[12]
Partnerships and international collaborations	Joint announcement on green finance initiatives and exploring collaborations with international entities	[13]
Green bonds	India's green bond market has grown consistently, with various entities issuing green bonds to fund sustainable projects, including renewable energy and sustainable infrastructure	[14]
Sustainable finance framework	In 2021, the Reserve Bank of India (RBI) introduced the sustainable finance framework, encouraging financial institutions to consider environmental and social aspects in financing decisions	[11]
National action plan on business and human rights	Launched in 2019, this plan emphasizes sustainable development and responsible business practices, recognizing the role of MSME and the need for financial support, including green finance, to promote sustainability	[15]
Government initiatives	The green climate fund, an international financing mechanism, supports climate-related projects in India, including those involving MSME. Programs like the credit guarantee fund trust for micro and small enterprises (CGTMSE) facilitate MSME lending, for instance, green projects	[16, 17]
Corporate social responsibility and nonfinancial incentives	Released in 2007, steps compel banks to stress sustainable development and environmental considerations, including aligning corporate social responsibility practices with sustainable development goals	[18]
National action plan on climate change (NAPCC)	Released in 2008, the NAPCC outlines a comprehensive policy framework for dealing climate change, including missions in sectors like solar energy, energy efficiency, and sustainable habitat	[19]
Climate change finance unit (CCFU)	Established in 2011 under the ministry of finance, CCFU mobilizes and manages climate finance, facilitating access to international funds and supporting climate change-related projects in India	[20]
Target of 450 GW renewable energy by 2030	Set up in 2011 under the ministry of finance, CCFU leverages and directs climate finance, facilitating access to overseas capital and financing Indian climate change projects. India had a high goal in 2021 of reaching 450 GW of renewable power installed capacity in 2030 such as solar, wind, biomass, and hydropower, with a commitment to cleaner and sustainable energy	[21]

Source(s) Authors' compilation

reassuring eco-friendly practices and financial investments that align with SDG 12 (Responsible Consumption and Production) [10]. Associating green finance with such SDGs not only encourages India's national sustainable development goal, but additionally puts it on the global map to achieve the United Nations' generalized sustainability goals. The following are the arrangement with India's Sustainable Development Goals:

(a) SDG 7: Clean and Affordable Energy: Green finance encourages investments in renewable energy projects [4]. India's goal to reach 450 GW of renewable energy capacity by 2030 [21] is consistent with SDG 7, which seeks to provide access to affordable, reliable, sustainable, and modern energy for everyone.
(b) SDG 9: Industry, Innovation and Infrastructure: Green finance encourages technological infrastructural development upgrades and innovations to enhance sustainability [6]. This is helpful in contributing to SDG 9 through inclusive and sustainable industrialization and innovation.
(c) SDG 11: Sustainable Cities and Communities: Green finance ensures sustainable infrastructure development [5]. This fits SDG 11, which aims to make cities and human settlements inclusive, safe, resilient, and sustainable.
(d) SDG 12: Responsible Consumption and Cleaner Production: Green finance promotes environmentally friendly production practices and green supply chains [5]. This is harmonizing to SDG 12 that encourages for sustainable consumption and production frameworks.
(e) SDG 13, Climate Action: Green finance has a pivotal responsibility in the hindrance of climate change and perception [4]. It unswervingly corresponds to SDG 13 that is about to obtain proximities against the climate change and its consequences (Fig. 2).

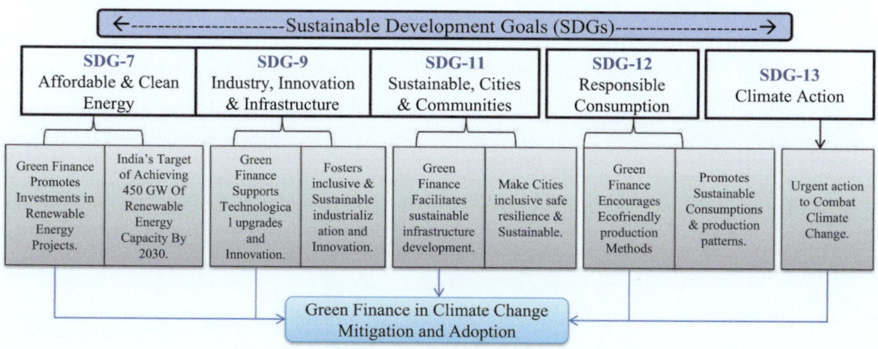

Fig. 2 India's sustainable development goals (SDGs)

3.4 *Potential to Contribute to a Greener and More Resilient Economy*

The hypothetical green finance that builds to assist a greener and more resilient financial sectors is substantial and varied. Green finance represents financial instruments and methodologies that seek to invest in projects and actions that have advantageous environmental products. Below are some of the most important ways green finance can assist in building a more sustainable and resilient economy:

(a) **Lower Carbon Footprint**: Green finance investments in renewable energy and energy efficiency lower greenhouse gas emissions, thus promoting a greener economy contexts [4].
(b) **Improved Energy Security:** Alteration to renewable energy sources through green finance advances energy security through reducing dependence on fossil fuels [21].
(c) **Improved Employment Opportunities**: Green finance projects, such as the renewable energy enterprises, postulate employment openings, which is imperative for economic growth [9].
(d) **Innovation and Technological Progression**: Green finance stimulates innovations in cleaner productions aside from sustainable practices [6]. It hastens technological progression and strengthens economic resilience.
(e) **Better Environmental Quality:** Green finance stimulates sustainable exercises resulting in superior environmental eminence and contributing to public health and digital well-being [3].
(f) **Access to International Green Markets**: Through the acceptance of green practices, Indian enterprises cross the threshold into global green markets and export green commodities and services [5].
(g) **Compliance with International Obligations**: India's entry into green finance is unswerving with global undertakings, such as the Paris Agreement, further accompanying its carriage on global sustainability [4] (Fig. 3).

Largely, green finance can safeguard amended environmental and economic performance, heading to a shift toward a greener, more buoyant, and sustainable

Fig. 3 Potential contributions to a greener and resilient economy

economy. The embracing of green finance ideologies is unswerving with global reactions to environmental encounters while stimulating economic growth and social well-being.

3.5 *Capacity Development Programs in Green Finance Initiatives for MSME*

Capacity development programs under green finance schemes have a crucial role in empowering MSME by providing requisite knowledge for adopting sustainable and environmental perspectives [29], as depicted in Table 2.

Table 2 Capacity development programs in green finance initiatives for MSME

Programs initiatives	Activities
Enhancing awareness and knowledge	Increase awareness on green finance advantages Offer critical knowledge on environmentally friendly technologies
Facilitating access to green finance	Supply MSME with data and capabilities for green finance access Inform MSME on funding needs and qualifications
Technical skills	Training for the installation of energy saving technologies Strengthen technical expertise for green production practices
Financial literacy	Strengthen financial literacy for green finance Offer capacity building in financial management
Fostering innovations	Support to green technology and practices for development of adaptively culture
Ensuring compliance	Educate MSME about compliance with environmental regulations and industry practices
Networking and collaboration	Encourage and built a collaborative network among MSME for green practice in industry
Competitiveness enhancement	For sustainable practices, MSME enable the market and tap to green market segments

Source(s) Authors' compilation

4 A Case Model of Green Finance Makeover for EcoFab Textiles

EcoFab Textiles, a small textile manufacturing unit based in Surat, Gujarat, had long struggled with outdated machinery and rising operational costs. Like many in the textile sector known for heavy resource consumption and pollution, the company recognized the urgent need to shift toward sustainable practices. However, accessing affordable finance for modernization remained a significant hurdle. In a strategic move, EcoFab Textiles partnered with a forward-thinking financial institution that provided them with a green loan specifically aimed at promoting environmentally responsible projects. The loan package supported comprehensive upgrade of their operations from acquiring energy-efficient machinery and installing rooftop solar panels to setting up water recycling systems and using eco-friendly dyes and chemicals.

This transformation had a twofold impact: On the business front, the new machinery improved production efficiency and lowered costs, while the green positioning helped attract environmentally conscious customers, expanding the company's market reach. On the environmental side, the changes led to a notable reduction in carbon emissions, water usage, and chemical discharge. EcoFab Textiles emerged as a benchmark for sustainable industrial practices. Its efforts were recognized by government bodies, earning awards and unlocking access to further incentives under green MSME initiatives.

"The story of EcoFab Textiles underscores the critical role that green finance can play in helping small enterprises embrace sustainability. It also highlights how ecological responsibility can go hand in hand with profitability and long-term growth. As more MSME follow suit, a ripple effect is set in motion leading to a cleaner, more resilient industrial.

5 Conclusion

Green finance primarily includes financial services and products that promote sustainable and green economic activities, which have tremendous potential to enhance India's crucial MSME sector and also promote sustainable development at the same time. Past few years, with the growing realization toward the need for sustainable development in India, the government has proactively taken steps to promote green finance in the MSME sector so as to boost economy and employment. The Indian government has launched a wide range of schemes, programs, and initiatives that seek out to encourage MSME to adopt environment-friendly practices and technical know-how. For instance, the National Clean Energy Fund has been established to provide crucial financial backing for renewable energy projects [21]. Simultaneously, the Technology Upgradation Fund Scheme offers financial assistance to MSME, enabling them to upgrade their technological capabilities [22]. Green finance serves as a pivotal enabler for MSME in India, facilitating access to cost-effective financing for investments in clean energy, energy-efficient technologies, and other sustainable practices Investments in sustainable practices, alternative energies, and energy-efficient technologies are crucial for businesses looking to reduce their environmental impact and operate more responsibly [30]. By empowering leisurelier approach to financing for green initiatives, such as clean power schemes and eco-friendly production procedures, green finance supports MSMEs to expand their sustainability metrics and affordability [31]. Moreover, it opens up new market prospects as consumers progressively prioritize eco-conscious products and services.

Businesses that contain green finance can not only augment their functional efficiency but also position themselves as front-runners in a shifting market countryside. By aligning with sprouting consumer preferences, MSMEs can fasten a larger market allocation and assemble long-term resilience. An example of this is EcoFab Textiles in Surat, Gujarat, India, which utilized green finance to renovate its operations and heighten revenue. EcoFab Textiles serves as a prototype for how companies can leverage green finance to capitalize in industrial equipment, technologies, and processes that align with sustainable practices. Through a strategic partnership with a forward-thinking banking institution, the company secured a green loan tailored to support environmentally friendly initiatives. As a result, EcoFab Textiles was able to invest in energy-efficient technologies and adopt sustainable practices, setting a positive example for other businesses to follow. Green finance emerged as a vital tool in strengthening India's MSME sector and advancing the India's sustainable development goals. Green finance empowers businesses to boost the competitiveness and growth, by providing affordable financing options for eco-friendly investments and helping MSME to improve the environmental practices,

6 Scope for Further Research

The chapter suggests some future scope suggesting the researchers could adopt the longitudinal point of views to compare the impact of green finance across the MSME sectors such as manufacturing, service, and agriculture. A detail study can be conducted on the effectiveness of the green financial products and services in improving the MSME performance and sustainability. Further, there is a scope to track the long-term effects of green finance adaption on operational efficiencies and environmental performance. The study may also focus on behavioral dimensions for MSME owners for environmental sustainability and their willingness to invest in green practices.

References

1. Singh K, Abrahamand R, Kolar P (2023) The sustainable FinTech revolution: building a greener future for finance: building a greener future for finance. IGI Global
2. Smith TM, Deshmukhand R, Roy S (2019) Green finance and micro, smalland medium enterprises (MSME) in India. In: Sustainability in a time of turmoil. Springer, Cham
3. The World Bank (2018) Energy efficiency in India: financing solutions and policy options. https://openknowledge.worldbank.org/handle/10986/30490
4. UNEP (2020) Green bonds and their role in climate mitigation
5. IISD (2021) Sustainable Infrastructure. https://www.iisd.org/topics/sustainableinfrastructure
6. Pereira E, Sousa T, Kinkyo T (2017) Green finance: a way to strengthen sustainable development in the European union. Sustainability 9(1):9. https://www.mdpi.com/20711050/9/1/9
7. Netscribes (2023) MSME financing market in India 2023. static.investindia.gov.in. Link: investindia.gov.in
8. World MSME Day 2023 (2023) How to empower MSME through financial innovation for sustainable development. Financial express. Published 25 June 2023. Accessed 9 Jan 2024. https://www.financialexpress.com/business/smemsmegreenworldmsmeday2023howtoempowerMSMEthroughfinancialinnovationtodrivesustainabledevelopment3140758/
9. Government of India (2021) Ministry of new and renewable energy. https://mnre.gov.in/
10. Rathi A, Akolkar A (2021) Green finance and ESG investing in India. In: Sustainable finance: challenges, success stories and opportunities. Springer, p 165181
11. Reserve Bank of India (RBI) (2021) Reserve Bank of India sustainable finance framework. Retrieved from https://www.rbi.org.in/Scripts/NotificationUser.aspx?Id=12189&Mode=0
12. NITI Aayog (2022) Report of the InterMinisterial committee on low carbon technologies formed under the India US. ReportCommitteeonLowCarbon Technologies.pdf (niti.gov.in)
13. Ministry of Environment, Forest and Climate Change (2023). Press Information Bureau (pib.gov.in)
14. Sustainable Growth Pillar of the Strategic Clean Energy Partnership Climate Bonds Initiative (2021) India. https://www.climatebonds.net/market/exposure/india
15. Government of India (2019) National action plan on business and human rights. https://www.mea.gov.in/Images/amb_bhr/MOFAIndia_National_Action_Plan_on_B usiness_and_Human_Rights.pdf
16. CGTMSE (n.d.) Credit guarantee fund trust for micro and small enterprises. https://www.cgtmse.in
17. Ministry of Environment Forest and Climate Change (2021) India third biennial update report to the united nations framework convention on climate change. India. Biennial update report (BUR). BUR3.|UNFCCC
18. Reserve Bank of India (2007) Guidelines on environmental and social risk management framework for banks and financial institutions. https://rbidocs.rbi.org.in/rdocs/PublicationReport/Pdfs/ESECGUIDELINES090513.pdf
19. Government of India (2008) National action plan on climate change
20. Government of India (2011) Climate change finance unit. https://ccfu.gov.in/
21. Ministry of New and Renewable Energy (2021). pib.gov.in/PressReleaseIframePage.aspx?PRID=1762960
22. Ministry of New & Renewable Energy (2022) MNRE annual report 2021
23. Srivastava S, Mathur HP (2012) Enhancing prosperity of India's micro and small enterprises through functional competencies. In: Micro finances: issues and challenges, pp 367–383. ISBN 9788183295154
24. Kumar A, Shrivastava RL (2020) Green finance: a comprehensive literature review and directions for future research. J Clean Prod 253:119892
25. Malhotra A, Choudhary N, Upadhyay A (2019) Sustainable finance in India: a regulatory perspective. J Adv Manag Res 16(2):145162
26. Choudhary A, Dwivedi YK (2021) Sustainability initiatives in the Indian MSME sector: a case study approach. Int J Inf Manag 57:102299

27. Das S, Banerjee S, Dey PK (2022) Technological innovations in sustainable finance: a bibliometric analysis and research agenda. Sustain Prod Consumption 28:13231336
28. SEBI (2022) Consultation paper on green and blue bonds as a mode of sustainable finance
29. Government of India (2021) Ministry of micro, small & medium enterprises. https://msme.gov.in/
30. CERC (2022) Draft central electricity regulatory commission (terms and conditions for renewable energy certificates for renewable energy generation) regulations
31. SBI (2023) Environmental, social and governance (ESG) financing framework report. ab4d507dc354d4d0b2e53cbbe1f75e0a (sbi.co.in)
32. Mukherjee, Chatterjee (2022) Carbon capture utilization and storage (CCUS)—policy framework and deployment mechanism in India. In Retrieved 17 Jan 2024, from https://www.niti.gov.in/sites/default/files/202211/CCUSReport.pdf
33. International Finance Corporation (2021) MSME sustainability guidebook. IFC ESG Guidebook

Navigating the Future: Exploring Intelligent Systems for Sustainable and Robust Finance

Priyanka Dhoot, Anuradha Patil, Abhishek Raidas, Palash Bairagi, and Samarjit Roy

Abstract AI, an integration of machine learning (ML) and deep learning (DL), is revolutionizing industries like automotive, healthcare, gaming, finance, and more. In finance, AI is being applied to asset management, algorithmic trading, credit underwriting, and blockchain-based finance. Its adoption is driven by the availability of data and affordable computing power. AI leverages big data to autonomously learn and improve performance without explicit human programming. This technology offers advantages in enhancing financial deepening and efficiency. However, there are concerns about the digital divide between advanced economies and developing nations. To address these concerns, policymakers need to understand the implications and risks of AI in finance. This chapter provides valuable insights into the advantages and risks associated with the adoption of AI techniques. It emphasizes the need for responsible AI use to safeguard financial stability, market integrity, competition, and consumer interests. To support responsible AI innovation, policy recommendations are proposed. These recommendations aim to strike a balance between fostering AI advancements and ensuring the protection of financial systems and consumers. By addressing potential risks, policymakers can promote the responsible use of AI and support its continued growth and development. In summary, AI is transforming the finance industry through its applications in various areas. Policymakers must carefully consider the implications and risks associated with AI adoption, whilst also promoting responsible use and supporting innovation in this rapidly evolving field.

P. Dhoot (✉) · A. Patil
School of Commerce and Management, D Y Patil International University, Akurdi, Pune, Maharashtra, India
e-mail: priyanka.dhoot@dypiu.ac.in

A. Raidas
Indira Institute of Management, Pune, Maharashtra, India

P. Bairagi
Department of Business Management, Central University of Odisha, Koraput, Odisha, India

S. Roy
School of Computing and Information Technology, Eastern International University, Becamex IDC Corp, Thu Dau Mot, Binh Duong Province, Vietnam

S. Roy et al. (eds.), *Humanized Intelligence in Sustainable Finance and Economic Behaviours*, Studies in Computational Intelligence 1226,
https://doi.org/10.1007/978-981-95-0599-9_5

Keywords Artificial intelligence · Machine learning · Financial services · FinTech

1 Introduction

The financial sector is undergoing a profound digital transformation, and at the heart of this revolution lies the integration of AI technology. Artificial intelligence has quickly become a pivotal force, empowering financial institutions to streamline operations, enhance decision-making processes, and deliver more personalized services to customers [1, 2]. As vast amounts of data continue to accumulate, AI's ability to analyse and interpret this information in real time has redefined risk management, enabling financial institutions to proactively respond to market shifts and make data-driven investment choices with unprecedented precision.

Moreover, AI-driven algorithms have revolutionized fraud detection, helping institutions detect and prevent fraudulent activities by spotting patterns and anomalies that may otherwise elude traditional systems. The seamless integration of AI in the financial sector promises not only increased efficiency but also the potential for democratizing access to financial services, making sophisticated wealth management solutions available to a broader spectrum of individuals and businesses.

Despite the transformative potential of AI, this technological evolution comes with its unique set of challenges and ethical considerations. As financial institutions rely more heavily on AI algorithms to make critical decisions, issues related to data privacy, transparency, and algorithmic bias demand careful attention. Striking the right balance between harnessing the capabilities of AI for growth and maintaining the trust of customers and regulatory bodies will be key to ensuring the sustainable and responsible development of AI technology in the financial sector. As the financial landscape continues to evolve, understanding the impact of AI and navigating its implications are paramount for stakeholders to stay at the forefront of this dynamic industry.

2 Foundations of AI in Transformative Finance

Artificial intelligence (AI), introduced by Stanford Professor John McCarthy in 1955, began with programming machines for clever behaviour but has evolved to focus on machine learning (ML), where computers learn from data [3]. Machine learning, as described by [4], improves performance with data exposure, making it invaluable in financial services for assessing credit risk, detecting fraud, optimizing investment portfolios, and predicting market trends [5]. This automation enhances processes and customer experiences significantly. Deep learning (DL), a specialized form of ML, uses multilayered neural networks for complex data processing, driving advancements in risk assessment, fraud detection, algorithmic trading, and sentiment analysis [6, 7]. Natural language processing (NLP) allows machines to understand and interact with human language, transforming financial applications from chatbots to sentiment analysis [8]. NLP automates customer support, analyses market sentiment, and aids in regulatory compliance, enabling better customer service, informed decisions, and efficient regulation adherence [9].

The depicted figure in Fig. 1 classifies the AI scenario by considering both the extent of AI adoption and the time needed to reach that level of adoption within the Indian context. One notable observation from the depicted use cases is the direct relationship between the adoption level and the timeframe. This connection is attributed to Indian financial services organizations embracing globally recognized and effective use cases.

Figure 2 showcases the adoption of AI in the Indian financial service industry which has experienced significant growth and transformation in recent years. Financial institutions in India are increasingly leveraging AI technologies across various domains such as customer service, risk management, fraud detection, and personalized financial advice. AI-powered chatbots and virtual assistants [10] are employed to enhance customer service, providing instant support and personalized recommendations. Moreover, AI algorithms analyse vast datasets for risk assessment, credit scoring, and fraud detection, enabling more accurate decision-making and risk mitigation strategies. Additionally, AI-driven tools aid in automating routine tasks, optimizing operational efficiency, and offering personalized financial guidance to customers. The evolving regulatory environment and increased digitization

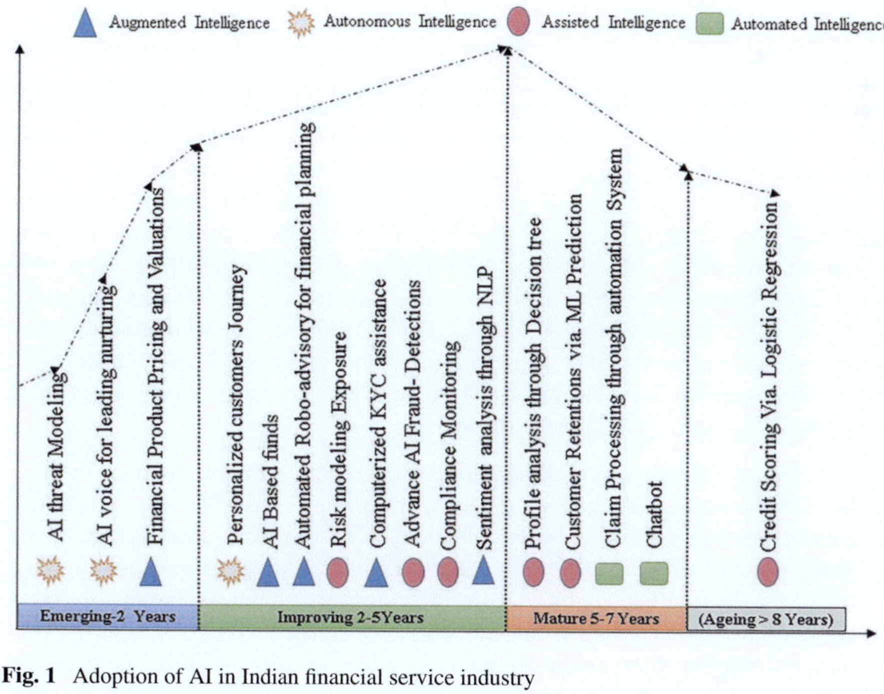

Fig. 1 Adoption of AI in Indian financial service industry

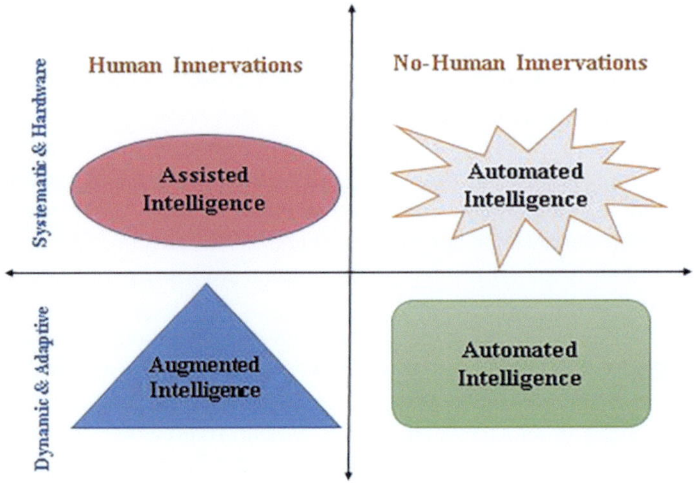

Fig. 2 Mapping AI adoption levels and timeframe

have spurred the rapid adoption of AI in the Indian financial sector, positioning it at the forefront of innovation and efficiency in delivering financial services.

2.1 Assisted Intelligence

Assisted intelligence involves technology that enhances human decision-making rather than replacing it, providing analytical tools and information to support human judgement. According to [11], assisted intelligence systems, such as decision support systems, have been instrumental in helping financial analysts interpret complex datasets and make more informed decisions. These systems utilize algorithms to analyse market trends and provide actionable insights, allowing financial professionals to leverage computational power whilst maintaining control over final decisions. This approach improves efficiency and accuracy in financial analyses [11]. Moreover, [12] emphasize that assisted intelligence augments human capabilities, particularly in data-rich environments where rapid analysis and response are critical. Application of assisted intelligence involves:

1. *Financial Exposure and Risk Modelling*: leveraging AI revolutionizes the way institutions assess and manage risks. AI algorithms, fed with vast and diverse datasets, enable precise analysis of financial exposures across various scenarios. These models can swiftly identify potential vulnerabilities, forecast market fluctuations, and simulate risk scenarios, offering insights into potential outcomes. By continuously learning from new data, AI enhances risk assessment accuracy, helping financial entities proactively adapt strategies to mitigate potential risks. This advanced modelling not only improves decision-making but also aids

in optimizing portfolio management, enabling institutions to navigate volatile markets with a deeper understanding of potential exposures and risks.

2. *Advance AI Fraud Detection*: Advanced AI-based fraud detection in the Indian financial services sector represents a transformative leap in identifying and preventing fraudulent activities. Using sophisticated machine learning algorithms, AI analyses massive volumes of transactional data in real time, detecting anomalies, unusual patterns, and suspicious behaviours that might indicate fraudulent activity. By continuously learning from new patterns and evolving tactics used by fraudsters, AI-driven systems can adapt and enhance their detection capabilities. This proactive approach helps financial institutions in India to stay ahead of emerging threats, safeguarding against various forms of fraud, including identity theft, payment fraud, and cyber-attacks, thereby bolstering the security and trust of their services for customers.
3. *Compliance Monitoring*: Compliance monitoring in the Indian financial services sector has undergone a significant evolution with the integration of advanced AI technologies. These systems employ AI-driven algorithms to scrutinize vast amounts of regulatory data, transactions, and records in real time, ensuring adherence to stringent compliance standards and regulations. AI not only automates routine monitoring tasks but also identifies potential compliance breaches, suspicious activities, or irregularities more efficiently than traditional methods. By continuously learning and adapting to regulatory changes, these AI-powered monitoring systems assist financial institutions in India in staying compliant, mitigating risks, and swiftly addressing potential compliance issues, thus reinforcing trust and stability within the industry.
4. *Customer Profiling using Decision Tree*: In the Indian financial service sector, customer profiling through decision trees has emerged as a powerful tool for segmentation and personalized service delivery. Decision tree algorithms analyse customer data points such as demographics, transaction history, credit behaviour, and preferences to categorize customers into distinct segments based on similarities and patterns. By employing these trees, financial institutions can tailor products, services, and marketing strategies to meet the specific needs of each segment. This approach not only aids in understanding customer behaviour and predicting preferences but also enhances customer experiences by offering targeted and relevant financial solutions, ultimately fostering stronger relationships and loyalty amongst diverse customer groups in the Indian market.
5. *Customer Retention and Churn Prediction*: In the financial sector, AI-driven tools are instrumental in customer retention and churn prediction by analysing vast datasets to identify patterns indicative of potential customer attrition. These AI algorithms assess various factors like transaction history, customer interactions, demographics, and behavioural patterns to create predictive models. By recognizing signals that precede churn, financial institutions can proactively intervene with personalized retention strategies. These strategies might include tailored offers, proactive customer service, or personalized recommendations aimed at addressing the specific needs or concerns of at-risk customers, thereby significantly reducing churn rates and enhancing overall customer retention. This

predictive approach empowered by AI not only helps in preserving customer relationships but also contributes to sustaining a loyal customer base within the dynamic landscape of the financial industry.

6. *Logistic Regression-Based Credit Scoring*: In Indian financial services, logistic regression-based credit scoring has become a cornerstone for assessing creditworthiness. Leveraging historical data on variables like credit history, income, debt-to-income ratio, and other relevant factors, logistic regression models calculate the probability of an individual defaulting on credit obligations. These models use a binary classification approach, categorizing applicants into creditworthy or non-creditworthy groups based on calculated probabilities. This method aids financial institutions in making informed decisions when extending credit, allowing them to manage risks effectively by identifying applicants likely to default, thereby optimizing lending practices and minimizing potential losses in the Indian financial landscape.

2.2 Automated Intelligence

Automated intelligence refers to systems that perform tasks without human intervention, including automated trading systems, fraud detection mechanisms, and robotic process automation (RPA). Chui et al. [13] highlighted the role of automated intelligence in enhancing operational efficiency and reducing the cost of financial transactions. Automated trading systems execute trades at high speeds and volumes based on predefined criteria, significantly outperforming human capabilities in terms of speed and consistency [13]. Additionally, [14] discussed how these systems help mitigate risks associated with human error and emotional decision-making, providing a more stable and predictable financial environment. The use of RPA in back-office operations has streamlined processes, reduced errors, and allowed human workers to focus on more strategic tasks [15]. Application of automated intelligence involves:

1. *Automated Claim Processing*: It has emerged as a transformative solution within Indian financial services, leveraging technology to streamline and expedite the claims management process. Through advanced algorithms and automated systems, claims are processed swiftly and accurately, reducing the need for manual intervention. These systems utilize data analysis, optical character recognition (OCR), and natural language processing (NLP) to extract information from submitted documents, verify claims against predefined criteria, and assess eligibility efficiently. By automating routine tasks, validating claims, and ensuring compliance, this technology not only accelerates the claims processing timeline but also enhances accuracy, ultimately leading to improved customer satisfaction and operational efficiency within the Indian financial sector.
2. *Use of Chatbots*: In the Indian financial service sector, chatbots have become integral in transforming customer interactions and support. These AI-powered conversational agents engage with customers in real time, offering personalized assistance, answering queries, and guiding users through various financial

processes. Chatbots are adept at handling a wide range of customer inquiries, from account queries to assisting in transactions, loan applications, or providing information about financial products. Their availability 24/7 ensures round-the-clock support, enhancing customer experiences by delivering immediate responses and relevant information. Additionally, they help financial institutions in India optimize their operations by automating routine tasks, reducing service waiting times, and freeing up human agents to focus on more complex queries, ultimately contributing to improved customer satisfaction and operational efficiency.

2.3 Augmented Intelligence

Augmented intelligence combines human intuition with advanced algorithms to enhance decision-making processes. This is particularly relevant in the financial services industry, where complex decision-making is required. Ransbotham et al. [16] indicate that augmented intelligence systems can process large volumes of data and provide insights that humans might overlook, thereby augmenting the decision-making capabilities of financial professionals. These systems are designed to work alongside humans, offering recommendations and highlighting potential risks or opportunities, thus improving decision quality [16]. Furthermore, Wilson and Daugherty [17] suggest that augmented intelligence fosters collaboration between humans and machines, leading to innovative solutions and enhanced performance in financial services. Applications of augmented intelligence involves:

1. *Product Pricing and Valuation*: In the realm of Indian financial services, product pricing and valuation are increasingly influenced by advanced methodologies, incorporating data-driven analytics and valuation models. These models employ diverse factors such as market trends, economic indicators, risk assessments, and customer behaviour to determine optimal pricing strategies and product valuations. With the aid of sophisticated algorithms and machine learning, financial institutions can more accurately assess the intrinsic value of products, tailor pricing structures to market dynamics, and optimize profitability. This data-driven approach enhances precision in pricing and valuation processes, enabling financial entities in India to remain competitive, adapt to market fluctuations, and offer products that align closely with customer expectations whilst ensuring financial sustainability.
2. *AI-Based Funds*: In the Indian financial services sector, AI-based funds have emerged as innovative investment vehicles leveraging artificial intelligence algorithms to manage portfolios. These funds utilize machine learning models and predictive analytics to analyse market trends, historical data, and various indicators to make investment decisions. AI algorithms continuously learn and adapt, identifying patterns and potential opportunities in the market, aiming to optimize investment strategies and deliver competitive returns. By automating investment decisions and adapting to changing market conditions in real time, AI-based funds offer a dynamic approach to portfolio management, attracting investors

seeking diversified and technologically driven investment options within India's financial landscape.

3. *Robo-Advisory for Financial Planning*: Robo-advisory services in the Indian financial landscape offer automated, algorithm-driven financial planning and investment advice to individuals. These platforms utilize AI and algorithms to analyse investors' financial goals, risk tolerance, and investment preferences. Based on these data, they recommend personalized investment strategies and portfolios that align with the individual's objectives. These services often offer diversified investment options, automated portfolio rebalancing, and real-time monitoring, providing cost-effective and accessible financial planning solutions to a broader audience. Robo-advisors in India aim to democratize financial planning by offering tailored advice without the need for extensive human intervention, catering to investors seeking convenient, low-cost, and technology-driven wealth management solutions.
4. *Computer Vision-Assisted KYC (Know Your Customer)*: In Indian financial services, advanced technology is introduced to streamline and enhance identity verification processes. Leveraging computer vision algorithms, this system verifies customer identities by analysing official documents such as IDs or passports. By extracting and verifying information from these documents through image recognition and analysis, it accelerates the KYC process, reducing manual intervention and enhancing accuracy. This technology aids financial institutions in India in meeting regulatory compliance requirements swiftly and efficiently whilst improving customer onboarding experiences by expediting the verification process.
5. *Credit Scoring using Neural Network*: Neural networks, a form of AI, process extensive datasets, including credit history, income, and other relevant factors, to predict credit risk. These networks utilize complex algorithms that mimic the human brain's learning process, enabling them to recognize intricate patterns within data and make more accurate credit assessments. By employing neural networks for credit scoring, financial institutions in India aim to enhance accuracy in evaluating borrowers, enabling better-informed lending decisions, and minimizing risks associated with extending credit, ultimately contributing to a more robust and efficient lending ecosystem.
6. *NLP Sentiment Analysis*: Natural language processing (NLP) sentiment analysis is revolutionizing how institutions gauge market sentiment and customer perceptions. NLP algorithms dissect vast amounts of textual data from sources like news articles, social media, and customer feedback to discern sentiments—positive, negative, or neutral—towards financial products, market trends, or institutions. By extracting nuanced insights and sentiments from unstructured data, NLP aids in understanding public opinion, market sentiments, and customer feedback. This analysis not only assists financial entities in India in making informed decisions but also helps in managing risks, predicting market movements, and tailoring customer-centric strategies, thereby enhancing their competitiveness and responsiveness to evolving market sentiments.

2.4 Autonomous Intelligence

Autonomous intelligence represents the most advanced form of AI, where systems operate independently of human intervention. In the financial services industry, autonomous intelligence is exemplified by technologies such as self-learning algorithms and AI-driven advisory services. Agrawal et al. [18] state that these systems can adapt to new data and evolving market conditions without explicit programming. Autonomous trading algorithms, for example, not only execute trades but also learn from each transaction to optimize future trading strategies [18]. Additionally, [19] discuss how autonomous systems continuously improve and adapt, making them highly valuable in dynamic financial markets where they can identify and exploit market inefficiencies more effectively than traditional methods. The potential for AI-driven financial advisory services to provide personalized, real-time investment advice is also significant, as highlighted by [20]. Application of autonomous intelligence involves:

1. *AI Threat Modelling*: AI threat modelling in Indian financial services involves the proactive identification and assessment of potential risks and vulnerabilities posed by artificial intelligence itself or its application in the sector. It encompasses analysing AI systems for potential weaknesses, vulnerabilities, or misuse that could lead to security breaches, data privacy concerns, or operational risks within financial institutions. By employing advanced techniques such as adversarial testing, scenario planning, and continuous monitoring, AI threat modelling aims to anticipate and mitigate risks associated with AI-powered systems. This approach enables financial entities in India to fortify their defence, ensuring robust security measures, safeguarding against potential threats, and fostering trust amongst stakeholders in an increasingly AI-driven financial ecosystem.
2. *Voice AI for lead nurturing*: Leveraging voice-based AI technologies, such as interactive voice response (IVR) systems and voice assistants, financial institutions can engage with potential leads by offering personalized information, assistance, and support. These systems analyse customer inquiries, preferences, and behaviour, enabling tailored responses and suggestions related to financial products or services. By utilizing voice AI for lead nurturing, financial entities in India enhance customer interactions, streamline queries, and provide timely and relevant information, fostering stronger connections and nurturing leads through personalized and efficient communication channels.
3. *Customer Journey Personalization*: It involves tailoring every stage of a customer's interaction with financial products or services to meet their specific needs and preferences. Leveraging advanced data analytics and AI-driven technologies, financial institutions in India analyse customer behaviours, transactions, and preferences to create personalized experiences. This personalization spans from marketing and acquisition to onboarding, usage, and ongoing support. By customizing interactions, product offerings, and communication channels,

these institutions aim to enhance customer satisfaction, foster loyalty, and optimize the overall customer experience in the dynamic and diverse market of India's financial services.

3 Review of Literature

Artificial intelligence (AI) and machine learning have significant effects on the financial sector. These include data-driven personalized product offerings, creative financial solutions, and the evolution of trading. It is especially important to assess the impact of these technologies and the methods and tools used to implement them [21]. AI-based solutions have notably improved the effectiveness and efficiency of fraud detection and risk management in the banking sector. The application of AI and machine learning in the financial sector presents significant ethical and regulatory challenges, but it also has the potential to significantly improve risk management and fraud detection in the industry. It is important to recognize that these technologies are not without limitations and potential biases. As noted in a Deloitte analysis (2020), for instance, worries have been expressed regarding the potential for AI-powered systems to reinforce or perpetuate pre-existing biases in financial decision-making. Concerns have also been raised concerning the interpretability and transparency of AI-powered systems, particularly with regard to regulatory compliance and accountability requirements [22].

Paper [23] examined the application of AI in the Indian banking industry with the goal of determining how Indian banks may take advantage of novel prospects and international technology norms whilst tackling the difficulties associated with adjusting to the new AI culture. According to the study, banks are experimenting with a range of applications, including enhancing chatbots for customer care, customizing offerings, and putting robots in place for self-service at dedicated Digital Banking Branches. These programmes are intended to lower fraud and security risks whilst increasing back-office productivity.

Arora et al. [24] investigated the role artificial intelligence (AI) plays in improving client experiences in the FinTech industry. The study evaluated a number of variables and sub-variables impacting consumers' opinions of FinTech services powered by AI. Service quality, commitment to trust, personalization, and perceived ease, commitment to relationships, perceived sacrifice, subjective norms, perceived usefulness, attitude, and vulnerability were amongst the variables that were looked at. According to the research, the main variables that have a big influence on how customers interact with FinTech services powered by artificial intelligence are perceived usefulness, perceived convenience, and service quality. Statements like "Using FinTech would enhance my efficacy in managing a portfolio (A2)", "My peer groups and friends influence the use of FinTech services (SN3)", and "Using FinTech would improve my efficacy in administering a portfolio (PU2)" were amongst the sub-criteria that were crucial for customers' experiences with FinTech services.

Kanwal and Misra [25] used a comprehensive multiple case study design to highlight AI's revolutionary significance as a key component in Indian P2P markets. In order to derive context-specific insights regarding AI's influence on the Indian P2P industry, the study carefully chosen six licensed non-banking financial company (NBFC)-P2P scenarios. Eighteen semi-structured interviews were included in the study, with six founders of NBFC-P2P and twelve FinTech and P2P industry specialists participating. The conclusions were confirmed by triangulating data using insights from several industry professionals. The findings were presented through distinct and context-specific results from the six NBFC-P2P scenarios. According to the findings, AI has reached a critical turning point in India's peer-to-peer markets.

Researcher has studied the significance of a number of variables from the viewpoint of employees at various organizational levels, including customer satisfaction, staff awareness of AI, and third-party partnerships. The author evaluated the potential of these new technologies to improve consumer happiness and looked into how important variables like third-party alliances and data security issues affected the uptake of AI technology. The research utilized logistic regression to evaluate the importance of the elements that were found and examined as they may have an impact on the adoption of AI-driven technologies in various departments. The results showed that artificial intelligence (AI) technology improves customer service and drastically lowers costs. The study discovered that the Indian banking and finance sector had a relatively low adoption rate of AI technology, despite its many benefits. The study has significance for organizations as well, offering a thorough 360-degree view of AI application and its effects, especially considering the paucity of research on this subject from lower organizational hierarchies [26].

The existing literature suggests that AI and ML could offer substantial benefits in the financial industry, particularly in areas like risk management and fraud detection. However, the implementation of these technologies requires careful consideration of relevant ethical and regulatory concerns. This study aims to offer a thorough overview of current research on the application of AI in the financial sector whilst highlighting potential areas for future exploration.

4 Research Questions

The existing study addresses the following research questions:

A. What are the existing adoptions of AI within the financial service sector?
B. How do various financial institutions engage in initiatives related to artificial intelligence?
C. What challenges arise from the deployment of AI in the Indian financial services industry?
D. How has AI transformed the Indian financial services industry?
E. What policy recommendations can be suggested to overcome existing challenges based on current research

These research questions will be tackled through an examination of pertinent literature in the field. The review will encompass studies and reports detailing the current state of AI and its application in the financial industry. In essence, this study aims to furnish an encompassing snapshot of the current status of AI deployment in the financial industry. Furthermore, it seeks to identify challenges and propose strategies to overcome challenges within this domain.

5 Objectives of the Chapter

The purpose of writing a chapter on "Navigating the Future: Exploring AI Technology and Its Implications in the Financial Sector" is to provide a comprehensive analysis of the impact of AI on the financial industry. The chapter aims to explore the potential benefits and challenges that AI technology brings to financial services. It serves as a valuable resource for policymakers, financial institutions, and stakeholders to understand the implications of adopting AI in the financial sector.

The chapter delves into the various applications of AI in finance, such as fraud detection, credit risk assessment, algorithmic trading, and customer service, amongst others. It highlights how AI can improve efficiency, enhance customer experiences, and streamline operations, leading to increased profitability and competitiveness for financial institutions.

However, the chapter also addresses the potential risks and challenges associated with the widespread adoption of AI in finance, such as data privacy concerns, algorithmic biases, regulatory compliance, and potential job displacement. It emphasizes the importance of responsible AI use, ethical considerations, and the need for safeguards to ensure fair and transparent practices.

Furthermore, the chapter proposes policy recommendations to guide the responsible integration of AI in the financial sector. These recommendations aim to strike a balance between promoting AI innovation, ensuring financial stability, protecting consumer interests, and addressing societal implications.

6 Case Study of AI-Induced Financial Institutions

Several financial institutions have embraced AI technology to enhance their services, improve efficiency, and offer innovative solutions to customers. Some notable examples include:

Figure 3 demonstrates how financial institutions are leveraging AI technology to innovate and improve customer experiences, risk management, and operational efficiency in the financial sector. As AI continues to advance, more institutions are expected to embrace its potential to transform the way financial services are delivered in India [27–29].

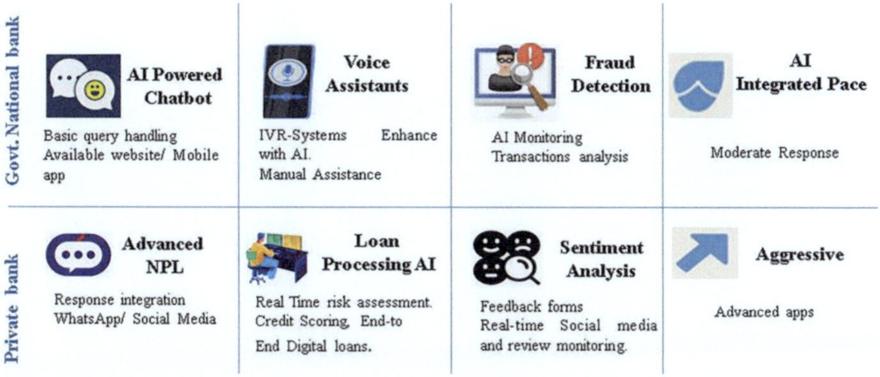

Fig. 3 Demonstration of financial institutions implementing AI

JP Morgan Chase: An AI Completed 360,000 Hours of Finance Work in Just Seconds.

JP Morgan Chase, a major U.S. bank, implemented an AI-based contract analysis system in 2017, significantly improving efficiency in reviewing and managing contracts. The system, demonstrated in Fig. 4, known as COiN (contract intelligence), employs unassisted AI to analyse legal documents [30], extracting key clauses, termination clauses, and non-disclosure agreements with a 95% accuracy rate. This initiative reduced the manual review time by approximately 360,000 h per year. JP Morgan recognized the potential of AI in 2016, establishing a Centre of Excellence within Intelligent Solutions to explore and implement diverse AI use cases, aligning with their commitment to cost reduction, operational efficiency, and improved customer experiences. COiN utilizes image recognition and a private cloud network to automate the document review process, extracting relevant attributes from business credit agreements within seconds. The success of COiN is attributed to JP Morgan's focus on innovation, building a strong internal team, and recruiting top AI talent globally [31]. The bank invests significantly in technology, allocating 9% of projected revenue in 2017 to research and development, double the industry average. The company collaborates with universities, research institutions, and start-ups, prioritizing ethics and values in their AI applications [32–34].

Looking ahead, JP Morgan aims to explore advanced applications of COiN for better predictive capabilities. Recognizing the need for workforce adaptation, the bank invests in education and job training programs. The success of JP Morgan's AI strategy serves as inspiration for other organizations to enhance their business strategies and embrace technological advancements. The broader implication for businesses is to streamline new technologies for competitiveness, as demonstrated by JP Morgan's adoption of AI in the financial sector.

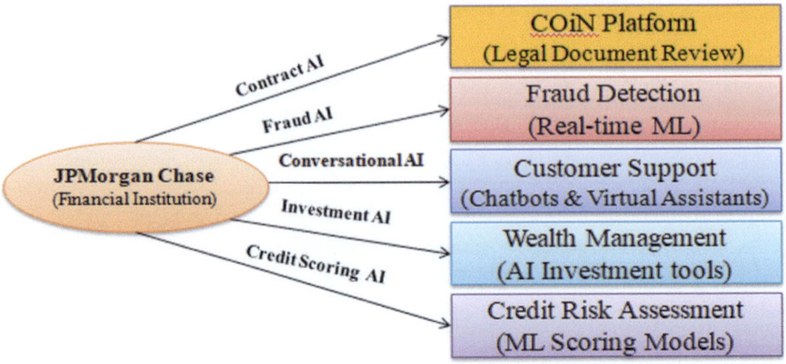

Fig. 4 Demonstration of JP Morgan's financial institution that leverages AI in diverge applications

7 Impact and Outcomes of AI Adoption

The adoption of AI in Indian financial services has significantly transformed the industry [35, 36]. AI-powered chatbots and virtual assistants [10] have revolutionized customer interactions by providing real-time assistance and promptly addressing queries, leading to higher customer satisfaction [37]. Additionally, AI algorithms have improved credit risk assessment and loan underwriting, enabling better lending decisions and reducing default rates, which facilitates easier access to credit and fosters economic growth [38].

AI has also revolutionized fraud detection and risk management. AI-driven fraud detection systems analyse vast amounts of transaction data in real time, quickly detecting suspicious activities and protecting customers [39]. Furthermore, AI-powered risk assessment models help financial institutions identify potential risks and make data-driven decisions, enhancing financial stability. AI-powered algorithmic trading improves market efficiency and liquidity, enhancing trading strategies and investment decisions [40]. As AI adoption expands, its potential to optimize operations, increase profitability, and drive innovation is crucial for India's economic growth and financial stability [41].

8 Mitigating Risks and Challenges

Mitigating risks and challenges in the adoption of AI in various domains, including finance, is critical to ensuring responsible and effective use of these technologies. Here are strategies for addressing key challenges in Fig. 5.

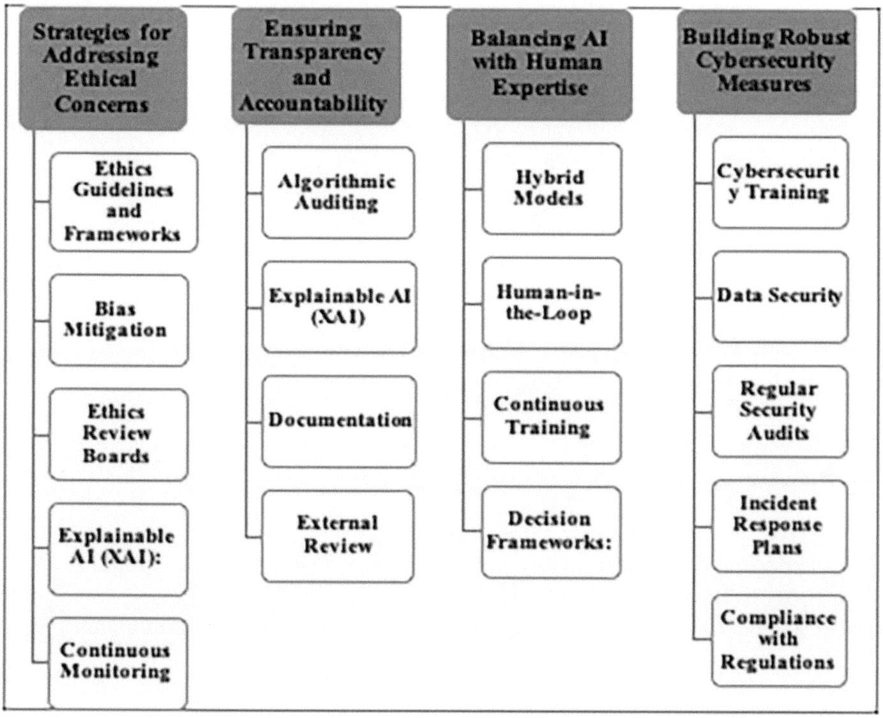

Fig. 5 Classified strategies for mitigating potential risks and challenges

8.1 *Strategies for Addressing Ethical Concerns*

1. *Ethics Guidelines and Frameworks*: Establish clear ethical guidelines and frameworks that govern the development and use of AI systems. These guidelines should emphasize fairness, transparency, accountability, and the avoidance of bias.
2. *Bias Mitigation*: Implement techniques to identify and mitigate bias in AI algorithms. This includes carefully curating training data, conducting bias audits, and using algorithmic approaches to reduce bias.
3. *Explainable AI (XAI)*: Employ explainable AI techniques to make AI decision-making processes transparent and interpretable. This helps build trust and allows for auditing AI systems for ethical compliance.
4. *Ethics Review Boards*: Establish ethics review boards or committees to evaluate and approve AI projects, especially those with potentially significant ethical implications.
5. *Continuous Monitoring*: Regularly monitor AI systems to identify and rectify any emerging ethical concerns. This includes ongoing audits and feedback loops to improve ethical performance.

8.2 Ensuring Transparency and Accountability

1. *Algorithmic Auditing*: Regularly audit AI algorithms to assess their performance, fairness, and adherence to ethical guidelines. Publish audit results when appropriate.
2. *Explainable AI (XAI)*: Use XAI techniques to make AI models more interpretable, allowing stakeholders to understand how decisions are made.
3. *Documentation*: Maintain thorough documentation of AI development processes, data sources, model architecture, and decision-making criteria. This documentation aids in accountability and transparency.
4. *External Review*: Consider external audits or third-party assessments of AI systems to ensure impartial evaluation and accountability.

8.3 Building Robust Cyber Security Measures

1. *Cyber security Training*: Train AI development teams on best practices in cyber security to build security into AI systems from the ground up.
2. *Data Security*: Implement strong data security measures, including encryption, access controls, and data anonymization to protect sensitive information used by AI systems.
3. *Regular Security Audits*: Conduct regular security audits and vulnerability assessments of AI systems to identify and patch potential weaknesses.
4. Incident Response Plans: Develop comprehensive incident response plans that outline actions to be taken in case of a cyber security breach or data leak related to AI systems.
5. *Compliance with Regulations*: Ensure that AI systems comply with relevant data protection and cyber security regulations, such as General Data Protection Regulation (GDPR), Health Insurance Portability and Accountability Act (HIPAA), and industry-specific standards.

By adopting these strategies, organizations can work to mitigate ethical, transparency, human–AI collaboration, and cyber security risks associated with AI deployment, promoting responsible and effective AI integration into their operations. It is important to note that AI governance and risk mitigation are ongoing processes that should evolve alongside advancements in AI technology and emerging threats.

9 Recommendations for Financial Institutions

In an era of rapid technological advancement and evolving customer expectations, financial institutions are facing unprecedented challenges and opportunities. To navigate this dynamic landscape successfully, it is crucial for these institutions to embrace digital transformation, prioritize cyber security, and harness the power of data analytics and AI. Moreover, fostering a customer-centric approach, coupled with a commitment to regulatory compliance and ethical AI, will be instrumental in maintaining trust and competitiveness.

9.1 Adoption Strategies and Considerations

Financial institutions should adopt a phased approach to AI implementation, starting with a clear definition of their strategic objectives and specific use cases. They must conduct a comprehensive risk assessment to address regulatory, ethical, and operational risks whilst ensuring compliance with financial regulations and data privacy laws [42–44]. Building a strong data governance framework, investing in talent acquisition and data infrastructure, and embracing cloud computing for scalability are crucial steps. Collaborations and partnerships with FinTech and established AI providers can enhance access to AI expertise and innovation. Lastly, prioritizing employee training and up skilling to foster a culture of continuous learning and responsible AI use is essential for successful AI adoption [45–48].

9.2 Building AI Capabilities and Infrastructure

Financial institutions should prioritize the development of robust AI capabilities and infrastructure by investing in talent acquisition, recruiting data scientists and machine learning experts, or partnering with external AI firms when necessary. Building modern data infrastructure and data management tools is pivotal, as high-quality data forms the foundation of AI. Leveraging cloud computing resources from providers like AWS, Azure, or Google Cloud can enhance scalability and flexibility. Establish a dedicated environment for AI model development and deployment, and implement systematic processes for model deployment and performance monitoring. This includes regular updates to keep AI models effective and compliant.

9.3 Collaboration and Partnerships with AI Providers

Financial institutions should actively seek collaboration and partnerships with AI providers to harness the full potential of AI in the financial sector. This involves conducting thorough assessments of AI vendors' capabilities, security measures, and regulatory compliance. Exploring strategic alliances with established AI solution providers and FinTech start-ups can offer access to cutting-edge technology and innovation. When forming partnerships, establish clear data sharing agreements to define data ownership, usage, and privacy rights, ensuring transparency and legal compliance. Collaborations with external organizations should align with the institution's strategic goals and enhance its ability to deliver innovative AI-driven solutions to customers whilst managing risks effectively.

9.4 *Training and up Skilling for Employees*

Financial institutions must prioritize training and up skilling programs for their employees to successfully integrate AI into their operations. These programs should cover AI concepts, tools, and best practices, enabling staff to understand the technology's potential and limitations. Furthermore, AI ethics education is essential to install responsible AI practices. Encourage a culture of continuous learning and innovation where employees are motivated to stay updated with AI advancements. Foster cross-functional teams that facilitate collaboration between AI experts and domain experts within the organization to ensure that AI solutions align with business objectives. Promote diversity within AI teams to enhance creativity and minimize biases in AI development. By investing in employee training and development, financial institutions can empower their workforce to adapt to AI-driven changes effectively and contribute to the responsible and ethical use of AI in finance.

10 Conclusion

In navigating the future of the financial sector, the exploration of AI technology has unveiled a landscape filled with immense promise and transformational potential. AI, with its ability to analyse vast datasets, make data-driven decisions, and enhance customer experiences, has already proven itself as a game changer in areas ranging from robo-advisors to risk management. However, this journey into the AI-powered future must be navigated with careful consideration of ethical, regulatory, and cyber security aspects. Financial institutions must commit to responsible AI adoption, ensuring transparency, fairness, and compliance with regulations. They must also embrace the opportunity to augment human expertise with AI, acknowledging that the most successful strategies often combine the strengths of both.

Moreover, a culture of continuous learning and innovation, along with partnerships with AI providers, will be integral to staying at the forefront of AI innovation in the financial sector. As we venture into this AI-powered era, the financial sector stands poised to evolve, offering more personalized, efficient, and secure services, ultimately benefiting both institutions and their customers. The path forward in the financial sector lies in harnessing AI's potential whilst upholding ethical values, promoting transparency, and fostering collaboration to shape a future that is not only technologically advanced but also sustainable and equitable.

11 Future Scope of Study

The progress and expansion of AI technology in finance is the future of this topic. Decentralized finance (DeFi) and blockchain technology will offer new options to integrate AI [49]. AI may improve DeFi protocol security and efficiency and automate and optimize blockchain transactions in the future study. The study might also generate AI-driven financial products for marginalized communities, encouraging financial inclusion and equity. AI, quantum computing, and advanced analytics can also transform financial modelling, risk assessment, and predictive analytics, providing more accurate and dependable financial insights.

AI adoption in finance has ethical, regulatory, and societal ramifications that the research might examine. As AI systems improve, justice, accountability, and openness will remain essential. Ethical AI governance frameworks, legislative methods across jurisdictions, and algorithmic bias mitigation best practices can be studied in the future. AI's effects on jobs and the workforce should also be studied. By investigating how AI-driven automation affects job responsibilities and skill needs, the study might inform educational and governmental initiatives to prepare the workforce for AI. To fully comprehend AI's transformational potential in the financial sector, this research will examine technology advances and the larger social and ethical environment.

References

1. Bahrammirzaee A (2010) A comparative survey of artificial intelligence in finance: artificial neural network. Expert system and hybrid intelligence system. Neural Comput Applic 1165–1195
2. Bharti Kumari JK (2021) System dynamics approach for adoption of artificial intelligence in finance. In: Saran VH, Misra RK (eds) Advances in systems engineering. Springer Nature Singapore Pte Ltd., pp 555–575. https://doi.org/10.1007/978-981-15-8025-3_54
3. Mitchell TM (1997) Machine learning. McGraw-Hill
4. Ng A (2018) Machine learning yearning: technical strategy for AI engineers, in the era of deep learning. Retrieved from https://www.deeplearning.ai/machine-learning-yearning/
5. Patel K (2018) Artificial Intelligence in Finance. IJSART 4(4):1–11
6. Goodfellow I, Bengio Y, Courville A (2016) Deep learning. MIT Press
7. LeCun Y, Bengio Y, Hinton G (2015) Deep learning. Nature 521(7553):436–444. https://doi.org/10.1038/nature14539
8. Jurafsky D, Martin JH (2020) Speech and language processing, 3rd edn. Pearson
9. Manning CD, Raghavan P, Schütze H (2008) Introduction to information retrieval. Cambridge University Press
10. Bhatt P (2018) Introducing 'Axis Aha!'—an AI-led virtual assistant to enhance. [Online]. Available at: https://www.axisbank.com/docs/default-source/press-releases/introducing-axis-aha!-an-ai-led-virtual-assistant-to-enhance-online-customer-experience.pdf
11. Davenport TH, Kirby J (2016) Only humans need apply: winners and losers in the age of smart machines. Harper Business
12. Brynjolfsson E, McAfee A (2014) The second machine age: work, progress, and prosperity in a time of brilliant technologies. W. W. Norton & Company

13. Chui M, Manyika J, Miremadi M (2016) Where machines could replace humans—and where they can't (yet). McKinsey Quarterly. Retrieved from https://www.mckinsey.com/business-functions/mckinsey-digital/our-insights/where-machines-could-replace-humans-and-where-they-cant-yet
14. Bostrom N, Yudkowsky E (2014) The ethics of artificial intelligence. In: Frankish K, Ramsey WM (eds) The Cambridge handbook of artificial intelligence. Cambridge University Press, pp 316–334
15. Willcocks LP, Lacity MC, Craig A (2015) The IT function and robotic process automation. London School of Economics and Political Science
16. Ransbotham S, Kiron D, Gerbert P, Reeves M (2017) Reshaping business with artificial intelligence. MIT Sloan Manag Rev 59(1). Retrieved from https://sloanreview.mit.edu/projects/reshaping-business-with-artificial-intelligence
17. Wilson HJ, Daugherty PR (2018) Collaborative Intelligence: humans and AI are joining forces. Harv Bus Rev 96(4):114–123
18. Agrawal A, Gans JS, Goldfarb A (2018) Prediction machines: the simple economics of artificial intelligence. Harvard Business Review Press
19. Arner DW, Barberis JN, Buckley RP (2017) FinTech, RegTech and the reconceptualization of financial regulation. Northwest J Int Law Bus 37(3):371–413
20. Firth N, Lawrence D (2018) AI and the future of financial advice. Financ Plann Res J 4(1):55–75
21. Martinello A, Jensen TL, Grenestam E, Meyer, BB (2022) AI and machine learning in the financial sector: five focus points. Economic Memo, No. 3, Danmarks Nationalbank, Copenhagen
22. https://www.pwc.in/research-insights/2022/ai-adoption-in-indian-financial-services-and-related-challenges.html
23. Chen S-H (2005) Computational intelligence in economics and finance: carrying on the legacy of Herbert Simon. Inf Sci Int J 121–131
24. Arora A, Gupta S, Devi C, Walia N (2023) Customer experiences in the era of artificial intelligence (AI) in context to FinTech: a fuzzy AHP approach. Benchmarking Int J 30(10):4342–4369
25. Kanwal A, Misra A (2022) Artificial intelligence in peer-to-peer lending in India: a cross-case analysis. Int J Emerg Mark 17(4):1085–1106. https://doi.org/10.1108/IJOEM-05-2021-0822
26. Deepthi B, Gupta P, Rai P, Arora H (2022) Assesssing the dynamics of AI driven technologies in Indian banking and financial sector. Vis J Bus Perspect 1–12
27. Halder R (2022) Analysing the impact of artificial intelligence in the financial sector. https://doi.org/10.13140/RG.2.2.18548.63362/1
28. Haswani K (2023) SBI embraces AI and ML technologies to transform banking operations. (Online). Available at: https://indianstartupnews.com/news/sbi-embraces-ai-and-ml-technologies-to-transform-banking-operations
29. Sharma D (2023) Kotak Mahindra Bank is experimenting with AI and blockchain. (Online). Available at: cio.economictimes.indiatimes.com/news/next-gen-technologies/kotak-mahindra-bank-is-experimenting-with-ai-and-blockchain-deepak-sharma-president-chief-digital-officer/104052705
30. Lin TCW (2016) Temple university legal studies research paper no. 2017-06. Brook J Corp Fin Com L 159. Available at SSRN: https://ssrn.com/abstract=2904664
31. Deepthi B, Gupta P, Rai P, Arora H (2022) Assessing the dynamics of AI driven technologies in Indian banking and financial sector. Vision 0(0). https://doi.org/10.1177/09722629221087371
32. Anon (2017). How chatbots are redefining banking experience? (Online). Available at: https://www.hdfcbank.com/personal/resources/learning-centre/digital-banking/chatbots-in-banking
33. Anon (2020) How AI is being used in Linding. (Online). Available at: https://www.bajajfinservmarkets.in/discover/personal-loan/loan-assessment-with-artificial-intelligence.html
34. Anon (n.d.) Artificial intelligence in loan assessment: how does it work? (Online). Available at: https://www.icicibank.com/blogs/personal-loan/artificial-intelligence-in-loan-assessment-how-does-it-work

35. Cao L (2018) AI in finance: a review. ACM Comput Surv 55(3):1–38, article no. 64. https://doi.org/10.1145/3502289
36. Mandala GN, Buddhi D, Arumugam M, Harbola S, Othman B, Almashaqbeh HA (2022) A critical review of applications of artificial intelligence (AI) and its powered technologies in the financial industry. In: 2022 2nd International conference on advance computing and innovative technologies in engineering (ICACITE), Greater Noida, India, pp 2362–2365. https://doi.org/10.1109/ICACITE53722.2022.9823776
37. Sharma N, Agarwal R (2021) Enhancing customer experience in Indian banking through AI-powered solutions. Banking Finan Rev 15(2):78–93
38. Kumar S, Gupta R, Agarwal P (2020) Credit risk assessment using AI: transforming lending in India. Int J Financ Stud 8(3):67–84
39. Rao P, Srivastava M (2019) Fraud detection in Indian banks using artificial intelligence. J Financ Crime Prev 14(4):301–315
40. Gupta R, Singh V (2022) Algorithmic trading and market efficiency: a study on Indian financial markets. J Financ Markets 18(1):23–37
41. Bansal A (2021) The impact of AI on the financial services industry in India. Financ Innovators J 12(2):45–59
42. Sharma VS (2023) Leveraging advanced AI capabilities to better financial services for our users. (Online). Available at: https://paytm.com/blog/investor-relations/leveraging-advanced-ai-capabilities-to-better-financial-services-for-our-users/#:~:text=Adding%20to%20it%2C%20Paytm%20is,payment%20technology%2C%20software%20and%20hardware
43. Tewari SB (2023) The revolutionary transformation of India's banking industry through artificial intelligence. In: 14th International conference on computing communication and networking technologies (ICCCN). Delhi, pp 1–5
44. Lin TCW (2019) Artificial intelligence, finance, and the law. 88 Fordham L Rev 531. Available at: https://ir.lawnet.fordham.edu/flr/vol88/iss2/6
45. Johnson K (2019) Artificial intelligence, machine learning, and bias in finance: toward responsible innovation. Fordham Law Rev 499–529
46. Kunwar M (2019) Artificial intelligence in finance: understanding how automation and machine learning is transforming the financial industry. Thesis, Centria University of Applied Sciences
47. Lin TCW (2016) Compliance, technology, and modern finance 11
48. Malali A, Gopalakrishnan S (2020) Application of artificial intelligence and its powered. IOSR J Humanit Soc Sci 25(4):55–60
49. McCarthy J (1955) A proposal for the Dartmouth summer research project on artificial intelligence. Retrieved from http://jmc.stanford.edu/articles/dartmouth/dartmouth.pdf

Financial Governance: A Roadmap for Economic and Environmental Sustainability

Arya Kumar

Abstract One way to guarantee that financial flows are compatible with resilience to climate and low-carbon growth is to reform financial institutions. Creating sustainable financial plans is a common initial step in this direction. The study interviews important Indian stakeholders to analyze the progress in implementing Indonesia's sustainable finance plan in 2019 and 2020. Based on their yearly sustainability assessments and commitment to producing financially sustainable strategies, the research concludes that financial institutions are highly compliant with the relevant procedures. However, there is a lot of variance and ambiguity in how financial companies define what counts as a "green" initiative, allowing specific financial organizations to participate in nothing more than symbolic gestures. The methods to encourage financial companies to take additional action and achieve accurate sustainability outcomes through existing regulatory oversight are still under consideration. More action to encourage climate financing by India's central bank is one solution this study offers for conquering barriers to the roadmap's adoption.

Keywords Sustainability finance · Carbon footprint · Governance · Financial roadmap · Climate change risk

1 Introduction

The Intergovernmental Panel on Climate Change [1] recommends quick, comprehensive, and historic transformation throughout all societal sectors to prevent disastrous climate changes. Modifications should accelerate the low-carbon economic revolution. By 2035, these initiatives might cost USD 93 trillion [2]. With financing from the government drying up, enterprises, institutional investors, financial institutions, and project developers must pay for low-carbon infrastructure, technology,

A. Kumar (✉)
School of Economics and Commerce, KIIT Deemed to Be University, Bhubaneswar, Odisha, India
e-mail: arya.kumarfcm@kiit.ac.in

S. Roy et al. (eds.), *Humanized Intelligence in Sustainable Finance and Economic Behaviours*, Studies in Computational Intelligence 1226,
https://doi.org/10.1007/978-981-95-0599-9_6

other climate finance, and partnerships. Due to its substantial financial resources and widespread connections to various industries, the evolution to a low-carbon economy and bringing about significant transformation can be expedited by the financial sector, which has great potential.

However, climate financing presents both risks and opportunities. Carney [3] suggests that by channeling trillions of dollars toward the change to a low-carbon economy, the financial sector can reap substantial benefits. Nonetheless, Durrani et al. [4] and Carney [3] caution that if the financial sector does not adequately address its environmental risks, the shift to a low-carbon economy may show an adverse properties on the financial structure. Most significantly, legislation and governance that best enable, incentivize, and compel financial institutions to act to hasten that transformation are needed to guide the financial system toward a speedy low-carbon transformation without disruption [5].

The Sustainable Financial Systems Roadmaps, which advise the financial industry on promoting sustainable and resilient economies, show that governments globally are either engaged in or preparing for an environmentally friendly finance sector. More Asia-Pacific financial and banking regulators are attempting to convince financial companies to apply governance standards, social and environmental, and investments opportunities [4].

There has to be more empirical study into building sustainable finance roadmaps in different nations, finding significant barriers to adequate climate finance oversight, and investigating solutions. Most climate finance research assumes a healthy capital market that can be diverted to low-carbon investment, as Anantharajah and Gunningham [6] note. Most Asian countries, like India, have a segmented capital market and underdeveloped financial infrastructure.

India's early adoption of sustainable finance makes it a perfect instance for an empirical study that cannot be repeated in countries only now exploring such challenges. The Government of India announced one of the first sustainable finance roadmaps in 2014 to scale up sustainable financing to promote strong and sustainable development. The influence of the sustainable financing plan on India's low-carbon growth is still unknown. This article aims to evaluate the advancements made in implementing the roadmap, determine among the most pressing governance challenges preventing its complete effective execution, and suggest ways Indian financial companies can accelerate the move to a more resilient and environmentally friendly economic model.

The paper is structured as follows: After this introduction, the study's theoretical structure and research methods are explained. The third section discusses Indian sustainable financing. The fourth section examines how far the sustainable finance plan process for implementation has progressed and identifies three significant barriers. In the fifth section, the results and findings are presented, while the sixth section provides conclusions and explores strategies to expedite sustainable financing in India.

1.1 Motivation

The goal of this chapter is to learn more about India's sustainable financial policies and environment. India is one among the world's greatest economies; thus, it plays a significant role in the fight against climate change. Recognizing India's fiscal sustainability efforts is crucial in light of the country's large GHG emissions and high climate change reduction targets. The purpose of this chapter is to provide insight into the potential and difficulties India confronts in obtaining private funding for sustainable development and making the transition to a low-carbon economy. The purpose of this chapter is to examine the nation's sustainable finance strategy and its development in order to determine the governance difficulties slowing efficient execution and to provide solutions that would speed up the move to a robust and sustainable economic system.

1.2 Contribution

This chapter adds to our knowledge of the difficulties inherent in Indian sustainable finance governance. This chapter examines India's sustainable finance strategy and its development to provide light on the country's attempts to move toward a low-carbon economy. To speed up the successful implementation of sustainable finance projects, it outlines critical governance concerns that must be solved. The results of this chapter are useful for policymakers, financial institutions, and scholars interested in climate financing and sustainable development because of the empirical inquiry done via interviews, which provides depth and context to the analysis. The overarching goal of this chapter is to provide actionable suggestions for fostering low-carbon growth and inspiring Indian financial institutions to take a larger role in creating an adaptable and sustainable economic model.

1.3 Chapter Organization

This chapter is categorized into several sections like introduction as first section, followed with methodology and analysis overview, third section deals with India's sustainable financial landscape and policies, and fourth states about the findings followed with discussions and conclusions.

Table 1 Contribution of researchers in similar context

Year	Researcher(s)	Discussed topics
2020	Durrani et al. [4]	Role of central banks in scaling up sustainable finance in Asia-Pacific
2020	Anantharajah and Gunningham [6]	Mobilizing private climate finance in emerging Asia
2020	Weikmans and Roberts [9]	International climate finance accounting and its challenges
2021	Gunningham [5]	The impact of crime, regulation, and climate finance
2023	Campiglio et al. [12]	Green central banking in emerging markets and developing countries

2 Methodology and Analysis Overview

2.1 *Overview of the Analysis*

2.1.1 Financing of Climate Management

The allocation of climate funds is an emerging area of study and policy. Recent academic publications by G20 [7] and UNEP-FI [8] have examined climate finance governance, assessing global efforts in climate funding [9, 10] analyzing the evolving framework for global climate finance. However, the research and theory surrounding the financing of climate management remain relatively limited.

Three essential components are needed to handle climate-related financial market disruptions [3]. The first step is correct information. Informational oversight, where the state requires the release of a business's social and environmental effects, could help investors comprehend the risks and rewards of financing in a specific corporate [5, 11]. Secondly, companies and financial organizations must identify, analyze, and deal with risks associated with climate change. Finally, governance structures have to be trustworthy and coherent. Table 1 presents about the various past researchers contributing in similar area.

2.1.2 Financial Governance

According to the socio-legal texts on the governance of the natural world, there have been three broad transitions in how the environment and, in recent years, the climates are governed: From various tiers of governance (such as the administrative state) to market economies (guided by neo-liberalism) to broader systems [13], the influence of different mechanisms can be observed. However, authorities aim at a moving goal (the third phase's difficulties are distinct from the first). Multiple mechanisms coordinated by numerous players, frequently from various industries, involving governments, corporations, and third parties may be best [14]. Complicated policy challenges need complex single-actor and single-instrument techniques.

2.1.3 Climate Change Risk

According to numerous recent research [4, 12, 15, 16], financial institutions may integrate climate financing and integrate risks related to climate change into their core policy structures. Financial institutions have been attacked for addressing the effects of climate change, but recent evidence reveals that undertaking so is essential to their primary goal of ensuring financial and price stability [4, 12, 15, 17]. They demonstrate climate change has the impact with direct, significant responsibility and shifting risks on the core financial and monetary stability mandates of central banks. They emphasize the importance of central banks integrating risks connected with change in climate, as well as governance, social, and environmental factors.

This article examines environmental finance governance challenges in India. It examines how India's finance regulator and central bank can promote low-carbon development and implement its sustainable finance plan.

2.2 Methods and Data Collection

As an element of a bigger investigation into how to finance the transition to renewable energy and combat climate change in the Pacific and Asia, this report is based on an empirical investigation conducted in Jakarta, India, from February 2019 to March 2020. It involved the participation of seven government officials and included a total of thirty-three semi-structured interviews, seventeen bank/financial organization officials, four non-governmental agency/research organization participants, and five international/multilateral organization members. Participants were asked about their sustainable finance activities in India. Interviewees were asked critical questions about executing the strategy. Interviews lasted 30 min.

This research examined laws, regulations, the sustainable financing strategy, and eight commercial banks' annual reports on sustainability. They are closer to sustainable financing than other banks. Qualitative data studies employed grounded theory, discourse analysis, and content analysis [18]. Trends and main ideas were identified [19]. By using both primary and secondary information to answer identical queries, triangulation increased the information's reliability, credibility, and predictability.

3 India's Sustainable Financial Landscape and Policies

The region's largest economy, India, emits a lot of GHGs [20]. Rising energy consumption is predicted to have a far more significant impact on the country's carbon footprint than reforestation and burning carbon-rich peat lands [21]. Demographic and economic expansion will boost emissions of carbon in the latter by 80% by 2030 [22]. Given the agreement of Paris, India established an ambitious climate change mitigation goal: a decrease of 29% in carbon emissions from the industry average by 2030 utilizing domestic money or 40% with assistance from abroad [23].

India will need USD 24.8 billion a decade to decrease the carbon it releases to its target. Since state funding and foreign development support are insufficient, India's financial sector is dominated by banks, making it crucial to restructure the sector in order to finance low-carbon activities. Private sector investment is also of great importance in this regard. Insurers and financial companies control 10.5%, 6.4%, and 2.5% of assets. Low-carbon construction initiatives nationwide need bank finance. Indian banks only lend a limited amount to green projects. Shareholders may engage in low-carbon initiatives with enough knowledge. Inadequate requirements for disclosure of environmental or risky situations over time aggravate it. The country needs more adaptability or mitigation initiatives.

Over the years, the government has been exploring alternative methods to attract private financing for low-carbon development. The primary responsibility lies with foreign initiatives that aim to incorporate sustainable finance and provide education to government and industry leaders. The IFSC created a sustainable finance plan in 2014. The International Financial Services Centers (IFSC) [24] provides an extensive plan for improving the availability and demand of sustainable financial services and oversees and works together with other government departments in order to carry out the sustainable finance goals.

Phase one (2014–2019) prepared financial institutions for integrating sustainable financing and created laws and regulations to govern it. IFSC started out concentrating on financial organizations' capacities. The IFSC established a sustainable finance regulation [24] to encourage environmentally friendly borrowing, improve financial institutions, and boost their ability to compete. In 2018, technical advice was provided to help financial institutions comply with sustainable finance requirements. IFSC [24] requires criteria for green bond issuance to promote sustainable finance tools.

The IFSC [24] rule compels financial companies to provide sustainable finance initiatives and post yearly sustainability reports to educate shareholders and the public about their sustainability efforts. The analysis should contain sustainable financial strategies, including green initiative investment, transparent funding, and risk management.

Furthermore, the IFSC has developed a pilot program dubbed "First Movers on Sustainable Banking," where eight commercial lenders representing forty-six percent of the country's financial assets will try to innovate in sustainable finance.

Green Banking. They improved sustainable oil palm finance regulations and generated reports on sustainability via capacity-building activities. These firms also created the "network on sustainable finance" to share insights [24].

4 Findings

4.1 Sustainable Finance Roadmap Implementation Status

India is making great strides in implementing a sustainable finance strategy. It is clear from an examination of interview information and reports that financial companies have shown a comparatively high degree of conformity with the requirements for sustainable financing set out in IFSC [24]. These requirements include submitting sustainable finance plans of action and publishing sustainability evaluations on a yearly basis, which provide updates on the implementation of the action plans. In 2019, sustainable finance action plans were filed by 80% of commercial banks [24]. However, there are no apparent repercussions for noncompliance with the sustainable finance rule due to the lack of disincentive measures.

Although yearly sustainability reports did not become necessary until 2018, most banks have already released them. All pioneering financial institutions have provided action plans for sustainable finance and released sustainability reports (Annex 1). Seven of eight early adopter financial institutions now have a written ESG policy [24]. Regarding developing sustainable finance solutions, some banks are more proactive than others. In an interview, a bank spokesperson explained in further detail how.

Our bank had a rough beginning in the sustainable finance project. It was feared that the sustainable finance platform would slow down our plans for corporate growth. However, we seized the opportunity to pioneer the field and establish the standard for the whole nation. Others will follow suit if a significant financial institution like ours is on board. We pioneered green bonds, which inspired imitators.

Eco-financial products are growing too. "Green bonds," environmentally friendly debt financing, have been offered by many central banks. In 2018, India issued $1.25 billion in sovereign sukuk green bonds, which are sharia-compliant green bond-type securities. The purpose of these bonds was to finance sustainable activities, such as waste management, renewable energy, green tourism, and energy efficiency. This development has led numerous banks to declare USD 3.2 million in green sukuk and USD 669 million in green bonds [24]. Another rising financial technique for environmentally friendly initiatives is blended finance, which promotes sustainable development through open-to-everyone, charitable, and private funds. In 2019, PT SMI invested four million dollars in environmental infrastructure initiatives.

Figure 1 shows the growing percentage of the borrowing portfolio dedicated to sustainable finance's main priorities. Sustainable agriculture is expected to receive about USD 108 million in financing in 2018. The oil palm business received 7.3% of USD 1466 million in loans during that period. The energy from renewable sources

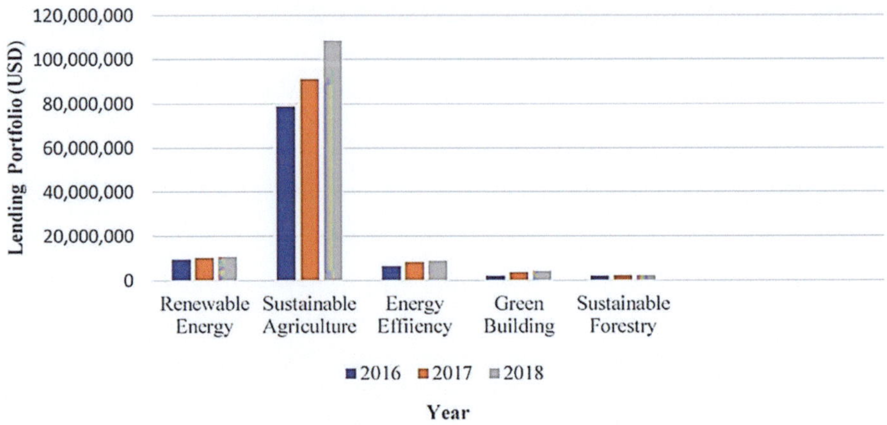

Fig. 1 Modified from IFSC's 2019: Financial sustainability plan, the environmentally friendly lending portfolio focusing on five key industry sectors

loan was USD 10.6 million, next to environmentally friendly farming. Palm oil exports 2017 were USD 23 billion, boosting India's economy [25]. Tax concessions were given to help entrepreneurs build oil palm farms and meet the government's high output goal [26, 27]. Thus, large banks' interest in funding the business is unsurprising.

In the initial stage of the Sustainable Finance plan, IFSC chose the palm oil sector as an experiment for integrating sustainability principles into first-mover banks' financing choices due to its importance and possible hazards to society and the environment. The financial sustainability NGO has published a financing policy for the oil palm business that considers environmental and social variables.

Indian economic growth relies on oil palm farming. This sector's environmental issues are well known. Since Indian central banks lend to oil palm companies, sustainability criteria must be mainstreamed. Thus, IFSC chose a plantation of oil palms by strengthening due diligence and helping banks discover financing mistakes.

All first mover bank officials said businesses seeking oil palm cultivation loans must demonstrate an environmentally friendly oil palm certification obtained from the RSPO or ISPO.

Bank India's involvement in developing green bonds has decreased after IFSC took over the financial sustainability plan. By joining the Network for Greening the Financial System (NFGS) [28] in November 2019, an international consortium of banking institutions and monetary regulators, it has positioned itself to show a more substantial part in future. The NFGS aims to enhance the financial sector's ability to prevent risks and channel funds toward sustainable and low-carbon investments.

4.2 Financial Sustainability Plan Problems with Governance

As shown below, notwithstanding early promising indicators of successful execution, a deeper review of the participant's information and secondary documentation shows severe and continuing obstacles in executing the sustainable finance plan.

4.2.1 Regulative Problems

Financial companies have struggled to create an environmentally friendly finance strategy and integrate it into their operations practices within present laws and regulations. Despite providing technical guidance to help financial companies cope with green finance legislation, 80% of financial institution officials' questioned said they need help understanding it. The regulation also lacks markers to measure adherence. Overall, an IFSC assessment suggests a high percentage of compliance by financial companies and banks to establish and implement the strategy.

Report compliance yearly [24], although it has to be seen if concrete sustainable benefits may be attained. The assessment of sustainable finance reveals that individuals within pioneering financial institutions exhibit a comprehensive understanding of a wide array of sustainable finance practices. Nevertheless, it is noteworthy that two out of eight financial institutions do not explicitly outline their primary sustainable operations, instead placing greater emphasis on the CSR and humanitarian dimensions of environmental friendly finance. Furthermore, only a solitary bank provides transparency regarding the regions it finances, thereby underscoring potential risks.

According to IFSC [24], the legal guidance, green initiatives/products are defined as those that adhere to the criteria of sustainable finance. This encompassing definition encompasses twelve distinct categories of environmentally friendly endeavors, including waste management, renewable energy, green tourism, and energy efficiency. In assessments, most financial institution officials said the green project definition was comprehensive, resulting in various interpretations of their sustainable finance holdings. For instance, a regional bank official advised that the financial institution included creating toll highways in its financially viable portfolio.

Financial sustainability includes toll roads. Since toll roads necessitate tree cutting, some may question whether this project is sustainable. The toll road might cut traffic and carbon emissions by speeding up automobiles.

Some situations do not need greater clarification. BNI, a state-owned bank developer in green finance, has filed a sustainability report clarifying its environmental plans, green banking rules and regulations, and internalizing sustainable practices.

However, banks sometimes have unclear, unique definitions of "green" assets. There are other causes besides the regulation's lousy drafting. Importantly, all financial institution officials questioned said sustainable finance measures are insufficient. IFSC [24] incentivizes financial sustainability conformity, particularly building capacity and rewards. They argued that such rewards were symbolic and insufficient to encourage their organizations to pursue and integrate financial sustainability into their business practices. One bank official proposed incentivizing developers of projects and financial institutions to increase green initiatives and assets.

Financial authorities' inadequate funds and technical expertise to supervise plan execution compound the issues above. During the period of the research, it was observed that a mere two government officials were assigned the responsibility of overseeing the plan and its corresponding regulations. This encompassed a vast network consisting of sixteen hundred rural banks, 119 commercial banks, as well as numerous non-bank financial bodies and financial markets [24]. Financial organizations' sustainability reporting may be ambitious, with few natural, sustainable achievements despite excellent compliance with procedures and inadequate control over substantial adherence. Halimatussadiah et al. [29] said that research on Indian domestic and foreign banks' environmentally friendly finance attitudes found that local banks lack sustainability finance capabilities, training, and knowledge of global best practices. Thus, these institutions' green finance portfolios are random, intermittent, and unrecorded, hindering sustainable finance's capabilities.

4.2.2 Inadequate Risk Management

Environmental change threatens the economy, but handling risks may reduce the risk [3]. Financial institutions and financial authorities globally are beginning to realize that ESG11 oversight of funding and borrowing decisions is essential to stability and economic and financial sustainability. ESG-compliant financial firms additionally do better financially [30, 31].

The incorporation of environmental and social factors into investment and lending choices has increasingly been increased in the nation's economic and financial laws and regulatory structures. The Banking Law 1998 requires financial institutions to conduct environmental impact assessments (EIAs, or AMDALs in India) for big or risky loans. Bank India advises banks to examine debtor ratings. The IFSC's technical suggestions on financial sustainability execution urge Indian banks to establish mechanisms that incorporate governance, social, and environmental concerns into handling risks [24].

Most banks must disclose their ESG policies on lending for extremely hazardous industries, and the IFSC's surveillance of the incorporation of ESG into institutions' risk mitigation strategies requires improvement. Most financial organizations need a sufficient EIA for their investment and lending choices. They frequently verify the borrower's EIA license without additional impact investigation. Several banks positively employ PROPER to evaluate an organization's environmental impact [32]. However, respondents stated that PROPER needed to be an excellent instrument to judge corporations' productivity owing to uncertain regulations and an extensive amount of red tape whereby a company can unfairly get a good grade. This corroborates previous research findings that highlight the inadequate instruments and capabilities of financial institutions within the country to effectively assess credit-related social and environmental risks [29]. It may distort risk-return assessments and slow financially viable industry growth.

Participants in the interview said that institutions seek RSPO and ISPO certification from oil palm cultivation loan applicants. Notwithstanding, it is imperative to acknowledge that the endorsement of corporate propositions might be contingent upon the presentation of a missive expressing the intent of aspiring borrowers to secure third-party accreditation from recognized entities such as RSPO or ISPO. Meanwhile, the bank shall remain vigilant supervision over the borrowers' operational efficacy. It is uncertain whether they would pull out of the financing arrangement without such a certification. Banks seldom analyze environmental and social variables.

Greening oil palm fields is still being determined. Regardless of its benefit to global economic growth, oil palm cultivation development has been blamed for conversion to forest and peat land clearing [33, 34], resulting in significant biodiversity loss, carbon emissions, and adverse social consequences like poor working conditions and land dispossession [35, 36]. Third-party certifications may not ensure the sustainable cultivation of oil palm [37, 38]. Hidayat et al. [38] proposed that ISPO, an Indian government-mandated third-party oil palm certification, has been criticized for not dealing with serious issues resulting from the industry's growth, such as social conflicts, loss of biodiversity, and deforestation.

4.2.3 Political Issue

Financial companies and regulatory authorities have different goals that make the execution of plans complex, according to this study. India's public banking/financial institutions need to employ sustainable finance. Interviews with bank officials showed a range of desire to popular it, notably from senior management. Financial organizations' long-term investment strategies are still restricted to overall growth. A government officer in the financial regulatory agency explained that establishing and carrying out more significant goals for success in the long-term financial plan might render financial companies less attractive to investors and have unforeseen adverse effects on the nation's economy. The person in charge advised to slow down while proposing greater financial sustainability ambitions. As authorities, we can block banks from financing coal mines and power facilities. We may, but whatever will occur to these industries' businesses? Who is going to supply financing if banks cannot? More ambition may hurt the economy.

A recent review of sustainable finance's sectorial composition reveals that India's national financial institutions have yet to formally pledge to transition from petroleum-based investments to energy from renewable sources financing. One bank in India reveals its conservation of energy and green energy investment portfolio, whereas most international banks in India declare lower emissions objectives. Regional financial institutions prefer coal power production over renewable as coal-powered generators are cost-effective and energy from renewable sources is risky.

Financial regulators' objectives for policy have changed due to management changes. The IFSC chairman championed sustainable finance in the first phase of the strategy. In 2019, a new IFSC leader changed objectives, jeopardizing the sustainable finance strategy. When the head of the organization has an alternate goal, integrating environmentally friendly finance is highly burdensome. We concentrate on virtual financial technologies. Policy talks reduced sustainability concerns. Innovation and environmental responsibility are the next generation of finance [39].

This adjustment prevented the subsequent stage of the sustainability plan from being produced during fieldwork. It is unclear if, where, and how much the plan in the second phase will incorporate more clear benchmarks and goals to speed up funding for carbon-free growth.

5 Discussion

IFSC, the financial sector regulator, develops and oversees regulatory structures under the environmentally friendly finance agenda. This research finds that financial companies follow procedures when creating sustainability strategies and filing yearly sustainability reports. However, regulatory obstacles hinder its adoption. The legislation lacks specific measures to monitor substantial adherence and basic incentive/disincentive systems for integrating sustainable finance. Financial organizations' definitions of green projects vary widely. Financial firms may have developed sustainable finance policies without concrete indications. They may also mislabel products as green.

The study concludes as follows: Due to significant resource limitations, the financial service regulator cannot offer regulatory supervision. The regulator fears that overregulating sustainable finance could harm financial organizations' competitiveness. The plan's latest reprioritization may hinder India's economic sustainability objective.

The limited substantial execution of the financial sustainability plan must be considered in light of the nation's political-economic difficulty. The nation's desire to cut emissions of carbon and development is in conflict.

Economic growths perceived requirement. This conflict makes it difficult to execute the plan since the latter outweighs the former. The existence of governmental statutes and regulations that lend support to the proliferation of high-carbon growth, exemplified by the domains of coal-based electricity generation and oil palm cultivation, undeniably manifests a significant upsurge in investments and financial allocations specifically targeted at these industries.

While considering reforming the investment and financing procedures in the oil palm cultivation industry by forcing new projects to fulfill an environmentally friendly norm, this proved to have limited influence on current funding and industry expansion. It will undoubtedly give rise to more oil palm plantings in forests. A recent study found that oil palm cultivation is still expanding in densely forested regions, especially in Kalimantan and Papua [27]. Financial institutions have not pledged to end petroleum investment. The more significant diseases associated with energy administration in India, including a complicated regulatory framework and mismatched regulations, have hindered green energy development [40].

The long-term viability of India's sustainable plan has yet to be discovered. Financial institutions' readiness to implement sustainable finance initiatives may determine the plan's viability without government involvement. Financial companies in the industrialized nations have become more conscious of their obligation to incorporate environmental risks into their company's making decisions and are shifting beyond high-carbon investments, conscious that environmentally friendly alternatives may assist these individuals in dealing with sustainability problems while enhancing their financial results [30, 31]. In India, financial firms seldom go beyond regulatory requirements to achieve sustainable finance. Self-monitoring will remain the same despite an attractive financial story and few brand benefits.

6 Conclusion

This paper contributes to the existing body of research on climate financing by delving into the intricacies of formulating and implementing a sustainable finance strategy within the context of a developing nation, such as India. Additionally, it highlights and elucidates three critical governance concerns associated with this endeavor. It shows that regardless of financial institutions' excessively noted adherence to the government's sustainable finance laws and regulations, despite adequate oversight from regulators and not certain disincentive and incentive processes, it is not clear the way this adherence can contribute to apparent long-term results. The government's strategic move casts doubt on the plan's viability.

The sustainable finance plan may be considered a necessary first step, but it failed to help India green its financial landscape, and there needs to be evidence that it will. How can we expand low-carbon finance?

The state, particularly the banking system and financial regulators, will need to define terms, require player involvement, and regulate financial markets [5]. IFSC encourages low-carbon financing and ensures critical stakeholders engage in financing for climate change in India. Nevertheless, the IFSC needs additional funding to implement sustainable finance laws. Improve resilience to refine sustainable finance legislation.

Transparency and handling risks are essential for sustainable financing. These metrics ought to monitor financial sustainability strategy implementation. To promote financial sustainability, specific reward and disincentive measures are needed. Employing distinct and targeted requirements, the financial regulator has the potential to incentivize and facilitate specific types of investments and offerings. This can be achieved through measures like imposing obligatory reserve rates on sustainable goods, or implementing disparate capital necessities that dynamically adjust ratios of capital adequacy in accordance with the attributes of financial institutions and the nature of their offerings. This represents a sophisticated approach to regulatory measures in the financial realm. By implementing such measures, an environment is created that fosters the growth and development of these specific investment types and offerings. Practical, informative requirements for disclosure for financial institutions and other banking on their reliance on carbon-intensive assets will improve the distribution of capital and assist investors interested in assessing the environmental hazards of investing in a company [12, 17].

The Reserve Bank of India (RBI), operating within a comprehensive legislative framework prioritizing price stability and inflation targeting, undertakes multifaceted responsibilities. While fulfilling its primary mandate, it concurrently assumes an implicit catalyst role in fostering sustainable finance, aligning seamlessly with India's overarching sustainable development agenda. To effectively discharge this pivotal role, the RBI wields an array of strategic options. Leveraging these options, the RBI can deploy incentivization mechanisms that promote investments characterized by low-carbon footprints, while seamlessly incorporating environmental, social, and governance (ESG) considerations into the bedrock frameworks governing policy implementation. A notable and viable strategy lies in the adoption of green macro prudential regulations, wherein lending limits are imposed on banks operating within sectors or entities entangled in carbon-intensive business models, thus facilitating the sustainable finance agenda. Additionally, carbon-intensive assets may be assigned higher risk weights in accordance with banks' capital requirements, while capital requirements themselves could be heightened in response to credit growth targeted toward capital investment. These measures have been extensively explored and discussed in studies conducted by Cecot [17].

To avoid financial sector risks related to climate change, Bank India might conduct a test and contingency design with numerous stakeholders. This method entails the creation of various simulation models that represent distinct low-carbon economic growth routes [4, 5, 17]. Enhancing capacity through informative disclosure, Bank of England's approach emphasizes collaboration with stakeholders, deriving potential benefits for Reserve Bank of India. Gunningham [5] highlights the importance of scenario assessment tools, facilitating adaptability testing to diverse climate-related events and outcomes. This approach empowers financial institutions to refine their methods and instruments while bolstering investors' ability to make informed decisions. This method has improved stakeholder comprehension and mainstreamed sustainable investing.

Given Bank India's commitment to strengthening climate change financing, it should leverage the governance of networks [41]. The likelihood of immediate development in the area of financial sustainability appears challenging, considering the partial knowledge and expertise of most Indian financial organizations in this domain, coupled with their current high-profit earnings from traditional operations. However, as highlighted by Durrani et al. [4], the Reserve Bank of India (RBI) has taken a significant step by joining the Network for Greening the Financial System (NFGS) with a dedicated focus on "macro-finance" and "scale-up sustainable finance." Acknowledging the status of financial regulators and central banks in addressing climate change with regard to financial issues, the participants of NFGS, as emphasized by NFGS [28], recognize the need for active stakeholder involvement to ensure effective governance of sustainable finance. To overcome the existing challenges, it is imperative for financial institutions to proactively integrate financial sustainability into their operations. They will recognize a need to expand their feasible financial abilities above their present responsibilities with no specific disincentive/incentive strategies from the government. Financial companies must be incentivized to embrace sustainable finance via building capacity and regulation changes.

Lastly, assisting the development of low-carbon sound finance administration requires several regulatory strategies and instruments organized by many players. Central banks and regulators of finance have complementary responsibilities in reforming the financial system [17]. The governing body must establish and execute an overall regulatory structure and climate change strategy to allow and support the financial sector's low-carbon growth.

The literature presented in this study on India's sustainable finance initiatives offers a valuable roadmap for the country's transition to a low-carbon economy. To effectively implement and reap the benefits of these insights in future endeavors, India should consider a multifaceted approach. Firstly, regulatory revisions are imperative, with a need for clear and comprehensive measures that promote adherence to sustainable finance principles and provide incentives for financial institutions to embrace them. Standardization of what qualifies as "green" projects is essential to ensure consistency in sustainable finance implementation. Furthermore, raising public awareness about the significance of sustainable finance can create a robust demand for eco-friendly investments and encourage financial institutions to expand their sustainable offerings. Government commitment is crucial, as setting ambitious sustainability goals and aligning economic growth with low-carbon development can pave the way for a more environmentally conscious economy. Lastly, fostering collaboration between financial regulators, institutions, and environmental organizations can create a shared commitment to sustainable finance and facilitate the exchange of best practices. By adopting these strategies, India can expedite its journey toward a resilient and eco-friendly economic model, benefiting both the environment and the nation's prosperity. Figure 2 flowchart illustrates the flow of the theme, from addressing the present challenges and opportunities in sustainable finance to ultimately supporting future endeavors. The examples provided within each section showcase real-world applications related to the respective factors and variables.

7 Scope for Further Research

India's sustainable financial structure and regulations provide attractive and diversified study opportunities. To measure the sustainable finance plan's influence on climate change and green investments, it must be comprehensive and impactful. This study should examine India's sustainable development, carbon emissions reduction, and low-carbon project finance. To identify banks and financial institutions' barriers to supporting green projects and create creative measures to promote and reward sustainable financing, the financial sector's transition must be examined. Evaluating India's sustainable finance regulatory system would reveal gaps and opportunities,

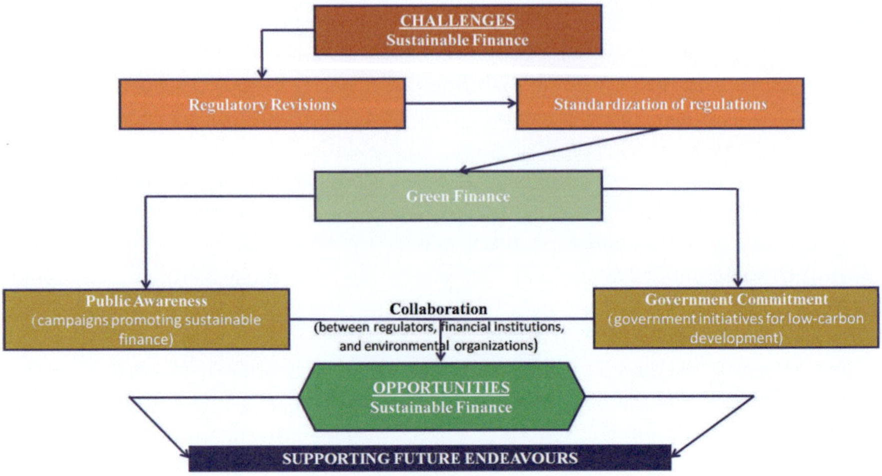

Fig. 2 Flowchart converting the challenges to opportunity for sustainable finance. *Source* Author creation

enabling more effective and incentivizing legislation to encourage financial businesses to adopt sustainability. To understand how environmental, social, and governance (ESG) factors affect lending decisions and risk exposure, Indian financial institutions must integrate ESG into risk management. To combine economic growth with climate change goals, India's sustainable finance strategy must be reviewed for long-term viability, paying special emphasis to environmental sustainability programs' synergies and drawbacks. To understand how international partnerships affect India's sustainable finance efforts, examine India's cooperation with overseas programs like the Network for Greening the Financial System (NGFS). Understanding how financial regulators, like the RBI, promote sustainable finance and finding areas for greater participation will encourage sustainable investments. Finally, sector-specific studies of oil palm output and renewable energy will help support sustainable initiatives and assess certifications and standards. These research strands will help India improve its sustainable financial strategy, expedite climate change mitigation, and create a greener, more resilient financial ecosystem.

References

1. Intergovernmental Panel on Climate Change (IPCC) (2018) Global warming of 1.5 °C. Retrieved 30 May 2023. http://report.ipcc.ch/sr15
2. UNESCAP (2017) Finance for climate action in Asia and the Pacific: a regional action agenda to access debt capital markets. Retrieved 28 May 2023, from https://www.unescap.org/
3. Carney M (2019) Fifty shades of green. Financ Dev 56(4):12–15

4. Durrani A, Rosmin M, Volz U (2020) The role of central banks in scaling up sustainable finance—what do monetary authorities in the Asia-Pacific region think? J Sustain Financ Invest 10(2):92–112
5. Gunningham N (2021) Crime, regulation and climate finance. In: Holley C, Phelan L, Shearing C (eds) Criminology and climate: insurance and the regulation of harmscapes. Routledge, pp 23–36
6. Anantharajah K, Gunningham N (2018) Mobilising private climate finance in emergence Asia. Retrieved 5April 2020, from http://www.climatefinanceinitiative.org/wp-content/uploads/2019/07/Working-Paper-1-Emerging-Asia.pdf
7. G20 (2018) Sustainable finance study group: Synthesis report. Retrieved 11 Nov 2018. www.unepinquiry.org/wp-content/uploads/2018/11/G20_Sustainable_Finance_Synthesis_Report_2018.pdf/
8. UNEP FI (2018) Making waves: aligning the financial system with sustainable development. Retrieved 12 Nov 2018. www.unepinquiry.org
9. Weikmans R, Roberts JT (2019) The international climate finance accounting muddle: is there hope on the horizon? Climate Dev 11(2):97–111. https://doi.org/10.1080/17565529.2017.1410087
10. Pickering J, Betzold C, Skovgaard J (2017) Managing fragmentation and complexity in the emerging system of international climate finance. Int Environ Agreements Polit Law Econ 17:1–16
11. Kumar A, Biswal SK (2019) Impulsive clustering and leverage effect of emerging stock market with special reference to Brazil, India, Indonesia and Pakistan. J Adv Res Dyn Control Syst 11:33–37
12. Campiglio E, Dafermos Y, Monnin P, Ryan-Collins J, Schotten G, Tanaka M (2018) Finance and climate change: what role for central banks and financial regulators. Nat Climate Chang 8(6):462–468
13. Gunningham N, Holley C (2016) Next-generation environmental regulation: law, regulation, and governance. Ann Rev Law Soc Sci 12:273–293. https://doi.org/10.1146/annurev-lawsocsci-110615-084651
14. Gunningham N, Sinclair D (2017) Smart regulation. Retrieved 11 Nov 2018, from http://press-files.anu.edu.au/downloads/press/n2304/pdf/ch08.pdf/
15. Dikau S, Volz U (2019) Bank mandates, sustainability objective and the promotion of green finance. SOAS working paper. Retrieved 21 May 2020. https://www.soas.ac.uk/economics/research/workingpapers/file145514.pdf/
16. Dikau S, Ryan-Collins J (2017) Green central banking in emerging market and developing country economies. https://eprints.soas.ac.uk/24876/1/Green-Central-Banking.pdf
17. Cecot M (2019) Climate emergency and central banks (12 Dec 2019). Retrieved 21 May 2020, from https://doi.org/10.2139/ssrn.3509012
18. Hajer M, Versteeg W (2005) A decade of discourse analysis of environmental politics: achievements, challenges, perspectives. J Environ Planning Policy Manage 7(3):175–184. https://doi.org/10.1080/15239080500339646
19. Tavory I, Timmermans S (2014) Abductive analysis: theorizing qualitative research. University of Chicago Press.
20. Edwards RB, Naylor RL, Higgins MM, Falcon WP (2020) Causes of Indonesia's forest fires. World Dev 127:104717
21. World Resources Institute (WRI) (2018) How can India achieve its climate mitigation goal? An analysis of potential emissions reduction from energy and land use policies. Retrieved 1 June 2023. http://www.wri.org/
22. IRENA (2017) Renewable energy prospects: India. Retrieved 2 June 2023. https://www.irena.org/
23. GoI (2016) Intended nationally determined contribution republic of India. Retrieved 2 Oct 2018. https://www4.unfccc.int/sites/submissions/INDC/Published%20Documents/India/1/INDC_REPUBLIC%20OF%20INDIA.pdf

24. International Financial Services Centres (IFSC) (2015) Securities Exchange Board of India (International Financial Services Centres) guidelines, 2015. Retrieved 1 June 2023
25. Reily M, Ekarina (2018) Rekor tertinggi, ekspor minyak sawit 2017 tembus US$ 22, 9 miliar. Katadata News. Retrieved 3 March 2023. https://katadata.co.id/berita/2018/01/30/rekor-tertinggi-ekspor-minyak-sawit-2017-tembus-us-229-miliar
26. Kumar A (2018) Prospects of GST on Indian economy-an econometric approach. Int J Manage IT Eng 8(7):147–164
27. Purnomo H, Okarda B, Dermawan A, Ilham QP, Pacheco P, Nurfatriani F, Suhendang E (2020) Reconciling oil palm economic development and environmental conservation in Indonesia: a value chain dynamic approach. Forest Policy Econ 111:102089
28. NFGS (2019) A call for action: climate change as a source of financial risks. Network for Greening the Financial System, Paris. Accessed 28 May 2023. https://www.banquefrance.fr/sites/default/files/media/2019/04/17/ngfs_first_comprehensive_report_-_17042019_0.pdf
29. Halimatussadiah A, Farahmita A, Machmud Z, Siregar AM, Iskandar SD, Sholihah NK (2018. Bankers' perception on the implementation of sustainable finance in India. In: International conference series on life cycle assessment: life cycle assessment as a metric to achieve sustainable development goals, vol 74. https://doi.org/10.1051/e3sconf//20127401002/
30. Friede G, Busch T, Bassen A (2015) ESG and financial performance: aggregated evidence from more than 2000 empirical studies. J Sustain Finance Invest 5(4):210–233
31. Nizam E, Adam N, Dewandaru G, Nagayev R, Nkoba MA (2019) The impact of social and environmental sustainability on financial performance: a global analysis of the banking sector. J Multinatl Financ Manag 49:35–53. https://doi.org/10.1016/j.mulfin.2019.01.002
32. Ramdlaningrum H, Mawesti D, Ningrum DR, Armintasari F (2019) Policy brief 16-long road to sustainable finance in Indonesia
33. Setiawan EN, Maryudi A, Purwanto RH, Lele G (2016) Opposing interests in the legalization of non-procedural forest conversion to oil palm in Central Kalimantan, Indonesia. Land Use Policy 58:472–481
34. Austin KG, Mosnier A, Pirker J, McCallum I, Fritz S, Kasibhatla PS (2017) Shifting patterns of oil palm driven deforestation in Indonesia and implications for zero-deforestation commitments. Land Use Policy 69:41–48
35. Dhiaulhaq A, De Bruyn T, Gritten D (2015) The use and effectiveness of mediation in forest and land conflict transformation in Southeast Asia: case studies from Cambodia, Indonesia and Thailand. Environ Sci Policy 45:132–145
36. Sinaga H (2020) Competitive pressure and labour rights: the Indian oil palm plantation and automobile sectors. Dissertation submitted to University of Kassel
37. Kusumaningtyas R (2017) External concerns on the RSPO and ISPO certification schemes. Profunda
38. Hidayat NK, Offermans A, Glasbergen P (2018) Sustainable palm oil as a public responsibility? On the governance capacity of Indonesian Standard for Sustainable Palm Oil (ISPO). Agric Hum Values 35:223–242. https://doi.org/10.1007/s10460-017-9816-6
39. Kumar A (2018) Impact of financial innovations on sustainable growth rate: a study on Nifty 50 listed banks in India. Int J Manag IT Eng 8(9):196–210
40. Guild J (2020) The political and institutional constraints on green finance in Indonesia. J Sustain Financ Invest 10(2):157–170. https://doi.org/10.1080/20430795.2019.1706312
41. Castells M (2011) The rise of the network society. John wiley & sons

Multicriteria Decision Making in Sustainable GRI-G4 Disclosure Framework

Mahesh Kumar, Palash Bairagi, and Mishra Padma Charan

Abstract This chapter aims to evaluate the standards and capacity of corporate sustainability reporting techniques used by Indian businesses following the Global Reporting Initiative (GRI-G4) guidelines. The report adheres to disclosure criteria that include economic, environmental, social, and governance aspects, which form the basis for ranking corporations based on their reporting performance. The study also explores the progress in sustainability reporting among GRI-compliant businesses in India using a multicriteria decision-making (MCDM) technique. To assess the quantity and quality of sustainability disclosures within the GRI-G4 framework, a content analysis method is employed. Corporate sustainability reporting (CSR) levels are measured using a binary coding scheme, with "1" indicating disclosure and "0" indicating non-disclosure. The quality of sustainability disclosures is rated on a four-point scale from "0" to "3." Criteria weights are determined using the MCDM technique, including entropy, and rankings are established using the TOPSIS technique (Technique for Order of Preference to Ideal Solution). This evaluation benefits various stakeholders, such as potential investors, asset managers, rating agencies, non-governmental organizations (NGOs), policymakers like the Securities and Exchange Board of India (SEBI) and the Ministry of Corporate Affairs, customers, academics, and students, by supporting informed decision making.

Keywords Sustainable finance · Sustainable disclosures · GRI-G4 framework · Tipple Bottom Line · MCDM

M. Kumar
School of Business, Dr. Vishwanath Karad MIT World Peace University, Pune, India

P. Bairagi (✉)
Department of Business Management, Central University of Odisha, Koraput, Odisha, India
e-mail: pbairagi@cuo.ac.in

M. Padma Charan
Department of Business Management, Central University of Odisha, Koraput, Odisha, India

S. Roy et al. (eds.), *Humanized Intelligence in Sustainable Finance and Economic Behaviours*, Studies in Computational Intelligence 1226,
https://doi.org/10.1007/978-981-95-0599-9_7

1 Introduction

The term "reporting" involves sharing details about events or activities within a certain period. Corporate reporting specifically refers to the disclosure of all corporate activities during a specified timeframe, acting as a communication tool between companies and their stakeholders. In today's business landscape, companies are seen not just as profit-making entities with financial impacts on shareholders, but as commercial enterprises affecting stakeholders, the environment, and society. As awareness has grown, stakeholders have increasingly demanded transparency about the environmental, sustainable, and social impacts of corporate activities.

A country's economic development depends on the sustainable growth of its corporations. Thus, in the modern business world, companies are required to produce sustainable reports that ensure they are not harming the environment or society and are conserving resources. This process is known as sustainable development. Organizational performance is now assessed based on the extent of sustainable growth achieved. Hence, modern corporations must develop their sustainability practices, emphasizing the importance of detailed reporting on their activities and their environmental, health, and societal impacts. Recently, the concept of "sustainability" has gained global importance, leading organizations to include sustainability in their reports. Recognizing sustainability's significance, many countries and the United Nations have established rules, regulations, and guidelines to standardize sustainability reporting. Beyond mandatory reporting, some corporations voluntarily provide additional disclosures out of ethical concerns, making reports more comprehensive and enabling informed stakeholder decisions.

The global perception of enterprises changed with the acceptance of the sustainable development notion by the "Brundtland Commission of the United Nations on Environment and Development in 1987" [1]. Companies are now seen as units encompassing social, economic, and environmental dimensions, essential for sustainable development (WCED, 1987). In promoting sustainability, accounting practitioners worldwide began viewing sustainability as encompassing environmental, ecological, and social issues. Consequently, "Corporate sustainability reporting (CSR)/Triple Bottom Line (TBL) reporting" has become crucial for organizations to increase transparency and accountability, presenting a clearer picture of sustainable development. Sustainability includes all activities from production to marketing and after-sales service. Stakeholders now expect more transparent reporting of financial and non-financial activities to assess it for organization sustainable development. Therefore, sustainability reporting can be considered the practice of "measuring, disclosing, and being accountable to internal and external stakeholders for organizational performance toward sustainable development" [2].

In an ever-changing environment with rising stakeholder expectations, companies must disclose their impacts on the environment and society. Organizational development and growth are linked to their contributions to overall social well-being. Recently, issues like climate change, societal health, educational development, and business sustainability have become central topics. As stakeholder expectations evolve, there is a demand for more accurate and transparent accounting and reporting of business activities. Business survival now depends on accounting for its social responsibilities. Governments worldwide have started scrutinizing corporate activities concerning sustainable development.

To measure sustainability performance, the concept of sustainability reporting was developed. It involves reporting the impact of business processes on society, the economy, and the environment. Over time, sustainability reporting has complemented traditional reporting practices, providing a true performance overview of an organization. However, to make sustainability reporting comprehensible to investors and stakeholders, common accepted norms are necessary. This facilitates the assessment of organizational performance and the impact of its activities on the environment, encouraging stakeholders to invest in organizations adopting sustainability reporting practices. Sustainability reporting has gained global traction as an essential tool for business firms to disclose their sustainable plans and performance, thereby gaining stakeholder confidence. By July 2024, approximately 14,000 organizations worldwide had adopted sustainability reporting based on the Global Reporting Initiative (GRI) framework reported by the global reporting organization (https://www.globalreporting.org/about-gri/).

2 Conceptual Development

2.1 Sustainability Reporting

Sustainable reporting is a form of non-financial reporting that increases the trust of stakeholders of a company and strengthens its relationship with them. A strong relationship with the stakeholders of a company is of utmost importance. Sustainable reporting enhances accountability and responsibility, which enhances transparent reporting of both the financial as well as non-financial aspects of a company, thereby providing complete information to the stakeholders.

Sustainability accounting incorporates the suggestions of external as well as internal stakeholders. Such suggestions help management to make better decisions in a cost-effective manner which increases the performance of an organization.

Sustainability reporting requires a clear definition of organizational mission & vision to explain the goals of an organization. It has a two-fold impact, first it helps with SWOT analysis of the company and second, it explains the efficiency and effectiveness of policies and the degree to which they are intermingled with the reporting practices of an organization. Besides this, it also covers the specific guidelines and legal compliance by a company. Sustainability accounting is governed by standards such as "GRI has G3, G3.1 & the latest G4." This aids in monitoring legal compliance & avoiding penalties & fines. It also facilitates intra industry as well as interindustry comparison of organizations. Companies doing sustainability reporting are often considered as innovative leaders who offer transparency to the stakeholders. Sustainability reporting also offers risk-management as it analyzes all the future-uncertainties' and probable solutions to mitigate them.

2.2 GRI-G4 Framework

The framework was initially introduced in 1997 through the collaboration of two independent bodies, Environmentally-Responsible-Economies (ERE) and the United Nations Environment Programme (UNEP). The formation of a multistakeholder Steering Committee in the year 1998, the GRI framework was prolonged and included the three key aspects such as economic aspects, social aspects, and governance aspects. The first set-up of GRI guidelines was released in June 2000, and the second version in early 2002. The third version of G3, which provided detailed instructions for sustainability reporting, was published in 2006. This was updated to G3.1 in 2011, and G4 was introduced in 2013. The latest version of these guidelines includes 91 indicators, 9 form economic performance, 34 form environmental performance, and 48 form social performance. According to KPMG survey reports, the GRI framework is widely accepted for reporting economic, social, and environmental performance [3–5]. The general disclosure of GRI framework for sustainability reporting has been unanimously agreed upon by the majority of investors and the organizations' in worldwide.

Several other frameworks complement the GRI framework in addressing sustainability reporting, such as the Dow Jones Sustainability Index (DJSI), the Sustainability Accounting Standards Board (SASB), the United Nations Global Compact (UNGC), and the International Integrated Reporting Council (IIRC) [1]. While these guidelines play an important role in sustainability reporting, their scope tends to be more focused or specific when compared to the comprehensive nature of the GRI framework.

2.3 Principles to Define Content of Report

- **Principle of Stakeholder-Inclusiveness**: A company needs to identify its stakeholders, and be rational in response to their interests and expectations.
- **Principle of Sustainability-Context**: A report should cover the organizational performance on the sustainability parameters.
- **Principle of Materiality**: A report should include significant economic, social, and environmental impacts which substantively affect the stakeholder's assessments and decisions

- **Principle of Completeness**: A report should sufficiently cover material aspects and their boundaries, reflecting important economic (Eco) social (S) and environmental impacts (Env.), to allow stakeholders to measure the performance of the organization.

2.4 Principles and Quality of Sustainable Reporting

These principles guide the quality of information shared through sustainable reporting. These principles are drafted in consideration with the transparency norms. Quality of report enables stakeholders to make better informed decisions and supports them to take appropriate action.

- **Principle of Balance**: Both positive and negative aspects should be covered in the report to present a true picture of organization.
- **Principle of Comparability**: Information presented in the report should be presented consistently. Such consistency enables stakeholders to compare figures and do analysis for making important decisions.
- **Principle of Accuracy**: Information depicted in a report should be sufficiently detailed and accurate so that the true position of the organizational performance may be presented.
- **Principle of Timeliness**: Reporting should be done on a timely basis for enabling stakeholders to take informed decisions.
- **Principle of Clarity**: Information presented in the report should be clear and should not be presented to create ambiguity
- **Principle of Reliability**: The report should be reliable and information given in the report should be complete and capable of being analyzed.

Companies committed to transparency and sustainability embrace the GRI guidelines, meticulously disclosing their company sustainability reports across four essential categories. First, governance (G) includes seven aspects, Second, economic (Eco) comprises four aspects, environment (Env.) covers twelve aspects, and social (S) encompasses twenty-nine aspects. The social category further breaks down into four subcategories: labor practices and decent work (LPDW), society, human rights and product responsibility (PR). Each aspect is defined by specific indicators, indicated in parentheses beside each aspect. In the GRI-G4 framework, a total of 149 indicators and 52 aspects are detailed across various categories.

This chapter is an attempt to investigate the sustainability reporting of Indian companies. The study moves ahead with the analysis of the adoption of the sustainability reporting standards in India and their alignment with the internationally defined standard for sustainability accounting and sustainability reporting. Further, performance of the companies of India has also been analyzed against the adoption of sustainability accounting and reporting framework. This research paves a way to research on this innovative aspect of the Indian corporate. Suggestions on the basis of analysis performed in this research also form the part of this study.

2.5 Motivation

The motivation behind widespread sustainability reporting has evolved, demanding that companies account for the environmental and societal impacts of their operations. Today, businesses are increasingly expected to contribute to societal well-being to achieve sustainable growth, alongside financial considerations. Issues such as climate change, community health, education, and business sustainability underscore the necessity for accurate and transparent reporting of accounting and business activities. Transparent and standardized reporting not only meets regulatory requirements but also enables comparative analysis among corporate peers and evaluation against benchmarks, facilitating corrective actions where necessary.

Sustainability reporting goes beyond regulatory compliance; it encourages investors to support organizations that diligently disclose sustainability reports alongside financial statements. Globally, the concept of sustainability reporting has gained prominence, leading to the establishment of sustainable development goals and reporting standards aimed at enhancing transparency in reporting and mitigating the environmental impacts of business activities. As of July 2024, more than 14,000 organizations, over 100 countries worldwide were reporting on sustainability parameters based on the Global Reporting Initiative (GRI) framework (https://www.globalreporting.org/, 2024). The above motivation highlights two key objectives to measures in this chapter are:

1. To measure the level and quality of sustainability disclosure practices by sample Indian companies following the GRI framework.
2. To measure the relative reporting performance and ranking sample Indian companies on sustainability disclosure practices following the GRI framework using the multicriteria decision-making (MCDM) technique.

3 Literature Review

Dimensional analysis of literature work regarding sustainability reporting, e.g., reporting level and quality and its relation with corporate performance is indispensable to know the existing research gaps. These identified gaps pave the way to further research over the subject matter. Several related research works regarding the dimensions mentioned can be categorized under the following headings.

3.1. Level and Quality of Sustainability Disclosure
3.2. Content Analysis and MCDM Techniques.

3.1 Level and Quality of Sustainability Disclosure

A researcher has conducted a thorough analysis of Australian energy retailers' sustainability reporting procedures by contrasting them with indicators of Global

Reporting Initiative (GRI) framework and found that there is significant difference between and within the three-category economic, environment, and social disclosure made by Australian energy retailers' sustainability report. Most of them have not addressed more than two-thirds of the GRI indicators. This study also demonstrates that positive disclosures predominate, and retailers who neglect to address indicators that call for information in numerical terms frequently report using declarative language without including any measurable data [5]. A researcher has conducted a study on small and medium size enterprises with exemplary application on how to prepare sustainability reports in accordance with GRI guidelines with proper utilization of resources for reporting activities. A consistency test is incorporated into the suggested approach to address the subjectivity, ambiguity, and vagueness that influence judgments. The findings give managers practical knowledge to align sustainability reporting, stakeholder responsibility, and CS strategic decision making [6]. A researcher has examined the ESG disclosure by Canada's biggest corporation after publication of UN Sustainable Development Goals and GRI standards. According to analysis, there are still issues that need to be resolved. For example, procedures and advancements frequently go unnoticed by investors because investors are not able to understand the report published by a third party on ESG and third-party reporters are always under pressure to reply to the information requested by the rating and reporting agencies [7].

Researchers have attempted to study the interfirm comparability performance by using the data reported in a sustainability report published by 17 automakers between 2014 and 2017. The study has covered the five primary indicators published by GRI standards on Climate. It reveals that it is impossible to draw any meaningful conclusions about how well the companies are performing in relation to one another. After careful assessment of the data obtained from the sustainability report, it was observed that there are four key issues that hinder rigorous credible ranking of the companies on climate performance [8]. Researchers have tried to study social sustainability and its measurement and reporting but are facing major problems due to non-availability of a common framework for it. The existing study covers social issues but not the standard approach. There is no single approach that can be universally applied in sustainable practices and financial reporting. It is often overlooked on social sustainability and its effects on stakeholders. So, adopting a 1:1 ratio of socially protected values to each stakeholder is the one way to address this issue and a four-level structure for the methodical exposition of social concerns can be suggested. Further subcategories can also be assigned to social indicators to mitigate the social issues [9]. Several researchers have also attempted to study Romanian companies which have published the GRI sustainability report on the GRI website and following the GRI indicators and understanding the ESG dimensions. In order to accomplish this goal, an analysis was conducted on the companies listed on the Global Reporting Initiative's official website (GRI). The research methodology is based on using a rating grid created from the literature review to assign scores for the reports' compliance levels [10]. Many have explored the corporate sustainability reports of firms listed on the Spanish Stock Index, focusing on the reporting standards employed and their alignment with the sustainable development goals (SDGs). This study examines these reports, the criteria they adhere to, and the correlation between the companies' performance on the index and their contribution to the SDGs [11, 12]. Some SDGs' performance and disclosure area also based on the following Spain's implementation of Law 11/2018 were used in the investigation [11]. It discovered that the GRI is the sustainability standard in use, that all companies display their verified sustainability reports, and that the percentage of ESG data is more than 75% [13]. Researchers have attempted to study the organization's institutional context that affects the compliance of sustainability reporting framework established by GRI. They also analyzed the panel data sample of 4380 companies, collected from the sustainability report published on the GRI framework, from 2006 to 2015. It was observed that state-owned enterprises, public institutions, large organizations, publicly traded companies, and organizations that conduct external audits of their reports have relatively high adherence ratings due to institutional norms and the pressure to achieve organizational legitimacy [11].

A study which provides decision-makers in SMEs with a new methodological methodology for assessing and ranking sustainability reporting (SR) standards, enabling to identify the most revent and impactful standards for enhancing transperancy, improve stakeholders trusts and aligning with long-term sustainable goals. It also can assist SMEs in resolving some of the issues they run into while starting SR. A set of decision criteria indicating the potential costs and benefits of SR adoption are identified in order to further investigate this choice. A sample of Spanish SMEs is used to test the methodological approach that is suggested, which is based on the best multicriteria and worst multicriteria of decision-making process. With GRI scoring highest among the SR options, the results demonstrate the importance of SR for the legitimacy and reputation of the company [14]. A study which investigates how sustainability policies affect the financial performance of 116 Swedish companies that are listed in 2019, offering insights into the relationship between environmental and social intiatives and corporate profitibilities. There are the researcher who suggested that the companies attain high ranking in corporate social responsibility, publish their sustainability report according to the GRI framework and participate in DJSI in order to improve financial performance [15]. Similarly, an empirical investigation highlighted on the relationship between corporate payouts and a firm's performance in corporate social responsibility (CSR). Findings demonstrate a positive relationship between the level and inclination of dividends, repurchases, and total payouts (dividends + repurchases) and the success of the firm in CSR. The correlation between CSR performance and corporate rewards is noteworthy, but it is limited to companies operating in areas with low levels of competition. CSR performance and corporate compensation do not appear to be significantly correlated in markets with high levels of competition [16].

In the present scenario, sustainability disclosure is closely linked with the quality of financial disclosure. Aligning with stakeholder theory, firms with a strong social reputation are viewed as trustworthy by both stakeholders and shareholders [17–19]. This finding supports the development of a cultural stakeholder theory, where societal values foster mutual relationships between firms and various stakeholders. By demonstrating trustworthiness through disclosures, firms can strengthen legitimacy. Enhance stakeholder confidence and builts long-term sustainable partnerships[17].

3.2 Content Analysis and MCDM Techniques

A multicriteria decision model can be applied to know the impact of Triple Bottom Line (TBL) dimensions on trader selection decisions in a largest developing economy. The result of the model is the comparative order for the companies or the firms' performance as potential traders or suppliers [20]. The outcomes of the research facilitated the decision-makers getting at an optimal decision considering TBL dimensions. "The proposed approach evaluates alternative suppliers, in terms of economic, social, and environmental dimensions by applying integrated TBL and ANP" [21]. A study which measured "Garanti Bank's corporate sustainability performance with prime dimension of governance, economic, social, and environment by applying the content analysis, entropy, and Technique for Order Preference by Similarity to Ideal Solution (TOPSIS) and entropy methods with in the period of 2010–2014, providing a comprehensive evaluation of Bank's sustainablity practices." It has also been found that sustainability performance increased with time and Economic and government determinants were found to have greatest effect on overall sustainability performance among all dimensions of MCDM. It is also found that social dimension has the highest number of disclosures, followed by environmental, governance, and economic dimensions, respectively [20]. A study contrastingly employed content analysis technique based on the GRI-G4 framework with a sample of 10 PSU banks, discovered that practices of sustainable reporting disclosure and transparency, which were not being adequately followed by most of these banks [22]. Another study applied MCDM techniques such as entropy and TOPSIS to measure the corporate sustainability performances of white goods industry in Turkey, and further assessed the sustainability performance of British Petroleum during 2010–2014, findings incremental improvements in the later years [23].

4 Methodological Development

4.1 Design and Analytical Framework

The chapter adopted a descriptive cum empirical design approach to culminate analytical development. The information extracted in the form of numbers is extracted from the available sustainability reports of a selected group of corporations using the content analysis technique. Corporate sustainability reporting (CSR) is measured by using binary coding methods, where each item receives a score of "1" if it is reported or a score of "0" otherwise. For the assessment of quality of sustainability disclosure, we employ four-point rating system, graded from 0 to 3. A score of 0 indicates "no disclosure", 1 represents partial disclosure, 2 denots full disclosure in narrative form, and 3 signifies complete disclosure with quantitative data [24, 25]. After obtaining the item-wise scores, the overall score for sustainability performance disclosure (SPD) is rigorously calculated using the following formula.

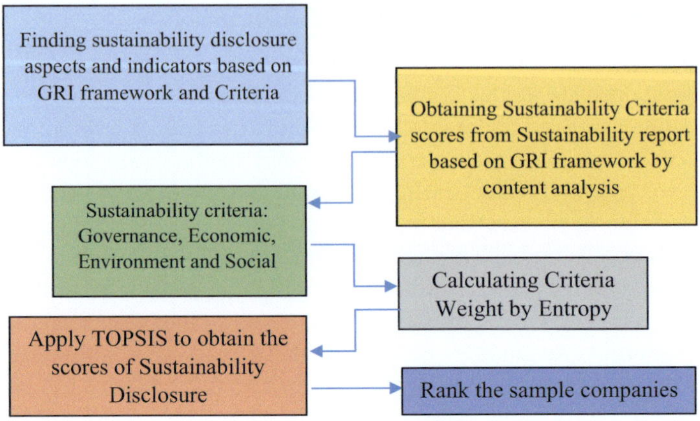

Fig. 1 MCDM technique framework for ranking of companies. *Source(s)* Author's own Creation

$$I_k = \sum_{i=0}^{n} \frac{X_{ik}}{N_k},$$

where

X_{ik} is the take value from 0 to 3 on the 4-point scale.
N_k is the maximum projected score for each category.

4.2 Procedure for Classifying Sample Companies on the Basis of Criteria and Alternatives

Figures 1 and 2 illustrate the framework utilized to achieve both objectives of the chapter. Initially, here we identify various criteria relevant to the study. Second, we gather and analyze data from sustainability reports published by the companies. The third step involves normalizing the data and applying additional methodologies outlined in the research methodology section to calculate criteria weights. Finally, TOPSIS (Technique for Order Preference by Similarity to Ideal Solution) is applied in conjunction with entropy in the fourth step [26].

4.3 Sampling Procedure and Sample Size

The secondary data gathered from the annual reports and sustainability reports posted on the websites of the corresponding companies will serve as the foundation for this study. This study consists of all listed businesses firms for the investigation. Due to the paucity of empirical data regarding corporate social responsibility (CSR) in

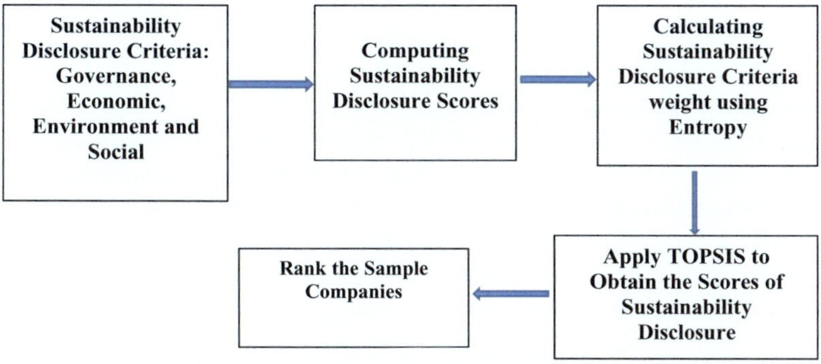

Fig. 2 Procedure for ranking the sample companies. *Source(s)* Author's own Creation

India, the sample was chosen based on disclosures posted on the GRI -2018 website. The following standards will be used to choose the listed companies are:

- The company or firm must be listed on NSE or BSE.
- The company is required to publish its sustainability reports on its own website or upload them to the GRI website.
- The company must follow the GRI rules and publish its reports consistently for a minimum of five years.

According to the GRI-G4 database GRI -2018, 116 Indian enterprises are now reporting on sustainability metrics. After a cursory examination, it was discovered that roughly 53 organizations have consistently reported on sustainability issues for the previous six years. However, financial companies are not included in this list since they follow the GRI-G4 report in addition to the "GRI-Financial Services Sector Supplement." Consequently, 53 is the sample size, with the exception of the financial sector, that has disclosed sustainability reports in accordance with GRI-G4 principles for the last six years, starting from 2017.

4.4 Selection of Companies for the Study

Step 1: The total number of Indian companies adhering to the GRI-G4 framework is 116.
Step 2: Excluding the financial sector, the total number of companies following GRI-G4+ is 8.
Step 3: Excluding companies whose reports were unavailable for six years amounts to 37.
Step 4: Excluding companies not listed on NSE/BSE results in 18.
Step 5: The total number of companies ultimately selected for the study is 53.

4.5 Data Source and Analysis Design

The data source comprises a total of 53 companies from various sectors, each reporting sustainability parameters in accordance with GRI-G4 guidelines over a span of six years ranging from 2017 to 2023. Secondary data were obtained from the prowess database (financial) and GRI database (non-financial) for this study. These datasets were compiled to meet the specific requirements outlined in this chapter. Secondary sources were chosen for their accuracy, lack of bias, and facilitation of replication in subsequent studies.

4.6 Multiple Criteria Decision-Making Methods (MCDM)

The process of making decisions from the viewpoint of several goals is addressed by multicriteria decision-making (MCDM) techniques. A decision-maker must choose between several factors, some of which can be measured and others of which cannot. This study utilized entropy to calculate criteria weights and Technique for Order Preference by Similarity to Ideal Solution (TOPSIS) for ranking companies.

4.6.1 Entropy

Entropy quantifies the potential information content of a message and holds significance in information theory [27]. It measures the actual variety and diversity of observations and analyzes the uneven frequency or probability of events that may occur. Moreover, it is considered a nonparametric measure of variety because it does not rely on any predefined or specific frequency or probability distribution [26]. The method of entropy measures in the following four consecutive steps.

Step 1: To construct a decision matrix of "n" alternatives i.e., $Y = \{Y_i | i = 1, 2, 3, \ldots n\}$ and "m" criteria i.e., $\text{X} = \{X_j | j = 1, 2, 3, \ldots m\}$ in MCDM problem and the decision matrix can be written as:

$$\begin{array}{cc} & X_1\ X_2\ X_3\ \cdots\ X_n \\ \text{DM} = & \begin{bmatrix} a_{11} & a_{12} & a_{13} & \cdots & a_{1m} \\ a_{21} & a_{22} & a_{23} & \cdots & a_{2m} \\ a_{31} & a_{32} & a_{33} & \cdots & a_{3m} \\ \vdots & \vdots & \vdots & \ddots & \vdots \\ a_{n1} & a_{n2} & a_{n3} & \cdots & a_{nm} \end{bmatrix} \\ & Y_1\ Y_2\ Y_3\ \cdots\ Y_n \end{array} \tag{1}$$

In which, "*i*" denoted as row elements and "*j*" denoted as column elements of a matrix.

Step 2: To normalize the decision matrix for each criterion of $X_j|j = 1, 2, 3, \ldots m$, which can be further denoted as NDM = $[X_{ij}]$. It can be express as:

$$x_{ij} = a_{ij} \Big/ \sum_{i=1}^{n} a_{ij} \tag{2}$$

where $i = 1, 2, 3\ldots,n$ and $j = 1, 2, 3,\ldots, m$.

Step 3: To calculate entropy from normalized decision matrix, which is denoted as B_j of j attributes can be express as:

$$B_j = -K\sum_{i=1}^{n} x_{ij} \times \log(x_{ij}), \tag{3}$$

where $i = 1, 2, 3, \ldots, n$ and $j = 1, 2, 3, \ldots, m$, "K" is constant and value of $K = 1/\log(m)$, "m" represents the number of criteria, where B_j ranges between 0 and 1 $(0 \leq B_j \leq 1)$.

Step 4: To calculate the degree of divergence D_j by deducting the weight of the Entropy (B_j) for each criterion of $X_j|j = 1, 2, \ldots, m$, can be express as:

$$D_j = 1 - B_j. \tag{4}$$

The higher the value of "D_j shows criteria X_j" is more important for the problem and the criteria weight can be calculate by this formula (5),

$$CW_j = \frac{D_j}{\sum_{j=1}^{m} D_j}, \text{ where } j = 1, 2, 3, \ldots, m. \tag{5}$$

4.6.2 TOPSIS

The Technique for Order Preference by Similarity to Ideal Solution (TOPSIS) method, Hwang and Yoon [27] elaborated, requires decision-making input to determine the weights for different criteria, as it does not assume equal importance for all criteria. The TOPSIS method yields a score ranging from 0 to 1. The following steps involved in the TOPSIS technique are:

Step 1: Hypothesis a decision matrix of "n" alternatives $(Y = \{i = 1, 2, \ldots, n\})$ and "m" alternatives $X = \{X_j|j = 1, 2, \ldots.., m\})$ in a MCDM problem with the equation expression as:

$$\text{DM} = \left[A_{ij}\right]_{n\times m},$$

where "i" denotes row and "j" denotes column in the matrix.

Step 2: To measure the normalization of decision matrix (NDM) for each criterion of $X_j(j = 1, 2 \ldots, m)$ as NDM $= \left[x_{ij}\right]$ in which we measure $\left[x_{ij}\right]$ as:

$$x_{ij} = \frac{A_{ij}}{\sqrt{\sum_{i=1}^{n} A_{ij}^2}}, \tag{6}$$

where$i = 1, 2, 3, \ldots, n$ and $j = 1, 2, 3, \ldots, m$.

Step 3: To determine the weighted normalized decision matrix (WNDM), each column in matrix NDM $= [x_{ij}]$ are multiplied by the corresponding weight of (CW_j) matrix in Eq. 5. The following matrix written as:

$$\text{WNDM} = [b_{ij}] = \left[\text{CW}_j\right] \times \left[x_{ij}\right], \tag{7}$$

where $i = 1, 2, 3, \ldots, n$ and $j = 1, 2, \ldots, m$.

Step 4: Further to identify the best and worst measure among the weighted normalized decision matrix which are extracted from step 4 can be measured by applying the following equations of BS and WS.

$$\text{Best solution (BS)} = \{(\text{Max}[b_{ij}]|j \in T_1), (\text{Min}[b_{ij}]|j \in T_2)\}. \tag{8}$$

Further the element of BS matrix represented as:

$$V_{+\text{ve}} = [b_1, b_2, b_3 \ldots b_n]$$

$$\text{Worst Solution (WS)} = \{\ (\text{Min}[b_{ij}]|j \in T_1), (\text{Max}[b_{ij}]|j \in T_2). \tag{9}$$

The element of WS matrix represented as

$$V_{-\text{ve}} = [b_1, b_2, b_3 \ldots b_n],$$

where $i = 1, 2, 3, \ldots, n$ and $j = 1, 2, 3, \ldots, m$. T_1 and T_2 are benefits and cost criteria, respectively.

Step 5: By using Euclidean distance formula, we measure the separation from best solution (BS) and worst solution (WS), which is define as:

$$S_i^+ = \sqrt{\left(\sum_{j=1}^{m}([b_{ij}] - V_{+\text{ve}})^2\right)} \tag{10}$$

$$S_i^- = \sqrt{\left(\sum_{j=1}^{m}([b_{ij}] - V_{-ve})^2\right)}, \tag{11}$$

where $i = 1, 2, 3, \ldots, n$ and $j = 1, 2, 3, \ldots, m$.

Step 6: Further to measure the coefficient of separation as "P_i, " which is defined as:

$$P_i = \frac{S_i^-}{S_i^+ + S_i^-}, \text{ where, } P_i \in [0, 1]\forall i = 1, 2, 3 \ldots, n. \tag{12}$$

Further, the preferred order can be determined based on the values of P_i, with the ranks chosen in descending order of these values."

5 Results and Discussion

5.1 Level/Scores of Sustainability Disclosures

The sustainability disclosure matrix for the collected data shown in Table 1, which represents four distance criteria of sustainable measures as governance, economic, environment, and social, along with 53 alternatives representing the companies under study. The sustainability disclosure matrix forms the basis for evaluating both the extent and quality of disclosures, as well as for ranking the companies. Two distinct matrices are analyzed: one assessing the level of disclosure and the other assessing the quality of disclosures. Consequently, the companies are ranked separately based on each criterion.

The results of the determinants used in this analysis include governance, economic, environment, and social criteria. The evaluation reveals that among these four GRI components, the companies generally exhibited stronger sustainability disclosures in terms of Governance with maximum score of 18,444 (81.5%), followed by economic with 2862 (78.3%), environmental with 10,812 (71.5%), and finally social with 15,264 (64.3%) reporting, which had the lowest level of disclosure."

5.2 Quality of Sustainability Disclosure

The quality of sustainability disclosure with cumulative score data shows in Table 2, compiled from the company's sustainability reports which was published on their websites or the GRI-G4 website. The quality of disclosure is assessed using a four-point scale (0–3) across four criteria of governance, economic, environment, and social, for the 53 companies, referred to as alternatives.

Table 1 Level of sustainability disclosure (cumulative score)

Companies	Governance	Economic	Environment	Social
Adani Power	346	37	155	117
Ambuja Cement	310	38	173	224
Bharti Airtel	222	36	62	85
BHEL	324	46	144	243
BPCL	324	49	163	202
Asian Paints Limited	314	46	160	204
Coal India Ltd	350	49	169	211
Gail India	309	47	171	224
GMR Energy	295	40	122	159
Havells	303	42	175	190
HCC	252	43	148	217
HCL	330	42	138	150
Hindalco Industries	305	44	160	213
HPCL	260	18	96	113
Arvind Limited	290	22	94	176
Indian Oil	334	54	189	242
ITC	344	45	171	203
Welspun India	200	44	158	254
Jindal Steel	298	43	102	115
JSW Steel	227	38	112	115
Jubilant Industries	282	48	158	213
Jubilient Life Science	280	47	184	249
Kansai Nerolac	330	51	198	233
Kirloskar Brthers Ltd	268	52	173	273
Mahindra Life Space Dv.	257	33	143	183
Mahindra & Mahindra Ltd.	268	47	157	185
Maruti_Suzuki	236	41	103	160
Hindustan Zinc Limited	298	47	167	226
NTPC	326	48	182	213
Petronet-LNG	215	36	146	152
RFC (Rastriya Chemical & Fertilizers)	279	49	119	166
RIL	326	44	162	174
Shree Cement Ltd.	342	34	135	187
Jain Irrigation Systems	201	34	109	132
Tata Motors	289	28	141	146
Tata Power	268	33	159	228
Tata Steel	332	39	175	176

(continued)

Table 1 (continued)

Companies	Governance	Economic	Environment	Social
TCS	315	53	170	202
Vodaphone-Idea	188	36	87	168
Wipro	289	45	149	164
Chambel Fertiliser & Chemical	249	38	128	146
The Indian Hotels Co. Ltd.	261	36	112	168
Infosys	348	35	138	220
JK Cements	229	38	147	160
Tata Chemicals	196	36	162	202
Tech-Mahindra	279	42	128	174
Dr. Reddy's India	307	48	134	175
Oil India Limited	181	53	130	142
ONGC	263	54	154	207
TVS Motors	259	49	124	143
ACC	338	50	186	221
Nalco	274	46	124	176
Mindtree	325	48	178	207
Total	15,035	2241	7724	9828
Maximum score	18,444	2862	10,812	15,264
Percentage (%)	81.5	78.3	71.44	64.38

Source(s) Author own calculation

The quality results of the sustainability disclosure matrix for the 53 selected Indian companies are evaluated based on the GRI standard reporting for governance, economic, environment, and social criteria. Among these four GRI components, the analysis reveals that all the companies studied exhibited low-quality sustainability disclosures. Specifically, the quality of disclosures was lowest in the economic criterion (41.77%), followed by governance (37.22%), environment (32.15%), and social (31.7%) reporting.

5.3 *Relative Performance of Sample Companies*

Evaluating sustainability performance begins with the calculation of item-wise scores. Each item within the assessment framework is assigned a score that reflects the level of disclosure or performance, typically based on specific criteria related to sustainability. These item-wise scores serve as the foundational data points from which the overall sustainability disclosure score is computed.

To calculate the overall disclosure score, we use the following formula:

Table 2 Quality of sustainability disclosure (cumulative score)

Companies	Governance	Economic	Environment	Social
Adani Power	417	65	178	181
Ambuja Cement	434	65	246	326
Bharti Airtel	283	56	90	138
BHEL	455	80	212	366
BPCL	464	78	253	334
Asian Paints Limited	424	68	226	298
Coal India Ltd.	458	79	239	318
Gail India	426	72	232	339
GMR Energy	463	71	203	309
Havells	367	73	214	258
HCC	363	65	201	337
HCL	446	68	184	233
Hindalco Industries	423	67	193	301
HPCL	420	73	168	221
Arvind Limited	387	46	137	271
Indian Oil	480	90	306	370
ITC	470	83	217	295
Welspun India	284	78	237	372
Jindal Steel	410	78	165	237
JSW Steel	426	77	210	312
Jubilant Industries	395	72	251	353
Jubilient Life Science	428	71	237	347
Kansai Nerolac	441	76	256	334
Kirloskar Brthers Ltd	395	81	229	397
Mahindra Life Space Dv.	413	74	198	299
Mahindra & Mahindra Ltd.	385	76	208	274
Maruti_Suzuki	336	53	142	201
Hindustan Zinc Limited	428	73	231	330
NTPC	456	86	265	329
Petronet-Lng	229	48	138	156
RFC (Rastriya Chemical & Fertilizers)	379	65	168	278
RIL	462	83	243	318
Shree Cement Ltd.	466	67	214	299
Jain Irrigation Systems	229	34	108	122
Tata Motors	414	67	207	240
Tata Power	416	65	214	338
Tata Steel	468	69	254	318

(continued)

Table 2 (continued)

Companies	Governance	Economic	Environment	Social
TCS	382	45	200	212
Vodaphone-Idea	279	66	120	258
Wipro	310	58	153	165
Chambel Fertiliser & Chemical	337	39	151	187
The Indian Hotels Co. Ltd.	279	58	113	184
Infosys	489	79	226	331
JK Cements	262	31	129	140
Tata Chemicals	326	60	225	307
Tech-Mahindra	351	73	155	218
Dr. Reddy's India	443	78	201	284
Oil India Limited	246	59	109	140
ONGC	347	69	170	281
TVS Motors	258	53	129	139
ACC	463	86	269	348
Nalco	341	66	154	246
Mindtree	443	75	252	328
Total	20,596	3587	10,430	14,517
Maximum score	55,332	8586	32,436	45,792
Percent (%)	37.22	41.77	32.15	31.7

Source(s) Author own calculation

$$I_k = \frac{\sum_{i=0}^{n} X_{ik}}{N_k}.$$

This formula sums up all the item-wise scores (denoted as X_{ik}) for a particular category k, and then divides this sum by N_k which represents the maximum expected score for that category. The sum $\sum_{i=0}^{n} X_{ik}$ includes all individual item scores within the category, providing a cumulative measure of performance. The variable 'N_k' is crucial because it establishes the highest possible score that can be achieved within each category. This normalization process ensures that the overall score is proportionate and comparable across different categories, regardless of the number of items or the maximum potential scores. Essentially, N_k acts as a benchmark for evaluating the relative completeness or thoroughness of the sustainability disclosures.

Each item-wise score X_{ik} is assigned a value on a four-point scale, ranging from 0 to 3. A score of "0" indicates no disclosure or performance, while a score of "3" reflects the highest level of disclosure or best performance. This scoring system allows for a nuanced evaluation of how well an organization is disclosing or relative performing of sample companies in Table 3.

Table 3 Relative performance of sample companies

Companies	Level	Rank_L	Quality	Rank_Q
Adani Power	0.733	31	0.597	38
Ambuja Cement	0.833	12	0.819	6
Bharti Airtel	0.453	53	0.433	52
BHEL	0.847	10	0.759	18
BPCL	0.826	15	0.863	1
Asian Paints Limited	0.810	17	0.777	16
Coal India Ltd.	0.871	4	0.836	4
Gail India	0.840	11	0.817	7
GMR Energy	0.689	35	0.478	47
Havells	0.794	20	0.789	15
HCC	0.738	27	0.533	43
HCL	0.738	28	0.712	24
Hindalco Industries	0.808	18	0.752	20
HPCL	0.545	50	0.674	30
Arvind Limited	0.651	40	0.643	34
Indian Oil	0.916	1	0.707	26
ITC	0.853	7	0.812	8
Welspun India	0.734	30	0.742	21
Jindal Steel	0.624	45	0.451	49
JSW Steel	0.550	49	0.600	37
Jubilant Industries	0.784	22	0.544	42
Jubilient Life Science	0.850	8	0.644	33
Kansai Nerolac	0.908	2	0.802	10
Kirloskar Brthers Ltd.	0.857	6	0.728	22
Mahindra Life Space Dv.	0.689	36	0.690	28
Mahindra & Mahindra Ltd.	0.735	29	0.721	23
Maruti_Suzuki	0.604	47	0.560	39
Hindustan Zinc Limited	0.826	16	0.812	9
NTPC	0.860	5	0.756	19
Petronet-Lng	0.614	46	0.437	51
RFC (Rastriya Chemical & Fertilizers)	0.686	37	0.680	29
RIL	0.790	21	0.709	25
Shree Cement Ltd.	0.781	23	0.800	11
Jain Irrigation Systems	0.532	52	0.377	53
Tata Motors	0.676	38	0.664	31
Tata Power	0.770	24	0.790	14
Tata Steel	0.808	19	0.848	3

(continued)

Table 3 (continued)

Companies	Level	Rank_L	Quality	Rank_Q
TCS	0.828	14	0.641	35
Vodaphone-Idea	0.536	51	0.553	40
Wipro	0.724	32	0.524	44
Chambel Fertiliser & Chemical	0.628	44	0.546	41
The Indian Hotels Co. Ltd.	0.645	41	0.485	46
Infosys	0.829	13	0.860	2
JK Cements	0.642	43	0.478	47
Tata Chemicals	0.663	39	0.702	27
Tech-Mahindra	0.697	33	0.820	5
Dr. Reddy's India	0.743	26	0.769	17
Oil India Limited	0.566	48	0.493	45
ONGC	0.758	25	0.663	32
TVS Motors	0.643	42	0.443	50
ACC	0.889	3	0.799	12
Nalco	0.694	34	0.617	36
Mindtree	0.848	9	0.794	13

Source(s) Author own calculation

Table 3 details the data regarding the level and quality of disclosure for each selected company individually. The analysis of the sustainability reporting and disclosures indicates that, among the companies chosen for this study, Indian Oil leads in the level of disclosure in their corporate social reporting. This is followed by Kansai Nerolac, ACC, Coal India Ltd., and NTPC. These companies are identified as the top five in terms of adherence to the required levels of CSR disclosures. In contrast, JSW Steel, HPCL, Vodafone India, Jain Irrigation Systems, and Bharti Airtel show the lowest levels of CSR disclosure among the selected companies.

In terms of the quality of sustainability reporting and disclosures, BPCL demonstrates the highest transparency in their corporate social reporting. They are followed by Infosys, Tata Steel, Coal India Ltd., and Tech-Mahindra. These companies rank among the top five concerning the quality of CSR disclosures. Conversely, Jindal Steel, TVS Motors, Petronet LNG, Bharti Airtel, and Jain Irrigation Systems exhibit the lowest quality of CSR disclosures among the selected companies. Notably, Coal India Ltd. maintains consistency in both the level and quality of its CSR reporting. Bharti Airtel and Jain Irrigation Systems also show a similar position in terms of both the level and quality of CSR disclosures.

5.4 Result of Rank-Based Level of Disclosure Using MCDM Technique

The result of rank-based disclosure of selected companies using MCDM techniques shows in Table 4. The divergence from the ideal best solution (+Ve) and the ideal worst solution (−Ve) among the alternatives is computed using the Euclidean distance formula, as outlined in step 5 of the Technique for Order Preference by Similarity to Ideal Solution (TOPSIS).

Table 4 represents the decision matrix for the level of corporate disclosures of 53 selected Indian companies, based on GRI standards reporting. The GRI criteria used for analysis in this matrix include governance (G), economic (E), environmental (Env), and social aspects (S). The analysis reveals that Indian Oil has the most transparent and comprehensive level of disclosure holding rank 1 with coefficient score of 0.88, followed by Kirloskar Brothers Ltd. (rank 2, coefficient score of 0.860), Kansai Nerolac (rank 3, coefficient score of 0.851), Jubilant Life Sciences (rank 4, coefficient score of 0.841), and ACC (rank 5, coefficient score of 0.802), respectively. In contrast, Jindal Steel, JSW Steel, Jain Irrigation Systems, HPCL, and Bharti Airtel show the least completeness and transparency in their sustainability reporting. Therefore, considering all four aspects of disclosure under study according to their respective weightages, Indian Oil ranks highest in transparency and completeness, followed by Kirloskar Brothers Ltd., Kansai Nerolac, Jubilant Life Sciences, and ACC, while Jindal Steel, JSW Steel, Jain Irrigation Systems, HPCL, and Bharti Airtel rank lowest in governance (G), economic (E), environmental (Env), and social (S) disclosures in their CSR reporting.

5.5 Findings of the Study

The following key findings of this chapter highlighted as:

- First, the chapter reveals that the level of sustainability disclosure across the four criteria—governance, economic, environment, and social—is 81.5%, 78.3%, 71.5%, and 64.3%, respectively, among the sample companies analyzed (refer to Table 1).
- Second, it further indicates that the quality of sustainability disclosure for the same criteria—governance, economic, environment, and social—is 37.22%, 41.77%, 32.15%, and 31.7%, respectively, for the companies selected for the study (refer to Table 2).
- Third, Relative Performance of Sample Companies on the Level of Disclosure: The analysis finds that Indian Oil exhibits the most transparent and comprehensive level of disclosure, followed by Kansai Nerolac, ACC, Coal India Ltd., and NTPC, respectively. Conversely, JSW Steel, HPCL, Vodafone Idea India, Jain Irrigation Systems, and Bharti Airtel demonstrate the lowest levels of completeness and transparency in their sustainability reporting (refer to Table 3).

Table 4 Rank-based level of disclosure of selected companies

Companies	$S1$	$S2$	$S1 + S2$	$S2/(S1 + S2)$	Rank
Adani Power	0.0399	0.0316	0.0715	0.4417	42
Ambuja Cement	0.0179	0.0466	0.0645	0.7230	14
Bharti Airtel	0.0583	0.0136	0.0719	0.1894	53
BHEL	0.0166	0.0485	0.0651	0.7446	9
BPCL	0.0192	0.0454	0.0646	0.7022	17
Asian Paints Limited	0.0199	0.0440	0.0638	0.6887	20
Coal India Ltd.	0.0166	0.0481	0.0647	0.7433	10
Gail India	0.0147	0.0488	0.0635	0.7688	6
GMR Energy	0.0347	0.0297	0.0644	0.4608	40
Havells	0.0224	0.0429	0.0652	0.6571	23
HCC	0.0215	0.0423	0.0639	0.6630	21
HCL	0.0337	0.0326	0.0663	0.4924	36
Hindalco Industries	0.0189	0.0445	0.0634	0.7021	18
HPCL	0.0530	0.0127	0.0657	0.1928	52
Arvind Limited	0.0422	0.0246	0.0669	0.3681	47
Indian Oil	0.0077	0.0569	0.0646	0.8810	1
ITC	0.0189	0.0458	0.0647	0.7082	15
Welspun India	0.0182	0.0501	0.0683	0.7331	12
Jindal Steel	0.0451	0.0241	0.0692	0.3487	49
JSW Steel	0.0455	0.0210	0.0665	0.3154	50
Jubilant Industries	0.0187	0.0451	0.0639	0.7070	16
Jubilient Life Science	0.0102	0.0542	0.0644	0.8416	4
Kansai Nerolac	0.0097	0.0559	0.0656	0.8520	3
Kirloskar Brthers Ltd.	0.0094	0.0582	0.0676	0.8604	2
Mahindra Life Space Dv.	0.0305	0.0333	0.0638	0.5215	33
Mahindra & Mahindra Ltd.	0.0245	0.0404	0.0649	0.6222	25
Maruti_Suzuki	0.0382	0.0268	0.0650	0.4126	46
Hindustan Zinc Limited	0.0151	0.0483	0.0634	0.7624	8
NTPC	0.0153	0.0495	0.0648	0.7635	7
Petronet-LNG	0.0356	0.0297	0.0653	0.4547	41
Rastriya Chemical & Fertilizers	0.0328	0.0339	0.0666	0.5085	35
RIL	0.0259	0.0398	0.0657	0.6055	27
Shree Cement Ltd.	0.0295	0.0350	0.0645	0.5423	31
Jain Irrigation Systems	0.0443	0.0201	0.0643	0.3121	51
Tata Motors	0.0384	0.0272	0.0656	0.4148	45

(continued)

Table 4 (continued)

Companies	$S1$	$S2$	$S1 + S2$	$S2/(S1 + S2)$	Rank
Tata Power	0.0222	0.0435	0.0656	0.6623	22
Tata Steel	0.0258	0.0409	0.0667	0.6127	26
TCS	0.0182	0.0477	0.0660	0.7235	13
Vodaphone-Idea	0.0418	0.0242	0.0661	0.3669	48
Wipro	0.0295	0.0361	0.0655	0.5503	30
Chambel Fertiliser & Chemical	0.0374	0.0270	0.0644	0.4195	44
The Indian Hotels Co. Ltd.	0.0361	0.0274	0.0636	0.4318	43
Infosys	0.0241	0.0413	0.0654	0.6321	24
JK Cements	0.0331	0.0317	0.0648	0.4886	37
Tata Chemicals	0.0264	0.0396	0.0659	0.6001	28
Tech-Mahindra	0.0310	0.0330	0.0640	0.5155	34
Dr. Reddy's India	0.0286	0.0370	0.0656	0.5639	29
Oil India Limited	0.0378	0.0337	0.0714	0.4712	38
ONGC	0.0203	0.0458	0.0661	0.6924	19
TVS Motors	0.0366	0.0315	0.0681	0.4621	39
ACC	0.0129	0.0521	0.0650	0.8021	5
NALCO	0.0306	0.0343	0.0650	0.5284	32
Mindtree	0.0169	0.0480	0.0649	0.7396	11

Source(s) Author own calculation

- Fourth, Relative Performance of Sample Companies on the Quality of Disclosure: BPCL stands out with the highest transparency in corporate social reporting, followed by Infosys, Tata Steel, Coal India Ltd., and Tech-Mahindra, ranking among the top five companies adhering to the required quality of CSR disclosures (refer to Table 3).
- Fifth, In contrast, Jindal Steel, TVS Motors, Petronet LNG, Bharti Airtel, and Jain Irrigation Systems provide the lowest quality of CSR disclosures among the selected companies. It is also notable that Coal India Ltd. is consistent in both the level and quality aspects of its CSR reporting. Bharti Airtel and Jain Irrigation Systems show similar standings regarding both the level and quality of their CSR disclosures (refer to Table 3).
- Sixth, Ranking Based on the Level of Disclosure Using MCDM Technique (Best): The findings indicate that Indian Oil leads in transparency and completeness, followed by Kirloskar Brothers Ltd., Kansai Nerolac, Jubilant Life Sciences, and ACC (refer to Table 4).
- Seven, Ranking Based on the Level of Disclosure Using MCDM Technique (Worst): On the other hand, Jindal Steel, JSW Steel, Jain Irrigation Systems,

HPCL, and Bharti Airtel rank the lowest in terms of the completeness and transparency of their sustainability reporting (refer to Table 4).

6 Conclusion

Elucidating the overall discussion and analysis conducted in this chapter concluded with drawing attention to the Global Reporting Initiative (GRI) framework, which provides a comprehensive and standardized approach to measuring and comparing sustainability disclosures across different companies. As noted by Richter et al. [28], the GRI framework facilitates an objective assessment of sustainability reporting, which is a significant advantage over other guidelines that may rely on more subjective evaluations. This objectivity allows for more accurate and meaningful comparisons of sustainability disclosures between companies, enhancing the reliability of the assessment process.

Furthermore, this chapter stands out by ranking Indian companies based on both the level and quality of their sustainability disclosures. By focusing on these two critical aspects, the research offers a detailed analysis of how well companies adhere to sustainability reporting standards and practices. The insights gained from this study provide a valuable resource for business leaders, company boards, and sustainability professionals, offering them a clearer understanding of how their organizations' reporting measures up against industry standards. This can aid in refining and improving their own sustainability reporting practices, ensuring greater transparency and effectiveness.

Moreover, the research serves as a guide for investors, asset managers, and ratings agencies by providing essential environmental, social, and governance (ESG) information. This information is crucial for assessing corporate sustainability performance and making informed investment decisions. The study highlights how companies communicate their sustainability efforts and contributions to sustainable development, offering potential investors a clearer picture of a company's commitment to ESG principles. Finally, this research contributes significantly to the body of knowledge on sustainability reporting by demonstrating the growing importance of adopting globally accepted standards. The findings underscore the value of the GRI framework in providing a consistent and objective basis for evaluating sustainability disclosures. This, in turn, supports better decision making for both investors and companies, reinforcing the importance of effective sustainability reporting in today's corporate landscape.

7 Limitation and Scope for Future Research

The insights gained from this chapter, have certain limitations and boundaries need to be considered. Firstly, the study is based on a sample of companies that have published sustainability reports using the GRI-G4 framework over the past six years (2017–2023). Financial companies were excluded from the study because they adhere to the GRI-G4 framework in conjunction with the GRI-Financial Services Sector Supplement. Additionally, the study covers a relatively short time frame of six years, which may limit the generalizability of the findings. Future scope of this research could benefit from a larger sample size and an extended time period to provide more comprehensive insights. Moreover, sustainability reporting parameters are continually evolving with advancements in reporting practices and corporate adaptability. This presents an opportunity for further research into the transparency of sustainability disclosures, particularly in light of new regulations, amendments, and voluntary disclosure frameworks adopted by companies.

References

1. Kumar M, Raj N, Singh RR (2023) Ranking Indian companies on sustainability disclosures using the GRI-G4 framework and MCDM techniques. Int J Sustain Dev Plan 18(9):2791–2799. https://doi.org/10.18280/ijsdp.180917
2. The Global Reporting Initiative (GRI) for independent not-for-profit organization (2023). Available: https://www.globalreporting.org/standards/globalsustainability-standards-board/
3. KPMG and Sustainability (2008) Count me in: the readers' take on sustainability reporting. KPMG Global Sustainability Services, Amstelveen
4. KPMG (2013) Indian corporate responsibility reporting survey. http://www.kpmg.com/IN/en/IssuesAndInsights/ArticlesPublications/Documents/India-CorporateResponsibility-Reporting-Survey-2013.pdf
5. Carrots and Sticks (2015) Sustainability reporting policies worldwide—today's best practice, tomorrow's trends. https://www.carrotsandsticks.net/media/4ldnklxn/carrots-sticks-2013.pdf
6. Mamun M (2023) Sustainability reporting of major electricity retailers in line with GRI: Australia evidence. J Account Organ Chang 19(3):474–493. https://doi.org/10.1108/JAOC-01-2022-0005
7. Calabrese A, Costa R, Levialdi Ghiron N, Menichini T (2019) Materiality analysis in sustainability reporting: a tool for directing corporate sustainability towards emerging economic, environmental and social opportunities. Technol Econ Dev Econ 25(5):1016–1038. https://doi.org/10.3846/tede.2019.10550
8. Cho CH, Bohr K, Choi TJ, Partridge K, Shah JM, Swierszcz A (2020) Advancing sustainability reporting in Canada: 2019 report on progress. Account Perspect 19(3):181–204. https://doi.org/10.1111/1911-3838.12232
9. Boiral O, Brotherton M, Rivaud L, Talbot D (2022) Comparing the uncomparable? An investigation of car manufacturers' climate performance. Bus Strat Environ 31(5):2213–2229. https://doi.org/10.1002/bse.3017
10. Richter F, Gawenko W, Götze U, Hinz M (2023) Toward a methodology for social sustainability assessment: a review of existing frameworks and a proposal for a catalog of criteria. Schmalenbach J Bus Res (SBUR) 75(4):587–626. https://doi.org/10.1007/s41471-023-00174-y

11. Gutiérrez-Ponce H (2023) Sustainability as a strategy base in Spanish firms: sustainability reports and performance on the sustainable development goals. Sustain Dev 31(4):3008–3023. https://doi.org/10.1002/sd.2566
12. Sierra García L, Bollas-Araya HM, García Benau MA (2022) Sustainable development goals and assurance of non-financial information reporting in Spain. Sustain Account Manag Policy J 13(4):878–898. https://doi.org/10.1108/SAMPJ-04-2021-0131
13. Marinescu A-O (2020) Analysis on the compliance of sustainability reports of Romanian companies with GRI conceptual framework. Audit Financiar 18(158):349–363. https://doi.org/10.20869/AUDITF/2020/158/011
14. Friske W, Nikolov AN, Morgan T (2023) Making the grade: an analysis of sustainability reporting standards and Global Reporting Initiative adherence ratings. Corp Soc Responsib Environ Manag 1. https://doi.org/10.1002/csr.2686
15. Rodríguez-Gutiérrez P, Guerrero-Baena MD, Luque-Vílchez M, Castilla-Polo F (2021) An approach to using the best-worst method for supporting sustainability reporting decision-making in SMEs. J Environ Planning Manage 64(14):2618–2640. https://doi.org/10.1080/09640568.2021.1876003
16. Pham DC, Do TNA, Doan TN, Nguyen TXH, Pham TKY (2021) The impact of sustainability practices on financial performance: empirical evidence from Sweden. Cogent Bus Manag 8(1):1–19. https://doi.org/10.1080/23311975.2021.1912526
17. Abeysekera I, Li F, Lu Y (2021) Financial disclosure quality and sustainability disclosure quality. A case in Chian. PLoS ONE 16(5):e0250884. https://doi.org/10.1371/journal.pone.0250884
18. Helfaya A, Whittington M (2019) Does designing environmental sustainability disclosure quality measures make a difference? Bus Strateg Environ 28(4):525–541. https://doi.org/10.1002/bse.2262
19. Trianaputri AR, Djakman CD (2019) Quality of sustainability disclosure among the Asean-5 countries and the role of stakeholders. Jurnal Akuntansi dan Keuangan Indonesia 16(2):4. https://doi.org/10.1002/csr.1330
20. Faisal MN, Al-Esmael B, Sharif KJ (2017) Supplier selection for a sustainable supply chain: triple bottom line (3BL) and analytic network process approach. Benchmarking Int J 24(7):1956–1976. https://doi.org/10.1108/BIJ-03-2016-0042
21. Sheikh S (2022) Corporate social responsibility and corporate payout policy: the impact of product market competition. Int J Manag Financ 18(1):49–74. https://doi.org/10.1108/IJMF-09-2020-0465
22. Aras G, Tezcan N, Kutlu Furtuna O, Hacioglu Kazak E (2017) Corporate sustainability measurement based on entropy weight and TOPSIS: a Turkish banking case study. Meditari Account Res 25(3):391–413. https://doi.org/10.1108/MEDAR-11-2016-0100
23. Kumar R, Pande N, Afreen S (2018) Developing a GRI-G4-based persuasive communication framework for sustainability reporting (SR): examining top 10 Indian banks. Int J Emerg Mark 13(1):136–161. https://doi.org/10.1108/IJoEM-01-2017-0015
24. Hummel K, Schlick C (2016) The relationship between sustainability performance and sustainability disclosure—reconciling voluntary disclosure theory and legitimacy theory. J Account Public Policy 35(5):455–476. https://doi.org/10.1016/j.jaccpubpol.2016.06.001
25. Laskar N, Gopal Maji S (2018) Disclosure of corporate sustainability performance and firm performance in Asia. Asian Rev Account 26(4):414–443. https://doi.org/10.1108/ARA-02-2017-0029
26. Krippendorff K (1986) Information theory: structural models for qualitative data (vol 62). Sage
27. Hwang CL, Yoon K (1981) Methods for multiple attribute decision making. In: Multiple attribute decision making: methods and applications a state-of-the-art survey, pp 58–191. Springer, Berlin, Heidelberg https://link.springer.com/chapter/10.1007/978-3-642-48318-9_3

28. Richter B, Marcondes G, Monteiro N, Costa SE, Loures E, Deschamps F, Lima E (2023) Industrial energy efficiency assessment and prioritization model-An approach based on multi-criteria method PROMETHEE. Int J Sustain Energ Planning Manag 37:41–60
29. Karaca S, Karaca A (2018) An evaluation of the performances of foreign capital trade companies in Bist and the sustainability index through the TOPSIS method. BRAND. Broad Res Account Negot Distrib 9(1):46–64. https://hdl.handle.net/11499/50927

Influence of Digital Wallets on Consumer Behaviors for Sustainable Economic Growth

G. Sunitha, Gowsya Shaik, and Sri Sai Chilukuri

Abstract Any software, electronic gadget, or internet service that enables electronic transactions for people or companies is referred to as a digital wallet. There is no longer a need for physical wallets because a digital wallet securely and compactly saves all of the users' payment information. The impact of digital wallets on consumer purchasing behavior is examined in this study, along with how these shifts affect sustainable economic growth. By enabling users to deposit funds and make payments via smart phones, digital wallets have emerged as a significant part of the digital payment process as an alternative to conventional payment methods. The study's goals included identifying the factors that support and hinder the utility of digital wallets, analyzing their impact on sustainable economic growth, and looking at how the use of digital wallets affects consumer purchasing patterns. Both descriptive and inferential statistics were utilized in the study's mixed method approach to assess the data. Customers who use their digital wallets were requested to complete a questionnaire in order to collect primary data. In order to assure diversity based on demographic parameters and consumption trends, the study employed a stratified random sample technique. Descriptive statistics was used to study the demographic factors whereas inferential tools like multiple regression, chi-square test, and correlation tests were used to check the relationship between the use of digital wallets, sustainable economic growth, consumer buying behavior, and challenges faced due to digital

Gowsya Shaik— Scopus id: 57201646214.

G. Sunitha
Department of Business Administration, Raja Bahadur Venkata Rama Reddy Institute of Technology, Hyderabad, Telangana, India

G. Shaik
Assistant Professor, Department of Management Studies, B.V. Raju Institute of Technology, Narsapur, Affiliated to JNTUH, Hyderabad 502313, Medak, Telangana, India
e-mail: gowsya2020@gmail.com; gowsya.sk@bvrit.ac.in

S. S. Chilukuri (✉)
Symbiosis School of Banking and Finance, Symbiosis International (Deemed) University, Pune, Maharashtra, India
e-mail: srisai103@gmail.com

S. Roy et al. (eds.), *Humanized Intelligence in Sustainable Finance and Economic Behaviours*, Studies in Computational Intelligence 1226,
https://doi.org/10.1007/978-981-95-0599-9_8

wallets. People who regularly use digital wallets were chosen as respondents for the study. A sample of 468 was gathered from important cities. The findings from the analysis explain that using a digital wallet considerably boosts consumer spending, mainly for e-commerce and all online payments. The study concluded by outlining how these digital wallets promote sustainable growth by enhancing financial inclusion and lowering transaction costs. However, wider use is hindered by cybersecurity issues and a lack of knowledge about digital wallets. The study's conclusion describes how digital wallets have a bigger influence on sustainable economic development and consumer purchasing behavior. It also implies that positive outcomes can be achieved by reducing adoption barriers and promoting increased financial knowledge.

Keywords Digital wallets · Consumer spending behavior · Sustainable economic growth · Digital payment systems · Payment methods

1 Introduction

One easy and safe way to conduct financial transactions with more change is using digital wallets [1]. Through the use of an electronic program known as digital wallets or e-wallets, consumers can make payments online by keeping their credit or debit card information, usually on their computers or smartphones [2]. The increasing use of smartphones and other digital devices in recent years has led to a growth in the use of digital wallets [3]. As digital wallets provide a practical and cost effective alternative to traditional banking, it has been seen recognition in both developing and developed countries [4]. Due to digital wallets, the consumer spending habits has been changing significantly [5]. Electronic payment systems like credit and debit cards were preferred in place of traditional payment systems by customers these days, and due to digital wallets, the payment method has revolutionized the way customers interact with business [6]. This change influenced the customer's preference to products and services and encourages smaller and regular purchases, in addition to other changes in consumer buying habits [7]. As consumers looking for a peaceful transaction experience found digital wallets to be an attractive alternative because of their security and ease of usage features along with rewards and loyalty programs which encourages customers to use digital wallets much more [8].

From the economy point of view, digital wallets have the ability to improve sustainable economic growth [9]. Digital wallets are the main reason for the step ahead for financial inclusion with the cashless transactions mainly for the people who do not access to banking sector which would encourage consumer spending habits by improving the growth of the economy [10]. As consumers are rapidly moving toward digital wallets, this has sparked innovation and financial opportunities which improved and developed new business models [11]. With the simple financial flows and transactions, these platforms started to grow by contributing for the development of a more well-organized and incorporated economy. The study's main objective is to evaluate the impact of digital wallets on consumer buying habits and promote sustainable economic growth [12]. This will aim to understand the elements which promote adoption and the ways in which digital wallets impact consumer choices and the broader economic indications. The study's evaluation process helps companies, policymakers, and financial institutions who wish to understand and capitalize on the capacity of digital wallets for influencing consumer behavior, demonstrated in Fig. 1, as well as sustainable economic growth [13].

In every business activity, financial transactions are carried out through both cash and online methods. Thanks in great part to technology like credit cards, debit cards, and other consumer-used payment methods, our nation has advanced in the current

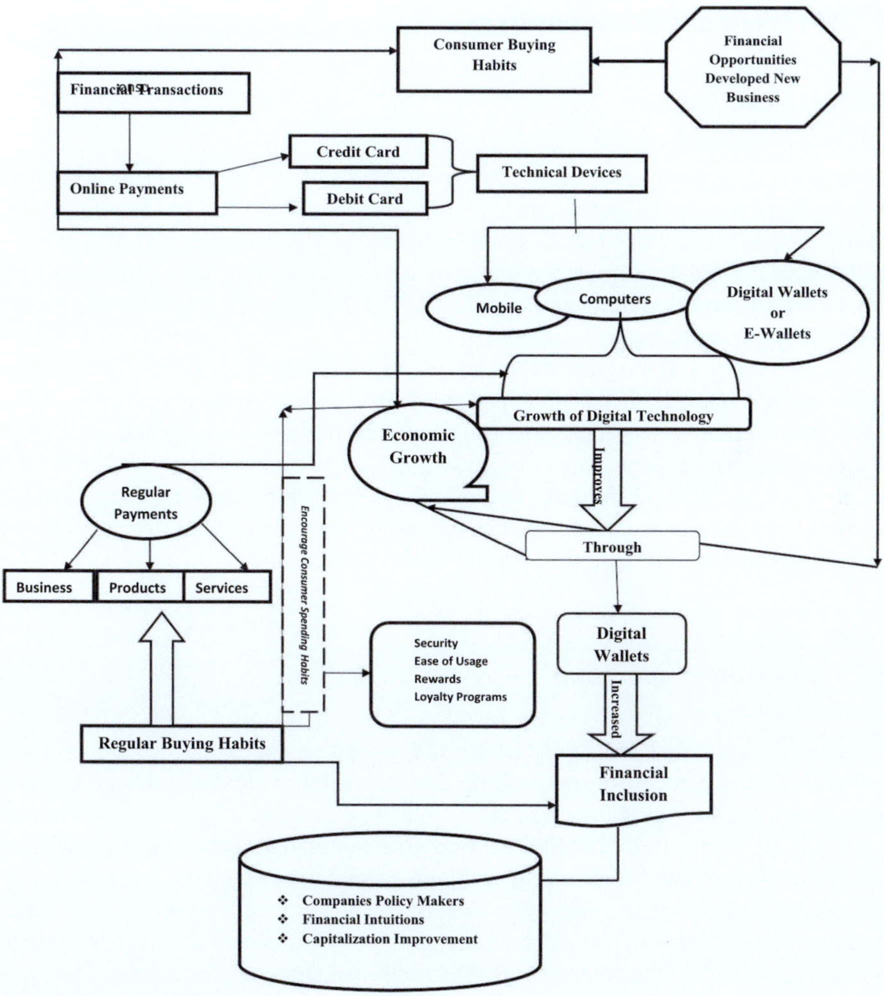

Fig. 1 Conceptual framework: digital wallets on consumer spending behavior and sustainable economic growth

global digital era. Digital wallets, technological gadgets, and mobile banking are all important in this regard. The adoption of digital wallets has been a major factor in our nation's digital technology progress. Financial inclusion will be attained when more people use these technologies, which will result in long-term economic growth. There will be a greater selection of financial products and services, which will strengthen the economy overall. Additionally, decision-making and firm policies will be improved, which will increase cash availability and facilitate easier access to financial institutions.

2 Theoretical Understandings

2.1 Digital Wallets

Digital wallets, sometimes referred to as e-wallets, are electronic programs or gadgets that let users utilize computers, smartphones, or other smart devices to safely store payment details and carry out financial transactions. Because they provide accessibility, speed, and ease in financial transactions, digital wallets have a revolutionary impact on consumer behavior and spending habits. By addressing unbanked communities, they promote financial inclusion while simultaneously encouraging higher consumer spending through cashback offers, rapid payments, and ease of access to funds. Additionally, by lowering dependency on currency, promoting transparency, aiding the formal sector, and supporting national digitalization objectives, digital wallets assist sustainable economic growth. They are essential to the goal of a cashless, inclusive, and digitally empowered economy because of their environmentally benign nature, which reduces the amount of paper used and encourages environmental sustainability through physical infrastructure.

2.2 Consumer Spending

The entire amount that people and households spend on goods and services for their own use is known as consumer spending. Because it has a direct impact on demand, production, and total economic growth, it is an essential part of a nation's economic activities. Given that digital wallets have completely changed how and when individuals spend their money, consumer spending is crucial in this context. Because digital wallets are quick, simple, and offer fast payment choices, they encourage more frequent and occasionally impulsive purchases. Additional features that encourage spending include cashback incentives, discounts, and smooth transactions. This shift in behavior encourages financial inclusion and boosts economic activity, particularly in the formal digital economy. Spending habits change as more customers use digital payment systems, which advance the larger objectives of sustainable economic growth by promoting greater transparency, effective resource allocation, and a transformation to a tech-driven, cashless economy.

2.3 Sustainable Economic Growth

A rate of economic expansion that satisfies current demands without jeopardizing the capacity of future generations to satisfy their own is known as sustainable economic growth. It comprises a comprehensive approach that ensures social inclusion, environmental protection, and efficient use of resources while fostering long-term economic prosperity. Sustainable growth becomes more important in this context as digital wallets increase financial inclusion, decrease cash-based transactions, and foster transparency, especially among marginalized groups. They promote the growth of a strong digital infrastructure, improve tax compliance, and formalize the economy by promoting digital transactions. Furthermore, the transition to contactless and paperless payments supports environmental sustainability. As a result, digital wallets not only affect consumer spending patterns but also significantly contribute to the development of an open, transparent, and ecologically responsible economic system.

3 Review of Literature

There are many different scholarly viewpoints that support the complex relationship between digital wallets and consumer spending patterns as well as sustainable economic growth. Digital payments have drastically changed consumer spending patterns by boosting purchase frequency, promoting impulsive purchases, and improving financial awareness through real-time expense tracking, claim Bhoopathy and Kanagaraj [5]. They also bring up important privacy and security issues, emphasizing the need to strike a careful balance between data protection, financial control, and ease. Sarna [9], who supports this change, highlights how India's demonetization policy sped up the uptake of digital wallets like Paytm, GPay, and MobiKwik, changing consumer behavior and propelling the country toward a cashless economy. The shift from cash to digital currency is noted, and stakeholders are given advice on how to take advantage of this trend by Kaakandikar [10], who investigates how the psychological, security, and ease of use of digital wallets affect customer purchasing behaviors.

Similarly, digital payments such as e-wallets, cards, and online banking have a substantial impact on consumer purchasing behavior and have wider societal ramifications, according to empirical findings from 100 users presented by Akilesh et al. [1]. Aji and Adawiyah [8] add psychological depth to this discussion by delving into the behavioral economics of e-wallet use and demonstrating how promotional incentives, ease of use, and the appearance of liquidity cause young adults to spend excessively, particularly when self-control is weakened. Last but not least, Sree Rajani et al. [13] emphasize how post-demonetization technology developments like UPI, AEPS, QR codes, and BHIM are changing digital banking and trade, with customers in places like Chittoor increasingly accepting them.

When taken as a whole, these studies make a strong case that digital wallets are not only changing consumer spending patterns by increasing convenience and psychological triggers, but they are also helping to advance sustainable economic growth by fostering a digital economy that is cashless, transparent, and inclusive.

4 Research Gap

Even while digital wallets are becoming more and more popular around the world, thorough studies analyzing their uptake rates among various consumer groups are still lacking, particularly in emerging nations [9]. Existing research frequently concentrates on broad adoption patterns without examining the ways in which age, income, and educational attainment affect consumer behavior. Furthermore, although research has examined how digital wallets affect consumer spending patterns, there are little empirical data connecting the use of digital wallets to certain measures of sustainable economic growth like retail sales, job creation, and GDP contribution [3]. Additionally, additional study is required to ascertain the relative significance of the elements that affect consumer selection for digital wallets: transaction costs, convenience, and security [14]. Customer choice is influenced by sociological and psychological aspects which was lacking in majority of researches which were published recently, concentrating only on technological developments.

Digital wallets offer benefits for sustainable economic growth such as promoting financial inclusion and lowering transaction costs which were not considered, and drawbacks like cyber security problems and regulatory barriers were recognized [8]. To cover up this gap, and in providing governments as well as businesses useful solutions, a comprehensive study is required which takes into account both difficulties and opportunities [10]. This study covers a thorough evaluation on digital wallets and their impact on consumer purchasing habits and their connection to sustainable economic growth and the variables which are influencing consumer preferences [7]. Suitable suggestions were also provided to overcome these obstacles and to optimize their economic potential.

5 Objectives

- To analyze the relationship between demographic variables (e.g., age, gender, income) and digital wallet adoption.
- To analyze the impact of digital wallets on consumer spending behavior.

- To study the relationship between digital wallet usage and sustainable economic growth indicators.
- To identify the factors influencing consumer preference toward digital wallets.
- To evaluate the challenges of digital wallets in driving sustainable economic growth.

6 Hypotheses

- H1: There is a significant association of demographic variables (e.g., age, gender, income) and digital wallet adoption
- H2: There is a significant impact of digital wallets on consumer spending behavior
- H3: There is a significant relationship between digital wallet usage and sustainable economic growth indicators
- H4: There is a significant impact of factors influencing consumer preference toward digital wallets
- H5: Challenges of digital wallets have a significant influence in driving sustainable economic growth

7 Scope of the Study

- The current study looks at how digital wallets affect sustainable economic growth and consumer purchasing patterns. The primary goal of the study is to examine the connection between usage of digital wallets and demographic variables.
- The research understands few important factors which influence consumer preference toward digital wallets along with the challenges of digital wallets and their influence in driving sustainable economic growth.
- The study considered both urban and semi-urban consumers in providing wider insights into unreliable responses from consumers.
- The present study helps banks, marketers, consumers, businesses, and policymakers by giving suggestions in integrating sustainable economic growth.

8 Research Methodology

The study combined quantitative and qualitative methods in a mixed-methods approach. To comprehend the financial ramifications, adoption patterns, and usage patterns of digital wallets, a descriptive study design is used. To collect primary data, standardized questionnaires and surveys are distributed to a broad spectrum of digital wallet users in urban and semi-urban areas. Stratified random sampling is used to achieve a sample size of 468 respondents in order to guarantee representation across demographics including age, gender, income, and education. To assess the stated objectives, statistical methods including multiple regression, correlation, chi-square testing, and descriptive analysis were employed. Reputable sources such as market research studies, financial institution reports, and government publications are used to collect secondary data.

A statistical method called regression analysis is employed in the study to examine the variables affecting consumer preference for digital wallets and the connection between shifts in consumer buying patterns and digital wallet usage. The challenges of digital wallets in driving sustainable economic growth were also evaluated using multiple regression analysis. The relationship between rising digital wallet usage and sustainable economic growth indicators is investigated using correlation analysis. The chi-square test was used to examine the connection between usage of digital wallets and demographic factors. Policymakers and industry experts are interviewed in-depth to obtain qualitative insights. The results provide a thorough grasp of how digital wallets affect consumer behavior and sustainable economic development, and they are triangulated to assure reliability.

9 Materials and Methods

Information from 468 consumers who utilize digital wallets is shown in Table 1. A brief description of the respondent's profile was given. The largest percentage of responders (168, or 36%) was in the 36–45 age range, followed by those in the 26–35 age range (134, or 29%). There were 154 (33%) females and 309 (66%) males, who made up the majority. Almost 148 (32%) of the respondents have completed their graduation followed by undergraduation 134 (29%). More than half of them were married 244 (52%), and 126 (27%) were unmarried. Majority of them were private employee 166 (36%) followed by self-employed 99 (21%). Maximum respondents were earning an income ranging 100,000–200,000 which were 170 (36%) followed by 0 to 100,000 which were 122 (26%).

Cronbach's alpha was used to analyze internal consistency in order to evaluate each component's dependability. With values ranging from 0 to 1, this statistical measure represents the average inter-item correlation between items in a scale. Stronger consistency between the elements is indicated by a higher alpha value. A low alpha could suggest that there are not enough pertinent items; this could be fixed by adding more pertinent questions. For multi-item scale dependability, a Cronbach's alpha of 0.7 or greater is typically regarded as satisfactory. As seen in Table 2, proper interpretation also entails comprehending the relationships between the test items.

High data dependability appropriate for additional analysis was indicated by the total reliability score of 0.969 for all chosen variables, which was higher than the suggested cutoff of 0.7. Separate assessments of each variable also showed values over 0.7, confirming their suitability for assessment. The standard deviations (1.055, 0.926, 1.050, and 0.995) show the variability or measurement error, whereas the mean

Table 1 Responders demographic profile

	Number	Percentage
Age		
< 25	74	15.8
26–35	134	28.6
36–45	168	35.9
46–55	50	10.7
≥ 56	42	9.0
Total	468	100.0
Gender		
Male	309	66.0
Female	154	32.9
Others	5	1.1
Total	468	100.0
Education		
Uneducated	14	3.0
Primary education	30	6.4
Secondary education	70	15.0
Undergraduation	134	28.6
Graduation	148	31.6
Others	72	15.4
Total	468	100.0
Marital status		
Married	244	52.1
Unmarried	126	26.9
Divorced	52	11.1
Widow	46	9.8
Total	468	100.0
Business type		
Student	24	5
Private employee	166	36
Government employee	85	18
Self employed	99	21
House wife	42	9
Others	52	12
Total	468	100
Income		
0–100,000	122	26.1
100,000–200,000	170	36.3
200,000–300,000	116	24.8

(continued)

Table 1 (continued)

	Number	Percentage
300,000–400,000	36	7.7
Above 400,000	24	5.1
Total	468	100.0

Source Authors gathered and calculated data

Table 2 Reliability test

	Mean	Standard deviation	Cronbach's alpha
Digital wallets	4.03	0.926	0.957
Consumer spending behavior	3.89	1.050	0.955
Sustainable economic growth indicators	3.95	0.995	0.957
Consumer preference for digital wallets	3.96	1.055	0.962

Cronbach's alpha: 0.969
N is 468
Source Authors gathered and calculated data

values (4.03, 3.89, 3.95, and 3.96) show the central trends of the variables. The corresponding Cronbach's alpha values, which further validated the data's consistency and dependability, were 0.957, 0.955, 0.957, and 0.962.

10 Hypothesis Testing

H1: There is a significant association of demographic variables (e.g., age, gender, income) and digital wallet adoption.

This test is very useful when dealing with nominal or ordinal categorical data. In this instance, marketing education is taken into consideration using ordinal. The alternative hypothesis in a chi-square test proposes an association between demographic characteristics and digital wallets, while the null hypothesis recommends no connection at all. The computed chi-square statistic is crucial in determining the test result. In this context, the computed statistic is higher than the critical value at 5% level of significance and establishes there is strong association between digital wallets and demographic traits. Therefore, the alternate hypothesis is failed to reject. The results interpreted are presented in Table 3. Digital wallets and gender do not correlate; however, when age and income are considered, there is a significant association. As demographics improve, especially for women, the use of digital wallets increases, and this implies that there is an unbreakable relationship between demographic characteristics and digital wallets.

Table 3 Chi-square test

	Value	Sig.
Chi-square test considering age		
Pearson chi-square	436.470	0.000
Phi	0.966	0.000
Cramer's V	0.483	0.000
Contingency coefficient	0.695	0.000
Chi-square test considering gender		
Pearson chi-square	62.232	0.734
Phi	0.365	0.734
Cramer's V	0.258	0.734
Contingency coefficient	0.343	0.734
Chi-square test considering income		
Pearson chi-square	386.739	0.000
Phi	0.909	0.000
Cramer's V	0.455	0.000
Contingency coefficient	0.673	0.000

Source Authors gathered and calculated data

H2: There is a significant impact of digital wallets on consumer spending behavior.

According to the findings of the multiple regression analysis, the R value was 0.921, the R-square was 0.848, and the adjusted R-square was 0.846 (Table 4). The p-value for the regression model was less than 0.05, despite the fact that these results were marginally below the threshold that was anticipated. This suggests that digital wallets have a considerable influence on consumer spending behavior, as evidenced by the statistically significant association. According to the coefficients, customer spending behavior is influenced by factors such as "Using a digital wallet is more convenient than traditional payment methods like cash or cards" and "Promotional offers (e.g., discounts, cashback) motivate me to use digital wallets for transactions." The factors like "I find digital wallets easy to use for my daily transactions," "I feel confident about the security features of digital wallets when making transactions," and "Using a digital wallet has encouraged me to spend more frequently than before" have no impact on consumer spending behavior. Out of all the elements that affect customer purchasing behavior, the outcome states that "Promotional offers (e.g., discounts, cashback) motivate me to use digital wallets for transactions." This shows that for every unit increase in "Promotional offers (e.g., discounts, cashback) motivate me to use digital wallets for transactions," consumer purchasing behavior increases by 0.742 units.

Table 4 Showing significant impact of digital wallets on consumer spending behavior

"*R*"	0.921	
"*R*-squared"	0.848	
"*F*"	515.984	
"Adjusted *R*-squared"	0.846	
"*P*-value (*F*)"	0.000	
"Durbin-Watson"	2.189	
Coefficients	***B***	**VIF**
(Constant)	0.265	
I find digital wallets easy to use for my daily transactions	0.026	5.139
Using a digital wallet is more convenient than traditional payment methods like cash or cards	0.171	5.107
I feel confident about the security features of digital wallets when making transactions	− 0.031	5.769
Using a digital wallet has encouraged me to spend more frequently than before	0.015	1.226
Promotional offers (e.g., discounts, cashback) motivate me to use digital wallets for transactions	**0.742**	23.903

Source Authors gathered and calculated data
'coefficients' represent significant impact of independent variables on dependent variable of the study. 'B' represents unstandardised beat coefficients and 'VIF' represents Variance Inflation Factor.

H3: There is a significant relationship between digital wallet usage and sustainable economic growth indicators.

The intensity and direction of the association between two continuous variables can be quantitatively measured using correlation coefficients, such as Pearson's correlation. The degree of a linear relationship is indicated by Pearson's correlation coefficient, or "*r*," which has values between − 1 and + 1. Strong correlations are indicated by values close to + 1 or − 1, where + 1 denotes a perfect positive linear relationship and − 1 a perfect negative one. There is no linear relationship when the coefficient is 0. Negative numbers indicate that one variable tends to decline as the other increases, but positive values indicate that the other tends to increase as well. The setting and topic of the investigation can affect how correlation strength is interpreted. There is no widely agreed-upon definition of what constitutes a "good" relationship; it depends on the research question and topic of study. While weaker correlations may be significant in some circumstances, stronger correlations may be expected in others. Table 5's significant and positive correlation coefficient indicates that the two variables are strongly correlated. This would suggest that consumer preference has grown as a result of the adoption of digital wallets, which is indicative of long-term economic growth (Table 6).

Table 5 Correlation demonstration

		I find digital wallets easy to use for my daily transactions	Using a digital wallet is more convenient than traditional payment methods like cash or cards	I feel confident about the security features of digital wallets when making transactions	Using a digital wallet has encouraged me to spend more frequently than before	Promotional offers (e.g., discounts, cashback) motivate me to use digital wallets for transactions	Sustainable economic growth indicators
I find digital wallets easy to use for my daily transactions	Pearson correlation	1	0.570^{**}	0.548^{**}	0.293^{**}	0.809^{**}	0.762^{**}
Using a digital wallet is more convenient than traditional payment methods like cash or cards	Pearson correlation	0.570^{**}	1	0.677^{**}	0.371^{**}	0.849^{**}	0.776^{**}
I feel confident about the security features of digital wallets when making transactions	Pearson correlation	0.548^{**}	0.677^{**}	1	0.392^{**}	0.854^{**}	0.734^{**}
Using a digital wallet has encouraged me to spend more frequently than before	Pearson correlation	0.293^{**}	0.371^{**}	0.392^{**}	1	0.421^{**}	0.359^{**}

(continued)

Table 5 (continued)

		I find digital wallets easy to use for my daily transactions	Using a digital wallet is more convenient than traditional payment methods like cash or cards	I feel confident about the security features of digital wallets when making transactions	Using a digital wallet has encouraged me to spend more frequently than before	Promotional offers (e.g., discounts, cashback) motivate me to use digital wallets for transactions	Sustainable economic growth indicators
Promotional offers (e.g., discounts, cashback) motivate me to use digital wallets for transactions	Pearson correlation	0.809**	0.849**	0.854**	0.421**	1	0.907**
Sustainable Economic Growth Indicators	Pearson correlation	0.762**	0.776**	0.734**	0.359**	0.907**	1

Sig. (2-tailed) is 0.000; N is 468
Source Authors gathered and calculated data

Table 6 Showing significant impact of digital wallets and sustainable economic growth indicators

R	897	
R-squared	0.805	
F	478.290	
Adjusted *R*-squared	0.803	
P-value (*F*)	0.000	
Durbin-Watson	2.032	
Coefficients	***B***	**VIF**
(Constant)	0.904	
I feel confident about the security and privacy of my personal and financial data when using a digital wallet	0.086	5.153
Using a digital wallet allows me to complete transactions more quickly than traditional payment methods	0.061	5.329
The rewards, cashback, or discounts offered by digital wallets influence my preference for using them	− 0.108	5.165
I find digital wallets more convenient than carrying cash or using credit/debit cards	**0.749**	22.700

Source Authors gathered and calculated data
'coefficients' represent significant impact of independent variables on dependent variable of the study. 'B' represents unstandardised beat coefficients and 'VIF' represents Variance Inflation Factor

H4: There is a significant impact of factors influencing consumer preference toward digital wallets.

The multiple regression explanations in Table 4 included "*R* value-0.897," "*R*-square value-0.805," and "adjusted *R*-square values-0.803," all of which were below the necessary level. The *P*-value was less than 0.05, which suggests that there are elements affecting consumers' inclination to utilize digital wallets. The coefficients show that consumer choice for digital wallets is influenced by the following factors: "I find digital wallets more convenient than carrying cash or using credit/debit cards," "I feel confident about the security and privacy of my personal and financial data when using a digital wallet," and "Using a digital wallet allows me to complete transactions more quickly than traditional payment methods." Consumer preference for digital wallets is unaffected by factors such as "The rewards, cashback, or discounts offered by digital wallets influence my preference for using them." According to the output, the statement "I find digital wallets more convenient than carrying cash or using credit/debit cards" had the biggest influence on consumers' choice for digital wallets out of all the criteria. This indicates that customer preference for digital wallets rises by 0.749 units for every unit increase in the statement, "I find digital wallets more convenient than carrying cash or using credit/debit cards." (Table 7).

Table 7 Showing significant impact of factors influencing consumer preference towards digital wallets

R	0.840	
R-squared	0.705	
F	220.620	
Adjusted *R*-squared	0.702	
P-value (*F*)	0.000	
Durbin-Watson	1.796	
Coefficients	***B***	**VIF**
(Constant)	– 0.038	
I feel that security concerns, such as data breaches and fraud, discourage widespread adoption of digital wallets	0.057	1.726
The complexity of using digital wallets poses a challenge for many consumers	0.245	1.666
Lack of digital literacy among consumers is a significant barrier to adopting digital wallets	0.171	1.734
Inadequate internet connectivity and smartphone access limit the use of digital wallets in certain areas	0.244	1.998
High transaction fees associated with digital wallet payments deter their usage	**0.260**	1.867

Source Authors gathered and calculated data
'coefficients' represent significant impact of independent variables on dependent variable of the study. 'B' represents unstandardised beat coefficients and 'VIF' represents Variance Inflation Factor

H5: Challenges of digital wallets have a significant influence in driving sustainable economic growth.

The multiple regression explanations in Table 4 included "*R* value-0.840," "*R*-square value-0.705," and "adjusted *R*-square values-0.702," all of which were below the necessary level. The *P*-value was less than 0.05; this suggests that issues with digital wallets significantly impacted sustainable economic expansion. Factors such as "high transaction fees associated with digital wallet payments deter their usage," "many consumers find digital wallets difficult to use," "inadequate internet connectivity and smartphone access limit the use of digital wallets in certain areas," and "lack of digital literacy among consumers is a significant barrier to adopting digital wallets" all have an effect on sustainable economic growth, according to the coefficients. The factors like "I feel that security concerns, such as data breaches and fraud, discourage widespread adoption of digital wallets" have no impact on sustainable economic growth. The result shows that, of all the factors, "High transaction fees associated with digital wallet payments deter their usage" has the biggest effect on sustainable economic growth, meaning that for every unit increase in this factor, sustainable economic growth rises by 0.260 units.

11 Findings

- The largest age group of respondents was 36–45 years, accounting for 168 individuals (36%), followed by the 26–35 age group at 134 (29%).
- Male respondents made up 309, or 66% of the total, while female respondents made up 154, or 33%.
- The majority of respondents was undergrads (134, 29%) and had completed their degrees (148, 32%).
- More than half of the respondents were married (244, 52%), while 126 (27%) were unmarried.
- The majority were private employees (166, 36%), followed by self-employed individuals (99, 21%).
- The highest proportion of respondents earned between ₹100,000 and ₹200,000 annually (170, 36%), followed by those earning up to ₹100,000 (122, 26%).
- With dependability ratings of 0.969 for all selected variables higher than 0.7, it is clear that the gathered data are more suitable and need to be selected for further examination.
- Demographic factors such as age and income showed a strong correlation with the usage of digital wallets, whereas gender had no effect.
- Using multiple regression analysis, the study examined how digital wallets affect consumer spending patterns and provided an explanation of how "promotional offers (such as discounts or cashback) motivate me to use digital wallets for transactions" was having a greater effect than other elements.
- The study evaluated the relationship between digital wallet usage and sustainable economic growth indicators using correlation, which represents that there was a strong relation among digital wallet and sustainable economic growth indicators.
- For the present study, multiple regression analysis was used to analyze the impact of factors influencing consumer preference toward digital wallets. Factor like "I find digital wallets more convenient than carrying cash or using credit/debit cards" was having more impact toward digital wallets.
- The challenges of digital wallets were having a significant impact toward sustainable economic growth. "High transaction fees associated with digital wallet payments deter their usage" was having more impact.

12 Recommendations

- As the majority of users fall in the 36–45 age groups, digital wallet providers should design campaigns and features tailored to this demographic, emphasizing convenience and ease of use.
- With only 33% of respondents being female, providers should explore strategies to increase adoption among women, such as offering features that cater to their specific financial and shopping needs.

- Since the majority of users are graduates, efforts should be made to reach less-educated individuals by simplifying the interface and offering educational resources on digital wallet usage.
- With a significant proportion of respondents earning between ₹100,000 and ₹200,000, providers can design cashback or discount offers that align with the spending habits of middle-income users.
- As promotional offers significantly influence consumer spending behavior, digital wallets should continue to invest in attractive cashback, discounts, and rewards to maintain user engagement.
- Highlighting the convenience of digital wallets over traditional payment methods can further boost consumer preference and adoption rates.
- To overcome the challenge of high transaction fees deterring usage, providers could consider lowering fees or offering fee waivers for frequent users.
- Because of the strong relationship between digital wallet usage and sustainable economic growth indicators, policies that promote digital payments could stimulate broader sustainable economic benefits.
- With a high Cronbach's alpha value (0.969), the data are robust, providing confidence for further strategic decision-making based on these findings.

13 Scope for Future Research

- Future studies can probe into identifying the growth in adoption of digital wallets and its influence on various sustainable economic indicators.
- Future studies can focus on examining the role of AI in digital security contributing sustainable financial development.

14 Conclusion

The study highlights the growing recognition of digital wallets as a disruptive factor in the financial sector. Digital wallets, a useful and efficient alternative to traditional payment methods, have improved the economy as a whole and prompted a shift in consumer purchasing habits. The results show that customers between the ages of 36 and 45 are the most likely to utilize digital wallets, with those between the ages of 26 and 35 coming in second. This suggests that a sizable percentage of consumers use digital payment services during their prime years for income and spending. The majority of responses were from men, but the user base may grow if initiatives were taken to encourage female adoption. A significant proportion of responders who were students and were graduates providing evidence those digital literacy initiatives are very important for users who are less educated, as education background were another important factor.

There is a significant influence of digital wallets over consumer purchases due to its quickness, easiness in using, and the attraction of promotional offers such as discounts and cashbacks. Out of all the characteristics evaluated, promotional incentives was showing a high impact and encouraged users in making more purchases. Top priority has to be given to customer-focused marketing efforts, and loyalty programs were implied by the businesses that provided payment options using digital wallets. The study further analyzed the substantial relationship among sustainable economic growth metrics and the use of digital wallets. As digital wallets make transitions easier and eliminate the need for currency, this has increased sustainable economic efficiency and financial inclusion. Yet a significant transaction costs, privacy, and security concerns were identified as very crucial barriers to wider use. However, by tracking these issues using new and advanced technology and implementing regulatory changes, the benefits from financing of digital payment systems can be elaborated.

From a wide perspective, the main importance of digital wallets in raising the transactions to a cashless economy and motivating sustainable economic growth was highlighted in this paper. Digital wallets are a very important part of eco-friendly sustainable economic growth, as they have the power to change consumer spending patterns and speed up the financial transactions. To eliminate the existing barriers, enhance digital infrastructure, and increase awareness, policymakers, businesses, and financial institutions have to collaborate to be completely realized to keep their promise. Finally, the study elaborates that digital wallets are significantly speeding up sustainable economic growth in addition to revolutionizing personal money management. By utilizing the strength and conquering challenges, interested parties can ensure the long-term success in digital revolution.

References

1. Akilesh KU, Hariharan G, Raju K, Jayasubramanian P (2023) A study on impact of digital payments on consumer's spending behaviour with special reference to Coimbatore City. Int J Res Anal Rev 10(1):70–76
2. Bansal S (2017) World Wide J Multidiscip Res Dev 3(9):10–12
3. Chaudhury S, Das SK (2021) Access to digital financial services: a study of rural India. J Financ Serv Mark 26(1):44–56
4. Budheshwar Prasad Singhraul YS (2018) Cashless economy: challenges and opportunities in India. Pac Bus Rev Int 10(9):54–63. Retrieved from www.pbr.co.in
5. Bhoopathy G, Kanagaraj P (2023) The impact of digital payments on consumer spending habits. Tuijin Jishu/J Propuls Technol 44(4):5373–5380
6. Sri Sai Chilukuri VV (2018) Modern banking services of scheduled commercial banks—an empirical study on customers satisfaction in Hyderabad City. J Advaced Res Dyn Control Syst 10(8):117–123
7. Williamson GD (2006) GE Money—America's enhance authentication in online banking. J Econ Crime Manag 4(2)
8. Aji HM, Adawiyah WR (2021) How e-wallets encourage excessive spending behavior among young adult consumers? J Asia Bus Stud 16(6):868–884

9. Sarna IS (2021) Impact of digital wallets on consumer behavior in India. Int J Adv Res 9(9):220–236
10. Kaakandikar R (2024) Unlocking spending trends: the behavioural impact of digital wallets on modern consumers. Econ Sci 20(1):127–143
11. Sunitha G, Venu Madhav V (2020) A study on role of fin- tech in banking services: Opportunities and Challenges. J Crit Rev 7(4):616–620. https://doi.org/10.31838/jcr.07.04.114
12. Johora FT, Hasan R, Akter J (2024) AI-powered fraud detection in banking: safeguarding financial transactions. Am J Manag Econ Innov 6(6)
13. Sree Rajani N, Bhargavi Reddy V, Susendiran S, Murali V (2022) A study on consumer behavior towards digital wallets—an empirical approach. IJFANS Int J Food Nutr Sci 11(3):1392–1402
14. Suwaidi MHA, Ahmed N (2021) Factors influencing the intention to adopt digital payment systems: evidence from the United Arab Emirates. J Financ Serv Mark 26(1):70–86

Behavioral FinTech: Nudging Consumers Toward Sustainable Financial Choices

Sahoo Kalyan Kumar, Ruchi Kushwah, Sheetal Jaldhari, and Mishra Padma Charan

Abstract The integration of behavioral economics and financial technology (FinTech) has provided new channels to encourage sustainable financial behavior. The rapid advancement of financial technology, or FinTech, has produced cutting-edge tools that use digital nudges and behavioral economics to promote sustainable financial practices. This study examines how customers' sustainable financial decisions (SFD) are affected by behavioral FinTech, specifically gamification, robo-advisors, and neobanks. Using a mixed-methods approach, we employ structural equation modeling (SEM) to analyze survey data from FinTech users and sentiment analysis of user comments to improve results. According to our findings, behavioral nudges have a significant mediating role in the meaningful facilitation of SFD by gamification, robo-advisors, and neobanks. Sentiment analysis finds environmental impact, transparency, and trust among top drivers of adoption, while data privacy and algorithmic bias remain concerns. The research gives actionable recommendations for FinTech companies like incorporating live sustainability metrics and policymakers to achieve ethical nudge design and accessible access. In bridging FinTech innovation with behavioral science, this research extends the debate about scalable, sustainable finance solutions and emphasizes the imperative for longitudinal and culturally representative studies to evaluate long-term behavioral effects.

Keywords Behavioral economics · FinTech · Nudging · Sustainable finance · Digital interventions · Consumer behavior

---

S. Kalyan Kumar · R. Kushwah · S. Jaldhari
Faculty of Management, Medicaps University, Indore, MP, India

M. Padma Charan (✉)
Department of Business Management, Central University of Odisha, Koraput, Odisha, India
e-mail: pcmishra@cuo.ac.in

S. Roy et al. (eds.), *Humanized Intelligence in Sustainable Finance and Economic Behaviours*, Studies in Computational Intelligence 1226,
https://doi.org/10.1007/978-981-95-0599-9_9

1 Introduction

The convergence of financial technology (FinTech) and sustainability has emerged as a pressing field of focus for scholars, practitioners, and policymakers in recent times. As international pressures around climate change, sustainable consumption, and ethical finance increase, the need for creative solutions that align economic incentives with environmental and social objectives has heightened. FinTech, with the use of digital technologies, behavioral economics, and data science, is well-positioned to affect financial habits in large numbers. New developments in the past, including gamification-based savings applications, ESG investment websites that utilize artificial intelligence, and carbon-aware neobanks, have tried to incorporate sustainability into the core of personal finance [21]. However, while they are becoming increasingly popular, doubts still linger as to whether these tools actually work in transforming consumer behavior toward truly sustainable ends.

Behavioral economics specifically the concepts of nudge theory and bounded rationality has played a formative part in developing FinTech solutions for nudging users toward more responsible financial practices [19]. Classic models of financial decision-making traditionally assumed rational agents on the part of consumers, but modern scholarship shows that there are biases, heuristics, and affective triggers, which shape spend, save, and invest habits [13]. FinTech platforms have been reacting by instilling behavioral feedback into their platforms: default choices, progress bars, goal reminders, and social comparisons are now widely used to "nudge" users gently but strongly. For instance, apps such as Aspiration give users cashback for purchases made at environmentally friendly merchants, while sites such as Betterment utilize robo-advisors to encourage ESG-oriented investment portfolios.

The incorporation of gamification as the use of game-like features in non-game environments has also increased the scope of FinTech to engage and inspire users. Evidence suggests that gamified FinTech instruments can boost user participation, goal commitment, and financial knowledge, particularly among younger populations [16, 17]. This increasing body of literature, however, also emphasizes important issues. There remains debate regarding whether such interventions actually spur long-term behavior change or whether they are simply based on superficial participation strategies. Also raised have been questions of the accessibility and inclusiveness of such tools, as they might not be equally beneficial to all demographic groups, with the potential for unequal benefits across customer segments.

Robo-advisors, another touchstone of behavioral FinTech, provide algorithmic investment advice that circumvents conventional human intermediaries. These online advisors purport to democratize access to sustainable investing by reducing entry barriers and advising users in accordance with their risk tolerance and ESG choices. Although there is evidence for their efficacy in alleviating decision fatigue and portfolio diversification [13], issues of transparency, algorithm dependence, and the possibility of manipulation persist. The challenge is how to keep such digital nudges ethical and context-aware—tuning personalization to regulatory needs and consumer protection.

Neobanks such as Tomorrow Bank and TreeCard have arisen as digital-native financial institutions that are consciously built with environmental objectives at their center. These platforms provide services like carbon tracking, climate giving, and green budgeting tools that invite consumers to align everyday purchases with sustainability objectives. Research indicates that these innovations are attractive to eco-friendly users, but the wider behavioral effect of these is unexamined [6]. In addition, the risk of greenwashing—a practice where firms overstate or distort sustainability claims—is a reputational risk and an ethical concern in this area. At the macroeconomic level, the growth of behavioral FinTech is not confined to the individual consumer but pervades institutional arrangements. Wan and Cui [22] underscore the importance of FinTech in mitigating risk in green supply chains and agriculture finance, particularly in emerging economies. Lee et al. [12] instead point out the convergence of ESG with AI in financial uses, noting the importance of transparency and responsible AI use on consumer-facing platforms. These wider developments highlight the need for interdisciplinary cooperation among technologists, behavioral scientists, and financial regulators.

In spite of these encouraging developments, current research has significant gaps. Firstly, most research has been based on Western or developed economies, with little empirical investigation in multi-cultural or socio-economic contexts. Secondly, longitudinal data are lacking to determine the sustainability of behavior changes brought about by digital nudges. Third, ethical concerns like data privacy, autonomy, and digital literacy are typically underemphasized in empirical studies even as they gain increasing importance. With improvements in the sophistication of behavioral nudging, issues related to manipulation, exclusion, and adverse side effects need to be taken into consideration [17, 21].

With this context in mind, the current study aims to investigate how behavioral FinTech devices affect sustainable financial decisions among consumers. In particular, this study will concentrate on three areas: (i) assessing the efficacy of gamification in encouraging green spending, (ii) determining the contribution of robo-advisors to the democratization of ESG investing, and (iii) measuring the efficacy of neobanks in encouraging long-term green financial behaviors. The integration of theoretical knowledge with practical case studies aims to enhance the understanding of behavioral FinTech as a catalyst for sustainable development.

The study will also provide actionable guidance to policymakers and FinTech businesses. By identifying the most important drivers of behavior to user takeup—e.g., trust deficits, psychological hooks, and structural constraints—the research will offer insight into how such instruments can be optimized to deliver sustainable financial outcomes. Besides, it will help advance the discussion on the ethical issues in the design and implementation of behavior nudges to guarantee that they achieve long-term positive social and environmental outcomes and minimize possible unwanted consequences.

2 Review of Literature

The intersection of behavioral economics and FinTech has emerged as a seminal interdisciplinary area of research, particularly in the context of encouraging sustainable financial behavior. Drawing from the pioneering work of Thaler and Sunstein [19], nudge theory proposes that individuals' money choices can be influenced by subtle design changes in the decision setting without removing freedom of choice. This theoretic model has had extended use in the FinTech arena, where online platforms employ tactics such as default options, social proof, gamification, and immediate feedback to nudge user behavior toward financially and environmentally sound directions [3, 4].

The greater significance of these sorts of behavioral interventions is that they have the potential to be scalable and flexible. FinTech innovations, from innovative mobile banking services to robo-advisors and decentralized finance platforms, are well suited to incorporate nudges into their user experience. This is beneficial in terms of mass personalized behavioral solutions that address individual financial goals and match wider sustainability goals. Singh et al. [18], in their empirical work, emphasized the role of behavioral interventions like cognitive framing, anchoring, and the default effect in influencing sustainable investment behavior. Such nudges are particularly effective when employed in real-time by digital financial products, confirming the place of FinTech to transform consumer understanding and response to financial information about sustainability.

In addition, Beshears et al. [3] theorize that standard economic models based on the rationality assumption are bound to fall short when it comes to explaining real money choices. Behavioral economics comes to the rescue by acknowledging biases such as present bias, loss aversion, and bounded rationality—each of which is highly influential in influencing financial behavior. The interaction of behavioral knowledge with FinTech not only allows for enhanced financial product design but also allows for consumer education and goal-setting through the utilization of intuitive user interfaces.

At a worldwide level, Kashif et al. [11] observed a significant association between the implementation of FinTech and sustainable finance in 89 countries. Through their study, they reveal that institutional quality acts as a moderator to reinforce the constructive impact of FinTech on promoting ethical financial practices and sustainability. Well-functioning institutions are the pillars that support regulatory mechanisms and customer trust, which are critical to the success of digital financial innovations. These macro-level findings are complemented by those of Nguyễn and Trần [15], who focused on ASEAN + 3 countries and found that FinTech innovation is a key driver of green bond market development. Their research highlights that FinTech platforms facilitate increased transparency, diminish transaction costs, and improve access to green investments, thereby performing a catalytic role in financing sustainable projects.

While digital nudges hold great promise, they also pose ethical concerns. Acquisti et al. [1] cautioned that behavioral interventions, if deployed without transparency, can infringe on the autonomy and privacy of users. This is especially pertinent in FinTech platforms that employ big data and AI-based algorithms to tailor financial nudges. Bounfour and Miyamoto [5] also point out that without ethical controls, such interventions may quickly become manipulative and produce unintended financial consequences for vulnerable users. They propose a behavioral-ethical approach to FinTech with informed consent, transparency, and user empowerment as the core pillars of ethical digital financial design.

In the Indian context, Gupta and Goswami [7] explored the intersection of behavioral economics and FinTech through an analysis of how psychological traits and demographic variables influence Socially Responsible Investing (SRI). Using Bayesian regression, they were able to identify strong gender-based differences in investment behavior, particularly among female investors who were most responsive to ESG-related information and behavioral nudge. They opine that products should be made gender-sensitive and that nudges must account for psychological as well as social heterogeneity in the users.

Mustafa and Rizwan [14] expanded on this learning through the focus on the trust function of FinTech-based crowdfunding platforms in Pakistan. Their structural equation modeling underlined the point that blockchain technology reinforces investor confidence, particularly when combined with behavioral cues like transparency, peer ratings, and social validation. By incorporating behavioral economics principles into digital trust design, these platforms are more likely to attract retail investors for worthy projects.

In addition, Tanchangya et al.'s [20] study offers useful insights into FinTech adoption in emerging economies. They point out that although FinTech platforms have behavioral benefits—e.g., nudges to budgeting, micro-investing, and savings—structural impediments such as digital illiteracy, unequal access to technology, and weak financial literacy have a tendency to restrict their impact. The authors advocate the combination of behavioral interventions with wider socio-economic development strategies so that FinTech-based sustainable finance is accessible to marginalized communities.

As FinTech keeps expanding, the application of environmental, social, and governance (ESG) standards has been introduced as a strong tool in assessing risk and innovative strategies. Gopal and Pitts [6] are strong advocates in the application of ESG metrics in the context of FinTech platforms in assessing and managing risk and ensuring long-term sustainability. By using geospatial data science, AI, and behavioral economics, their cross-disciplinary process develops predictive models for sustainability-based financial products. Not only does this minimize information asymmetry among issuers and investors, but it also enhances regulatory compliance and investor trust.

At the capital supply side, crowdfunding platforms and decentralized finance (DeFi) have greened access to capital for environmental projects. They are inclined to leverage popularity of initiatives, user reviews, and milestone metrics as behavior triggers in order to heighten interest and contribution. Though such features enable engagement, they also produce new behavioral risks such as herd behavior and choice fatigue. Writing emphasizes the importance of balancing behavior design with mitigation of manipulation or disinformation.

The influence of FinTech in influencing behavior is not just at the level of personal financial decision-making but also at the level of corporate financial behavior. Corporates are now making greater use of FinTech data analytics to measure investors' sentiment, quantify ESG performance, and forecast sustainability-related risk. Business intelligence derived from big data allows businesses to set strategy in line with stakeholders' expectations and environmental norms.

The World Bank [23] identifies the expanding role of FinTech tools directed by behavior within global sustainable finance. It is in favor of establishing regulatory sandboxes for piloting behavioral nudges within a controlled environment. The sandboxes allow FinTech start-ups to test innovative remedies while being held in line with ethical and compliance standards. The World Bank also identifies the need for interdisciplinary collaboration of behavioral scientists, data scientists, regulators, and FinTech entrepreneurs to come up with a balanced and inclusive financial environment.

In addition to policy recommendations, other scholars also advocate for an interdisciplinary research agenda to examine the long-term behavioral effect of digital nudges. As users increasingly rely on algorithmic recommendations, the risk of behavioral habituation or dependency is higher. Experiments must be performed to evaluate the long-term consequences of exposure on financial literacy, autonomy, and decision-making resilience. There is similarly a paucity of information about cultural and socio-economic background factors affecting the effectiveness of behavioral interventions in digital finance.

One of the future research avenues is exploring the impact of machine learning and artificial intelligence on behaviorally targeted nudges. AI-powered FinTech apps have the ability to dynamically change nudges in real-time as per user actions, creating a feedback loop to continuously refine experience. However, it raises new concerns about data governance, algorithmic bias, and decision-making transparency. Scholars recommend the creation of ethical AI architectures that guarantee user rights and consent.

Furthermore, the concept of "smart nudging" is gaining traction, whereby digital interventions are not just context-aware but also ethically informed. This approach incorporates fairness, equity, and long-term impact into nudge design, such that behavioral economics is serving the larger goals of sustainable development and not short-term compliance or profit maximization.

Finally, the literature is inclined toward the notion that FinTech combined with behavioral knowledge and ethical thinking is the best approach to attaining the objectives. Additionally, the term "smart nudging" is becoming more popular. It means using digital tools in such a way that not only fits the user's context but is also ethically appropriate. The nudges are designed with fairness, equality, and long-term impacts in consideration, making sure they promote sustainable development objectives and not merely lead individuals to make short-term commitments or profits.

On the basis of this thorough literature review, a conceptual diagram has been developed illustrating the interplay between FinTech tools, behavioral nudges, and

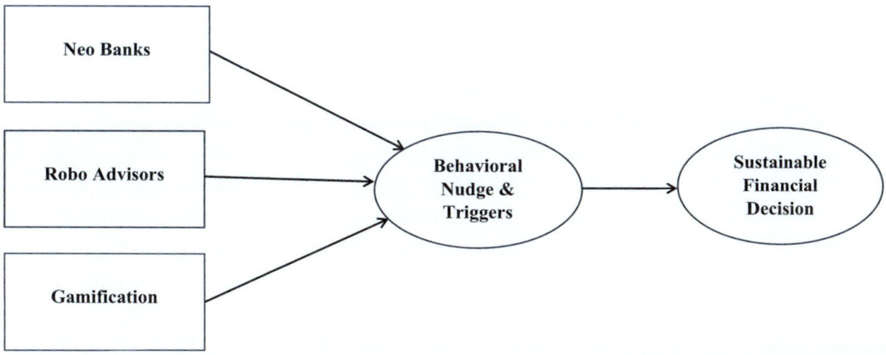

Fig. 1 Conceptual framework for the study

sustainable financial decision-making (Fig. 1). The diagram features major behavioral drivers such as psychological triggers and nudges influencing user engagement and adoption of these tools. According to this framework, structural equation modeling (SEM) has been developed to test the proposed relations and ascertain the mediating functions of these behavioral variables on the effectiveness of FinTech tools to facilitate sustainable financial habits.

3 Research Methodology

The research is quantitative, and the aim is to assess the effectiveness of behavioral FinTech products like gamification, robo-advisors, and neobanks in inducing sustainable financial decision-making (SFD). The research aims for a positivist approach, where attention is drawn toward measuring associations among FinTech solutions and user behaviors, aiming for the discovery of the mediating role of psychological triggers and behavioral nudges in inducing adoption.

A cross-sectional descriptive design is employed, where data can be collected at one point in time to examine the relationships between the independent variables (FinTech tools) and dependent variable (sustainable financial decision-making). This type of design yields evidence of FinTech tool adoption and usage trends as well as determinants for users' sustainable financial behavior.

The population for the study is 500 users who have used at least one of the below FinTech tools: gamification tools, robo-advisors, and neobanks. To ensure the sample is representative of the entire population of users of the tools, stratified random sampling approach is employed. The users of FinTech are grouped into separate strata according to their usage of various types of FinTech tools. Each level comprises users using gamification features, robo-advisors, or neobanks individually. Random samples are then drawn from each level such that all groups of users are represented in the final sample quite evenly. This probability sampling is useful in the enhancement of generalizability of the study results to a greater population of the FinTech user base. Sampling for this research was done by collecting data based on a structured survey questionnaire and was administered online across different platforms of FinTech and social networks. The survey was constructed to gather information regarding FinTech tool usage, environmentally friendly financial decision-making mindsets, and the behavioral nudges and psychological cues that are shaping adoption. The survey was based on a previous literature review of behavioral finance, sustainable investing, and user behavior on online finance websites. It was pre-tested in terms of context and clarity by being administered to an initial sample of 50 participants before data collection went on at scale.

The data collected were structured and examined using structural equation modeling (SEM), a statistical analysis technique well adapted to exploring intricate interactions between observed variables and latent factors. SEM was applied to examine the hypothesized hypotheses and analyze the mediating influence of behavioral nudges and psychological triggers on the interaction between FinTech tools and sustainable financial decision-making. The analysis was achieved using the support of AMOS (Analysis of Moment Structures), an extendable used set of software used in SEM to estimate indirect and direct effects and understand the mechanisms in which FinTech tools affect the behavior of the users.

4 Research Objectives

i. To evaluate the effectiveness of behavioral FinTech tools (gamification, robo-advisors, neobanks) in promoting sustainable financial decisions.
ii. To identify key behavioral drivers (e.g., nudges, psychological triggers) that influence user adoption.
iii. To provide practical recommendations for FinTech firms and policymakers to utilize sustainable finance solutions.

5 Data Analysis and Findings

In order to meet the first research goal—testing the effectiveness of behavioral FinTech technologies like gamification, robo-advisors, and neobanks in fostering sustainable financial choices—structural equation modeling (SEM) was employed. SEM was used because it has the ability to test both the measurement as well as structural aspects of the model simultaneously and hence capture latent constructs and their relationships fully (Fig. 2).

The model included three exogenous variables (gamification, robo-advisors, neobanks), a mediator (nudges and triggers), and the endogenous outcome variable (sustainable financial decisions).

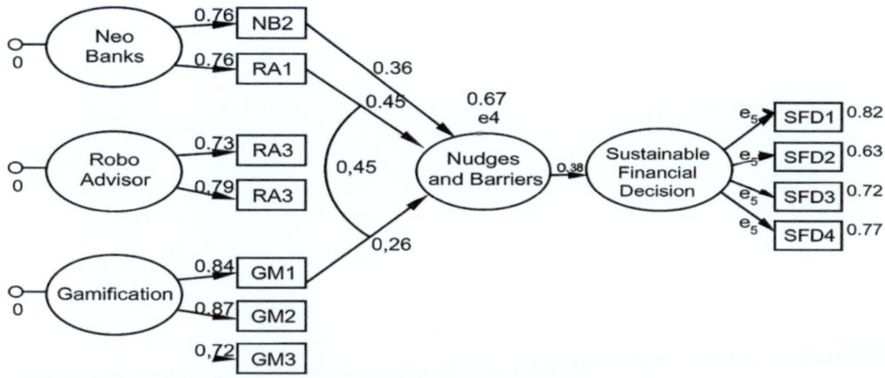

Fig. 2 Structural equation model for the study

The internal consistency of the constructs was checked using Cronbach's Alpha and composite reliability (CR) measures. All the constructs had high internal consistency, with Cronbach's Alpha measures between 0.82 and 0.89 and CR measures between 0.84 and 0.91, both higher than the suggested threshold of 0.70 [8] (Refer to Table 1).

Convergent validity was verified by checking Average Variance Extracted (AVE) values, which were all higher than the standard threshold of 0.50. AVE values were between 0.61 and 0.74, signifying that every construct accounts for over half the variance of its indicators. The findings, therefore, confirm that the measurement model satisfies the threshold of reliability and convergent validity requirements (Table 2).

The structural equation modeling (SEM) fit is good according to the results. The χ^2/df ratio of 1.44 is significantly less than the suggested threshold of 3, indicating a reasonable level of model parsimony. Also, important incremental fit measures like CFI (0.963) and TLI (0.957) are well above the conventional cut-off of 0.90, establishing a good model fit. RMSEA at 0.041 and SRMR at 0.036 are both well below the acceptable ranges (< 0.06 and < 0.08 respectively), which means there is little residual error between the observed and the predicted covariance matrices. The GFI (0.902)

Table 1 Reliability and validity measurements

Construct	Cronbach's alpha (α)	Composite reliability (CR)	Average variance extracted (AVE)
Neobanks	0.75	0.85	0.65
Robo-advisors	0.78	0.86	0.68
Gamification	0.82	0.89	0.73
Nudges	0.74	0.84	0.57
Sustainable financial decision	0.79	0.87	0.63

Source Primary Data

Table 2 Structural model fit indices

Model fit indices	Obtained value	Standard value
χ^2/df	1.44	< 3
CFI	0.963	> 0.90
TLI	0.957	> 0.90
RMSEA	0.041	< 0.06
SRMR	0.036	< 0.08
GFI	0.902	> 0.90
AGFI	0.881	> 0.80

Source Primary Data

and AGFI (0.881) indices also confirm the goodness of the model structure. The fit indices all together exhibit that the given model connecting neobanks, robo-advisors, and gamification to behavioral nudges and sustainable financial decision-making is a good fit for the data that has been observed.

Mediation analysis was used to explore the role of nudges in the association between the FinTech variables and sustainable financial decision-making. The mediation analysis results are displayed in Table 3.

The mediation analysis found that nudges showed partial mediation for the relationship between sustainable financial decision-making and FinTech innovations. Neobanks partly mediated this process, which also had both a significant direct (0.15, $p < 0.05$) as well as indirect effect (0.22, $p < 0.01$), proving that neobanks drive financial decision-making directly, while they influence this process indirectly by nudging. In contrast, robo-advisors and gamification were found to exert full mediation, with the direct effects being non-significant (0.08, $p > 0.05$ and 0.10, $p > 0.05$, respectively), but highly significant indirect effects (0.29, $p < 0.001$ and 0.25, $p < 0.001$), which suggests that these constructs affect decision-making completely through the mediating power of nudges.

To address the third goal giving actionable suggestions to FinTech designers and policymakers, results of both SEM and sentiment analysis were triangulated. The collective findings indicated that FinTech design aimed at behavioral change should

Table 3 Mediation analysis summary

Path	Direct effect	Indirect effect	Total effect	Mediation type
Neobanks → Sustainable financial decision	0.15	0.22	0.37	Partial mediation
Robo-advisors → Sustainable financial decision	0.08	0.29	0.37	Full mediation
Gamification → Sustainable financial decision	0.10	0.25	0.35	Full mediation

Source Primary data

Table 4 Sentiment analysis summary

Theme	Sentiment polarity	Frequency	Illustrative keywords
Trust and transparency	Positive	87	"secure," "transparent," "reliable"
Ease of use	Positive	75	"simple," "user-friendly," "intuitive"
Environmental impact	Positive	64	"green," "sustainable," "eco-conscious"
Data privacy concerns	Negative	49	"privacy," "tracking," "unauthorized"
Algorithmic bias	Negative	38	"biased," "unfair," "lack of clarity"
Lack of personalization	Negative	41	"generic," "impersonal," "not tailored"

Source Primary Data

prioritize personalization, transparency, and sustainability communication. Suggestions include having real-time carbon tracking functionality in neobanks, increasing algorithmic fairness in robo-advisors, and having goal-based rewards in gamification platforms. These user-focusing functionalities will most likely raise engagement and trust levels, hence supporting sustainable financial behavior (Table 4).

This methodological design allows for a comprehensive and analytically sound basis on which to tackle the research goals. Through the integration of cutting-edge quantitative modeling with qualitative text analysis, the research presents actionable findings on how behavioral design in FinTech can drive sustainable financial impacts.

6 Discussions

This research explores the behavioral pathways via which novel FinTech innovations like gamification, robo-advisors, and neobanks influence sustainable financial decision-making (SFD). Applying structural equation modeling (SEM), the analysis not just verifies the assumed paths between FinTech enablers and sustainable results but also identifies the mediating effect of behavioral nudges and triggers. These results provide useful insights into how digital financial services can influence sustainable consumer behavior, consistent with the wider goals of responsible finance and ESG integration.

FinTech Enablers and Sustainable Financial Behavior: SEM findings reveal gamification has a noteworthy impact on sustainable financial choices ($\beta = 0.34$, $p < 0.01$). This finding is consistent with previous behavioral finance literature that highlights the importance of game-centric mechanics—like setting targets, rewarding, and feedback cycles in promoting positive financial behaviors. The behavioral design incorporated into gamified instruments encourages ongoing participation, which can instill prudent spending and saving behavior that is consistent with environmental and social objectives.

Also, robo-advisors positively influence SFD ($\beta = 0.28$, $p < 0.05$), which indicates that algorithmic, automated investment advice—particularly when using ESG metrics—can lead to decision-making that leads to sustainability objectives. Automated nudges, green asset rebalancing in the portfolio, and low entry barriers are reasons for this positive influence on sustainable financial choices.

Neobanks, which incorporate sustainability-oriented features like carbon footprint monitoring, budgeting reminders, and impact statements, were observed to have a moderate but considerable influence on sustainable financial behavior ($\beta = 0.21$, $p < 0.05$). This is in line with the increasing view of digital-first banks as not only transactional platforms, but as value-oriented ecosystems that align with user ethics and environmental values.

Role of Behavioral Nudges and Triggers as Mediators: One of the key findings in this research is the discovery of behavioral nudges and triggers as influential mediators of FinTech enablers to SFD. These behavioral interventions, including reminders, suggestions, or prompts are central to facilitating FinTech features into lasting financial behaviors. For instance, nudges such as investment reminders in environmental assets or warnings on the carbon intensity of purchases strategically direct users toward choices that fit sustainability objectives. Mediation analysis validates that behavioral nudges and triggers strongly reinforce the mechanisms between digital interventions and environmentally friendly financial outcomes.

In addition to these quantitative results, a sentiment analysis of open-ended survey responses and online feedback was carried out. By utilizing natural language processing (NLP), repeated emotional themes were determined with positive sentiment being in dominance in user feedback, which is in particular with regards to the behavioral nudges and triggers within the platforms. Still, issues regarding privacy and standardization were observed as well, reflecting areas where FinTech developers need to ensure the ethical use of nudges and continue user participation.

Theoretically, the findings add to the developing literature at the nexus of behavioral finance, digital nudging, and sustainable consumption. By empirically confirming the mediating role of behavioral nudges and triggers, this research fills gaps among technology design, psychological theory, and financial sustainability.

In practice, the conclusions imply that FinTech companies ought to concentrate on creating nudges and triggers that stimulate sustainable financial conduct. This might involve individualized reminders, goal-setting capabilities, and alerts that recognize the environmental effects of financial decisions. Policymakers can utilize the findings in order to back regulatory systems that stimulate the creation of environmentally friendly and ethically sound FinTech products and services, eventually leading to a more sustainable financial system.

7 Conclusion

This study was able to achieve its three main objectives. One, it was able to affirm that behavioral FinTech tools—gamification, robo-advisors, and neobanks—positively influence sustainable financial choices. Gamification, particularly through mechanisms like Ant Forest, has been shown to encourage environmentally friendly consumption patterns by combining game mechanics with environmental goals [24]. Robo-advisors, in presenting ESG-focused investment schemes, are reported to be in line with proof of such sites facilitating sustainable investment [9]. Neobanks, through features such as automatic carbon footprint tracking, promote environmentally friendly financial behaviors, as demonstrated by recent FinTech advancements [10]. Second, the study concluded that behaviorally nudging and psychologically priming are primary drivers of user adoption on FinTech apps. Nudges in digital environments, such as framing and default options, have been found to induce effectively energy saving behavior [2]. Besides, personalized nudges delivered through large language models have been discovered to possess potential in enhancing conservation intentions, demonstrating the potential of personalized behavioral interventions on digital platforms.

Finally, the study provides recommendations to FinTech firms and policymakers. To FinTech firms, the integration of nudges of behavior into their platforms such as personalized reminders and goal-setting instruments can stimulate user engagement as well as promote sustainable financial behavior. Policymakers can utilize the findings to help inform policies that facilitate the development of green and morally sound FinTech solutions to create a financial system in support of sustainability on an expanded level.

At a societal level, the wider adoption of sustainable financial practices through the use of FinTech technologies can facilitate more responsible consumption and investment patterns and thereby further environmental and social well-being over the longer term. By infusing behavioral knowledge into the process of financial decision-making, FinTech platforms not only stand to benefit in terms of user uptake but also are well placed to drive the global sustainability agenda.

References

1. Acquisti A, Brandimarte L, Loewenstein G (2015) Privacy and human behavior in the age of information. Science 347(6221):509–514. https://doi.org/10.1126/science.aaa1465
2. Berger M, Greinacher E, Wolf L (2022) Digital nudging to promote energy conservation behavior–Framing and default rules in a smart home app. ECIS 2022 Research Papers. 92 https://aisel.aisnet.org/ecis2022_rp/92.
3. Beshears J, Choi JJ, Laibson D, Madrian BC (2018) Behavioral household finance. In: Bernheim BD, DellaVigna S, Laibson S (eds) Handbook of behavioral economics: applications and foundations 1, vol 1. Elsevier, pp 177–276. https://doi.org/10.1016/bs.hesbe.2018.07.004
4. Björk V, Shinkle GA (2022) Digital nudges and their ethical implications in FinTech: an integrative review. J Bus Ethics. https://doi.org/10.1007/s10551-022-05199-7

5. Bounfour A, Miyamoto K (2023) Fintech nudges for sustainability: a behavioral lens on green consumer finance. OECD Digit Econ Papers. https://doi.org/10.1787/12345678-en
6. Gopal S, Pitts J (2025) ESG integration: unveiling risk and driving innovation in sustainable finance. In: The FinTech revolution: bridging geospatial data science, AI, and sustainability). Springer Nature Switzerland, Cham, pp 35–81. https://doi.org/10.1007/978-3-031-74418-1_2
7. Gupta R, Goswami A (2024) Behavioral biases and gender in socially responsible investing: a Bayesian approach. J Behav Exp Financ. https://doi.org/10.1016/j.jbef.2024.xxxx
8. Hair JF, Black WC, Babin BJ, Anderson RE (2010) Multivariate data analysis, 7th (ed) Pearson, London
9. Investopedia (2019) Robo-advisors and sustainable investing. Investopedia. https://www.investopedia.com/robo-advisors-and-sustainable-investing
10. Investopedia (2023) Neobanks and sustainable financial behaviors. Investopedia. https://www.investopedia.com/neobanks-and-sustainable-financial-behaviors
11. Kashif M, Rehman MA, Zhang Y, Khan SAR (2023) FinTech, institutional quality, and sustainable finance: evidence from 89 countries. Technol Forecast Soc Chang 191:122487. https://doi.org/10.1016/j.techfore.2023.122487
12. Lee SU, Perera H, Liu Y, Xia B, Lu Q, Zhu L, Cairns J, Nottage M (2024) Integrating ESG and AI: a comprehensive responsible AI assessment framework. arXiv preprint. https://arxiv.org/abs/2408.00965
13. Lisauskiene N (2024) Robo-advisors help investors overcome biases and improve decision-making. The Financial Analyst. Retrieved from https://thefinancialanalyst.net/2024/11/05/robo-advisors-help-investors-overcome-biases-and-improve-decision-making/
14. Mustafa K, Rizwan M (2022) Crowdfunding, blockchain, and investor trust: a structural equation modeling approach in Pakistan. Sustainability 14(15):9387. https://doi.org/10.3390/su14159387
15. Nguyễn LT, Trần MH (2024) Fintech and green bond development in ASEAN+3: empirical insights and policy recommendations. Asia-Pacific Finan Markets. https://doi.org/10.1007/s10690-024-00423-7
16. Pour Mohammad Bagher L, Safarabadi NS (2024) Increasing the diversity of investment portfolios with integration of gamified components in the FinTech applications lifecycle. arXiv preprint. https://arxiv.org/abs/2402.17754
17. Sarwar S (2025) Engagement through gamification: the role of FinTech in enhancing user motivation for achieving financial goals. J Appl Financ Res 24(2), Article 002. https://doi.org/10.26652/jafr/24.02.002
18. Singh R, Singh A, Singh M (2024) Nudging sustainable financial behavior: cognitive biases, interventions, and policy implications. J Financ Serv Mark 29(1):55–68. https://doi.org/10.1057/s41264-023-00223-1
19. Thaler RH, Sunstein CR (2008) Nudge: improving decisions about health, wealth, and happiness. Yale University Press
20. Tanchangya T, Raihan A, Mia MR, Tafsirun U, Naher K, Islam N, Rashid F, Sarker SR (2025) Financ Risk Manage Rev 11(1):37–71. https://doi.org/10.18488/89.v11i1.4076
21. Trotta A, Rania F, Strano E (2024) Exploring the linkages between FinTech and ESG: a bibliometric perspective. Res Int Bus Financ 60:102200. https://doi.org/10.1016/j.ribaf.2023.102200
22. Wan Q, Cui J (2024) Dynamic evolutionary game analysis of how FinTech in banking mitigates risks in agricultural supply chain finance. arXiv preprint. https://arxiv.org/abs/2411.07604
23. World Bank (2023) FinTech and sustainability: leveraging behavioral tools for green finance. World Bank Group
24. Zhang Y, Zhang Y, Zhang Y (2023) The impact of gamification motivation on green consumption behavior—an empirical study based on ant forest. Sustainability 15(1):512. https://doi.org/10.3390/su15010512

Agronomics: A Sustainable Economics for Consumer Electronics in Agriculture

Satabdwi Sarkar, Anirbit Sengupta, Paramita Dey, Abhijit Das, Debashis De, and Nurzaman Ahmed

Abstract Agricultural economics is the study of the way goods produced via farming can be traded and used. Even while it might not be responsible for the majority of a nation's GDP these days, agriculture is always essential to its development since it provides the nation with its own food. Growing, selling, and trading the products produced on a farm is a business as farming is not just for personal consumption. The system that examines components of agriculture trade is, by definition, the agricultural economy. It began as a means of monitoring the yield that farmers were receiving from their crops in terms of the condition of the soil in which they were being cultivated. Nowadays, agricultural economics influences food production in addition to crop output. Due to the supply of many of the commodities used to create other goods and services for other economic sectors, agriculture contributes to the whole economy. An economy is the exchange and distribution of resources, and economics examines resource allocation. Thus, the economy related to agriculture is similar to the economy overall. The study of agricultural economics examines the exchange of goods like fiber and crops, as well as the effects of environmental

S. Sarkar (✉)
Department of Computer Science and Engineering, RCC Institute of Information Technology, Kolkata, India
e-mail: satabdwi.rcciit@gmail.com

S. Sarkar · P. Dey
Department of Information Technology, Government College of Engineering & Ceramic Technology, Kolkata, India
e-mail: dey.paramita77@gmail.com

A. Sengupta
Computer Science and Engineering—AI, Brainware University, Barasat, West Bengal, India

A. Das
Department of Information Technology, RCC Institute of Information Technology, Kolkata, India

D. De
Department of Computer Science and Engineering, Maulana Abul Kalam Azad University of Technology, Nadia, West Bengal, India

N. Ahmed
Donald Danforth Plant Science Center, St. Louis, MO, USA

S. Roy et al. (eds.), *Humanized Intelligence in Sustainable Finance and Economic Behaviours*, Studies in Computational Intelligence 1226,
https://doi.org/10.1007/978-981-95-0599-9_10

externalities crucial to preventing any negative effects on the economy, as agriculture relies heavily on them.

Keywords Agriculture 5.0 · Sustainability · Economics · Consumer electronics · Precision agriculture

1 Introduction

Since the termination of World War II, agronomy has transformed in a significant manner. This type of changes has created the farming process to grow more foods with help of fewer farmers at lower prices. Different types of IoT services are in the form of agriculture. The farmers are gating different types of services like cultivation of crops, pricing, fertilizers, diseases identification method and provides solutions based on the suggestions regarding modern techniques for cultivation for the involvement of cloud in agriculture and IoT, as our agriculture field is updating day by day. With the technological advancement and help of industrial equipment, the cultivation process has been changed drastically. Now here we have explored the part of Internet of Things (IoT) in agricultural segment. Cloud computing to dew computing can contribute to the advancement of scientific computational productivity through automation. We are being empowered to use the different features like the integration of sensors/actuators, digital transmission, minimal consumption of power, scalability using different IoT applications.

For the last four decades, there has been a search for the reason behind these excessive costs and provides inventive alternatives. These days, our food production systems are accepting and supporting the sustainable agriculture movement more and more. Three primary aims of sustainable agriculture are social equality, financial prosperity, and environmental health (Fig. 1). While many different ideologies, laws, and methods have assisted achieve these purposes, utmost definitions of sustainable agriculture share a few basic concepts and perceptions.

Li et al. in their paper use the spatial analysis method to estimate the agricultural green GDP of many developing countries and their regional units based on energy theory. It also determines the spatial dispersal features and reasons that influence the commercial growth of green agronomy in many developing countries [1].

1.1 Basics of Sustainability

Four key pillars are necessary for sustainability. The pillars for sustainability are Human, Social, Economic, and Environmental (HSEE). Environmental sustainability is the capability to meet current necessities without losing the accessibility of

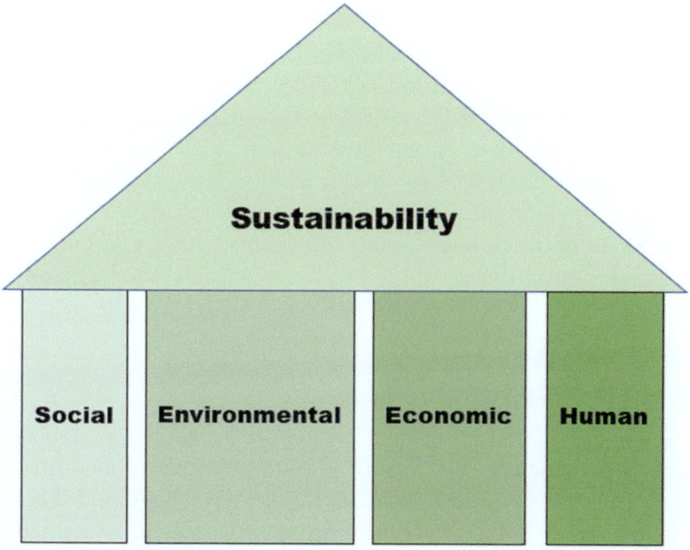

Fig. 1 Pillars of sustainability

resources for upcoming generations. This can be achieved via the execution of appropriate practices and laws that gradually conserve and defense the natural environment. As we discussed about plans, projects, and actions that are related to preserve a specific resource, we're talking about sustainability. As the four pillars of sustainability, it truly refers to four different domains: personal, social, economic, and environmental through most descriptions of agriculture that is sustainable [2, 3].

The different programs that include availability of facilities, nourishment, education and health system investments, knowledge and skill growth, and nutrition categorized in the all-purpose type of human sustainability. Depending on the versatile availability of resources and spaces, it is important to maintain a balance between reaching general economic welfare by enhancing health through helping nonstop progress. The organization that upholds human capital-respecting corporate values through different operations in the business world to be a part of society. Industries can have a beneficial or bad influence on groups all around the world by the ways they source raw materials. The growth of human potential and abilities to sustenance an organization's operations, confirm its advancement in social and public welfare is entirely encompassed in the concept of human sustainability.

1.2 Social Sustainability

Social sustainability devotes in and generates the amenities that type our society's framework to protect social capital. In terms of cultures, societies, and globalization, this knowledge permits us to an additional complete perception of the world. It indicates that our actions might affect other people and the entire planet through consideration and defending and the next generation. Social sustainability's aim is to reserve and boost the quality of society through the employment of ideas like cohesiveness, reciprocity, honesty, and the value of interpersonal relationships. It also promotes and supports the laws, knowledge, and a common understanding of rights and equivalence. The thought of maintainable growth as drawn in the Sustainable Development Goals of the United Nations is combined into social sustainability. The idea of sustainable growth concentrates on enhancing social and financial conditions.

1.3 Environmental Sustainability

The goal of environmental sustainability is to improve human well-being by preserving normal resources, such as minerals, air, water, and land. The population requirements are satisfied and are considered environmentally sustainable, without imperiling the requirements of future generations, initiatives, and programs. In terms of environmental sustainability, according to the four pillars of sustainability principle, issues must be resolved regarding each of the four pillars of sustainability. As there may be instances of overlapping, it is crucial to determine for the four traits as each of them has its own special qualities, which helps the green business to focus on. Companies must decide some course of action strategically for integrating selected methods successfully into their plans and procedures.

According to the four pillars of sustainability principle, problems necessarily are resolved regarding each of the four pillars of sustainability before being maintained. Although there may be instances where these overlap, it is vital to control which kind of green business to focus on because each of the four has its particular distinct traits. Companies must decide on a strategic course of action to successfully integrate the selected method into their policies and processes [4].

1.4 Economic Sustainability

Hawking et al. have discussed in their paper that the defenders of the model concede that a significant lacuna in contemporary accounting procedures is the exclusion of the cost of environmental harm from market prices [2]. Benn et al. described in their paper that the ecological and social mechanisms of this model are only moderately incorporated, as recognized by a more modern perspective on economics. Natural capital, or ecological systems, and social capital, or interpersonal ties, are both included in new economics, which questions the capitalist axiom that uninterrupted expansion is desirable and greater is better, particularly if it puts the ecological and human systems at risk [4].

1.5 Human Sustainability

The aim of human sustainability is to reserve and progress society's human capital. Programs contains access to amenities, nutrition, education and health system savings, knowledge and skill growth, and nourishment fall under the overall category of human sustainability. As so many natural resources and spaces are available, it is important to strike a balance between achieving universal economic well-being, to enhance health, and endorsing continuous growth. An organization that operates in the business world will see itself as a part of society and uphold human capital-respecting corporate values. According to Benn et al., human sustainability highlights the value of all parties who are either directly or indirectly involved in the production of goods, the delivery of services, or the involvement of larger stakeholders. Businesses and the ways they source raw resources can have a valuable or bad impact on communities all around the world. The development of human potential and abilities to support an organization's operations, ensure its sustainability, and advance societal and community well-being is all included in the concept of human sustainability [5].

2 Architecture of Sustainable Agriculture

The sustainable architecture of Fig. 2 comprises of four different blocks: Edge and cloud computing block, application block, user block, and processing block. A Sengupta et al. have designed FarmFox to collect the on-field real-time data [6]. The real-time data collected is saved in EPROM, and after every hour, it is sent to the cloud for storage. After the collection, the data analysis is processed using this data to check whether it is in line with normal data or there is any difference. The analysis which determines it is the trend analysis. If there is any mismatch during the trend analysis, a signal is triggered to indicate the disruption. Using the real-time sensed data sensed by the soil moisture and soil temperature, the farmers of agriculture enthusiast can determine the soil health condition helps in crop management. To manage water usage field can be managed using two real-time data, i.e., soil temperature and turbidity. The PH sensor sensed data plays an important role in determining the presence of heavy metals ion, organic compound, etc. The decision-making block is the most integral block. The most important block is the edge/ cloud block, where the sensed data is kept in the cloud for future utilization. Cleansing and processing of data are action which is done in edge/cloud block [7].

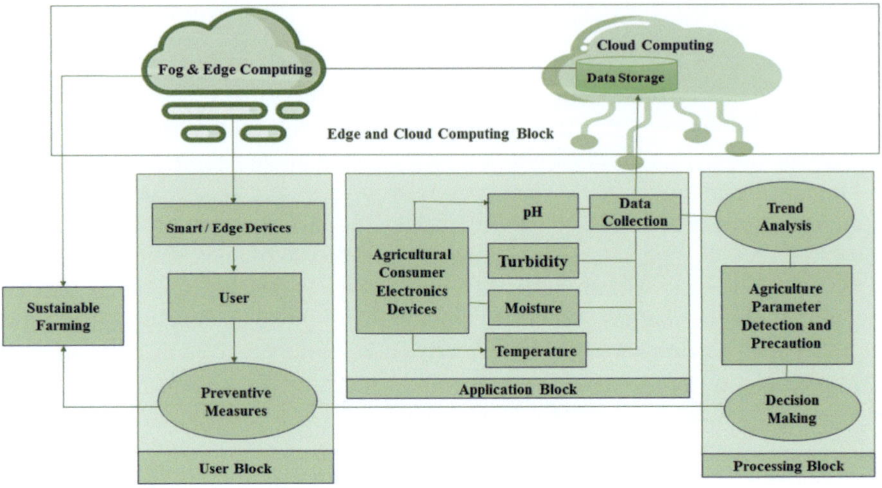

Fig. 2 Architecture for sustainable agriculture

2.1 Sustainable Economics in Agriculture

These progresses have significant costs and had many positive effects that have reduced many risks in farming. Prominent among these are topsoil depletion, groundwater contamination, air pollution, greenhouse gas emissions, the decline of family farms are due to the negligence of the living and working conditions of farm laborers. The new threats to human health and safety are due to the spread of new pathogens, economic concentration in food and agricultural industries, and disintegration of rural communities.

2.2 Types of Sustainable Agriculture

Economic Impact in Agriculture: The main objective of any developing nation is to become a high-income nation. Other crucial development goals are like ensuring food security and increasing nutrition, and where the agriculture is the key to change financial prudence to grasp the target. Therefore, agricultural transformation needs to happen if we are to eradicate malnutrition and hunger while also boosting economic growth [4].

The term "economic transformation," also known as "structural transformation," refers to a nation's change in the proportion of its sectors and technologies to its GDP as a whole: from cultivation to manufacturing and economic activity to high-income services, and from traditional to modern technology. Agriculture needs to be upgraded for this process to succeed.

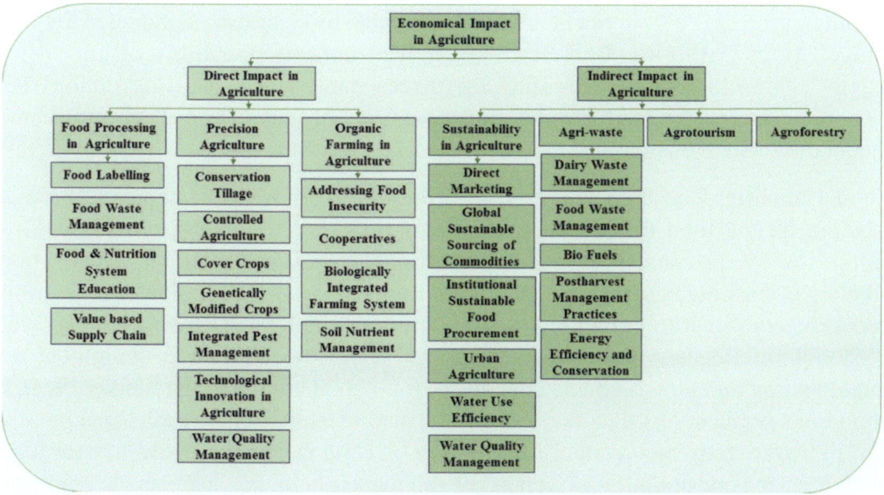

Fig. 3 Hierarchical diagram of economic impact in agriculture

This economic impact can be classified in two categories: Direct impact and indirect impact (Fig. 3).

Direct impact can be classified in three categories like food processing in agriculture, precision agriculture, and organic farming; these branches are also related with each other. Indirect impact can be classified as sustainability in agriculture, agri-waste, agro-tourism, and agro-forestry.

Now basically, food processing is the most important type of this agriculture industry.

Food Processing: Employment in the agricultural and food systems has always been linked to low pay and little access to healthcare. Agricultural laborers and farm laborers are well-known for putting in long hours under often hazardous circumstances. People who produce and serve food for a living are frequently at risk for chronic diseases linked to diet, malnutrition, hazardous living and working circumstances, and insufficient access to healthcare. The demand from consumers for sustainable products that address social and environmental issues as well as growing attention from policy and regulation has caused players in the food supply chain to move more and more toward models of production and operations that give workers fair compensation and good working conditions.

Food workers and jobs in the food business are defined as production-oriented farm laborers, workers in slaughterhouses and processing plants, workers in distribution or warehouses, workers in retail establishments, and workers in service industries.

Wages in a large portion of the food industry have remained low due to consumer expectations for affordable, abundant food. The majority of the lowest-paid occupations in the USA are in the food service industry, including dishwashers, baristas, cooks for fast food restaurants, and those who prepare and serve food [5, 8].

Studies on the employment and labor in the food system conducted by the University of California, Berkeley, have looked into the possibility of improving supply chain efficiencies, providing more reasonably priced housing options for those working in the food system, and the effects of technological advancements that could replace manual farm labor with mechanization.

Food Labeling: Consumers can relate to the information that is depicted through labeling on the food that is out for consumption. These labeling provides information on the product's ingredients, nutritional value, and allergen status. Food labeling is considered to be the one of the primitive marketing solutions which influences the consumer to purchase the food based on quality and other preferences of consumption. Because of this, there is an increasing amount of study being done on food labeling and how it affects customers, food companies and the market [9, 10]. Numerous publications that have been published recently, for example, specifically aim to better study how certain information on food packaging could impact food purchases and consumption as well as overall dietary behavior and provide credence to this [11–14].

Food Nutrition Management: A lot of people use the term "food system," but not many systematic frameworks fully capture the scale and organization of the nutrition and food system. To find current models of farming, food production, health, nutrition, and environmental systems, a variety of methods were used, including bibliographic searches, focus groups, expert interviews, and a modified Delphi procedure. To create an integrated conceptual model of the nutritional and food system, these models were analyzed, categorized, and combined. Most models in use concentrated on a single system segment or one discipline, and very few provided a general description of the system. The integrated model that was created here comprised nine stages (production, processing, distribution, acquisition, preparation, consumption, digestion, transport, and metabolism) and three subsystems (producer, consumer, and nutrition) [15, 16].

Food Waste Management: The most important aspect of the food supply chain is food waste that happens at multiple points of food supply chain. The steps that food undergoes from farm cultivation and harvest to reach the consumer make up the food supply chain. Agricultural production (harvest), managing after harvest and storage, processing (slaughter, manufacturing bread from wheat, etc.), distribution (markets and merchants), and consumption are some of these steps.

Precision Agriculture: Environmentally friendly crop production input management is possible with the use of precision agriculture (PA). The on-field data collected helps in P.A to determine the utilization of fertilizer and seeds keeping a close look at the chemical parameters of the soil. The determination of the use of fertilizer and other components is purely data based (information based on real-time data analysis). For long-term sustainability of farming, precision agriculture is one of the major techniques used by the farmers and agriculture enthusiast to understand. Implementation of P.A in agriculture helps the farmers to be more specific to the use of inputs that increases the crop yield. P.A also helps the farmers to increase the crop yield by pointing out the areas like insect's attack, imbalance nutrition of crops, soil health. The use of P.A is limited to certain areas which is also affecting the areas under orthodox framing. The sensitivity analysis's findings indicate that, for a variety of N application-level constraints, PA is a somewhat more profitable option than complete field supervision. These limitations may be imposed by laws or the landowner's view of environmental responsibility. In the example, even if nitrogen is limited to not more than half of the suggested uniform rate, farm profitability is maintained by the variable rate [17–20].

Conservation Tillage: The aim of conservation of agronomy is an agricultural management strategy and is to lessen the regularity or severity of plowing processes to support specific environmental and economic advantages.

These include a decline in greenhouse gas releases and carbon dioxide emissions, a reduction in the procedure of farm equipment and apparatus, and a general decrease in labor and fuel expenses. Additionally, it has been established that conservation tillage techniques increase soil health, lessen surplus, and control the amount of erosion. An agronomic system's sustainability can be improved by a well-designed and completely combined with conservation tillage method, which has several promising ecological and economic profits.

Plant residue removal, weed management, loosening compacted surface soil, and planting preparation have all traditionally been accomplished with tillage techniques. Even though tillage can be an essential part of a successful and lucrative agro-ecological business, reducing mechanical activities and soil disturbance in a field can have advantages like lessening soil erosion and the pollution it causes in the air and water, cutting down on production costs and fuel, and lessening subsurface soil compaction from tractor passes [21].

Controlled Environment Agriculture: Plant residue removal, weed management, loosening compacted surface soil, and planting preparation have all traditionally been accomplished with tillage techniques. While tillage can play a significant role in an agro-ecological venture's performance and profitability, a field's ability to minimize mechanical operations and disturbance to the soil can affect the growing plants and their products—including flowers and vegetables. Agriculture inside of buildings with controlled environments, like greenhouses, vertical farms, and growth chambers is known as controlled environment agriculture, or CEA.

We can efficiently and ecologically friendly grow crops with high value at maximum productivity by adopting CEA. Hydroponics and aeroponics are two technologically driven growing techniques used in the field of CEA.

Hydroponic Farming: Plant residue removal, weed management, loosening compacted surface soil, and planting preparation have all traditionally been accomplished with tillage techniques. While tillage can play a significant role in an agro-ecological venture's success and profitability, a field's-controlled environment can be enhanced by avoiding mechanical processes and soil disturbance. Plants produced hydroponically are cultivated in water instead of dirt. To guarantee healthy plants that yield their utmost potential, nutrients are supplied to the water. We must change the way we farm as the world's population concentrates in metropolitan areas, fertile land and water increasingly scarcer and food insecurity rises.

The existing food system needs to be more economical, environmentally friendly, and sustainable for both humans and the earth. Hydroponics and vertical farming have the potential to enhance the system and address current issues, covering the way for a more sustainable and healthful upcoming.

Aeroponic: Plants cultivated aero-phonically have their roots open to the atmosphere. After that, water and vitamin solutions are frequently misted over the roots, allowing them to absorb them. With the roots exposed to the greenhouse air, aeroponics is a technique for giving plants the necessary nutrients and water. Using a sprayer to create tiny water droplets, fogging or misting technology delivers water to the exposed roots. Because the water droplets remain held in the air for a longer amount of time, the plant canopy becomes drier and the rate of evaporation increases [6].

Cover Crops: Any non-cash crop farmed as a supplement to the main cash crop falls under this broad definition. These crops may enhance soil fertility and organic substance, lessen erosion, boost soil structure, reassure water infiltration, and regulate insect and disease eruptions. Cover cropping has several returns, but alike any management strategy, there are costs and drawbacks that necessarily to be considered. In many cases, cover crops can result in increased agricultural productivity and a reduced dependency on fossil fuels.

Genetically Modified Crops: Six nations grew genetically modified (GM) crops in 1996. There were twenty-five countries growing genetically modified crops in 2009. In 2019, biotech crops will have been commercialized for the 24th year, and there are now 29 nations that use GM crops. Genetically modified corn that is resistant to insect larvae is one example [22]. DNA insertion into an organism's genome is a technique known as genetic modification. New DNA is implanted into a plant's cells to generate a genetically improved plant. Following that, these cells are cultured in tissue culture, where they develop into plants. These plants will produce seeds with unique DNA. The gene guns approach is the most frequently used technique of implantation. Agrobacterium, microinjection, and electroporation are the other genetic engineering methods. These remain the three prime categories of genomic changes.

1. Transgenic—Genes from different animals have been introduced into plants.
2. Cisgenic—Genes from identical species or near connected classes are used to make plants.
3. Subgeneric—Change a plant's genetic arrangement without accumulating genes from different plants.

Integrated Pest Management (IPM): The Integrated Pest Management (IPM) process can be used to manage pests in urban, agronomic, or usual areas. The focus of the approach is to implement a combination of supervision procedures that are designated to minimize the amount of ecological deprivation and lessen the impact

Fig. 4 Integrated pest management

of chemical inputs on humans and non-target creatures. IPM is a serious element of the sustainability of several agro-ecological systems.

Based on scientific studies on pests and pest management, the IPM approach was developed. Any organism that harms residences or other structures, gets in the way of valuable plants in a farming environment, or has an adverse effect on the health or welfare of people or animals is regarded as a pest. Insects, animals, plants, and disease-causing bacteria, viruses, or fungi can all be considered pests. IPM prioritizes long-term defense. By investigating the environmental elements that influence a pest, the treatment approach can be adjusted to provide negative circumstances and lower the likelihood of recurrence (Fig. 4).

Integrated Pest Management (IPM) "programs" are designed to combine cultural, biological, mechanical, and chemical control strategies with the goal of making agricultural systems more sustainable. An IPM program for a particular crop determines the pest damage levels that may be realistically stood before chemical management is needed to keep pest populations regulated and maintain a viable crop; this does not mean that IPM automatically eliminates the need of pesticides as a means of control. Farmers may be able to reduce the effects of pesticides on non-target species and the environment while lowering the frequency and related costs of treatments by applying pesticides at the right time and concentration [23].

3 Technological Innovation in Agriculture

(1) Bee vectoring technologies, (2) Precision agriculture, (3) Indoor vertical farming, (4) Livestock farming technology, (5) Laser scarecrows.

The value of honey bees to the US crop producing industry is $20 billion. Because bees are vital to human survival, agricultural equipment is becoming more innovative in order to safeguard bees and optimize their pollination potential [24–26] (Fig. 5).

Different data from precision agriculture is used in management decisions to enhance farmland and farming production in several important parts, such as:

- Resource use efficiency.
- Sustainability.
- Profitability.
- Productivity.
- Quality.

Farm food is grown indoor vertical farming in a regulated and enclosed space, where it is piled one on top of the other. The method increases crop output in confined locations by using growing shelves that are installed vertically. Plants are grown hydroponically, which is a gardening technique that uses water and nutrient solutions.

In aeroponics, the crop roots are suspended in the air and are intermittently sprayed with nutrients and water by emitters.

Agronomists can achieve healthier and higher yields by controlling elements like temperatures, light, water, and infrequently, carbon dioxide levels, in indoor vertical farms.

Livestock Farming Technology: New developments in livestock expertise give farmers data-driven visions that simplify farm operations, enhance animal care, and increase output.

Several advances are redefining cattle husbandry, a few of which are as follows:

Dairy farms with automated systems milk cows without the need for human involvement. The milk quality is also monitored by farmers thanks to the milk sensors.

Waste is removed by automatic cleaning systems, resulting in cleaner, healthier settings.

Acoustic pulse technology (APT) is the non-antibiotic treatment offered by Armenta for bovine mastitis, a condition that affects cows and causes losses of more

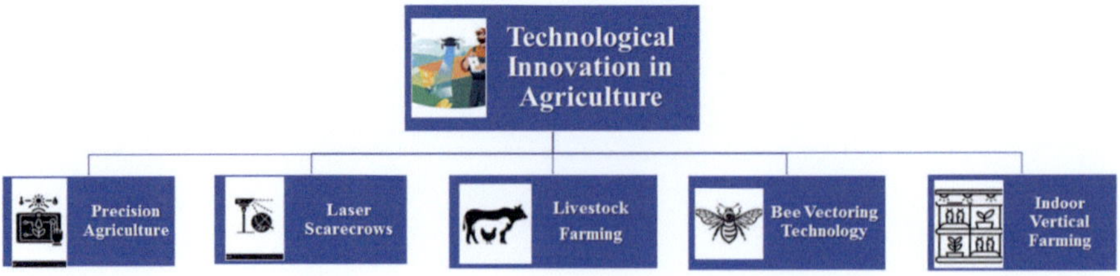

Fig. 5 Hierarchical diagram of technological innovation in agriculture

than $6 billion a year in the USA and Europe. Animals receive meal mixes that are precisely matched to their requirements in the proper quantity thanks to automated feeder systems.

Faromatics uses big data, artificial intelligence, and robots to improve farm efficiency and animal care [27].

Although they are amazing animals, birds also pose challenges to farmers worldwide. Indian farmers create "SCARECROWS," man-sized effigies meant to deter birds from eating farmland-destroying seeds. Since the beginning of agriculture, there has been a constant battle between farmers and birds. Within 48 h after harvest, pests like crows, blackbirds, and starlings can destroy up to 75% of crops, causing a large financial loss. Because they eat recently dispersed seeds, uproot budding corn, pull ripe grapes from the vine, and nibble at grains and vegetables before harvest, crows have gained the reputation of being pest birds [28].

A big issue facing most of the humanity today is hunger. Reduced agricultural output is one of the main causes. The farming population in many third-world countries has endured ongoing hardship. The lower classes of farmers suffer because they are unaware of the techniques that modern technology has made widely accessible to everyone and because they are unable to use them themselves. The limited area of the land that they own privately makes it more challenging to use the equipment that could have otherwise replaced their present degree of expertise. Paradoxically, if better machinery were to replace farmers, this may result in up to 10% unemployment—that is, 14.4 million individuals without other relevant skill sets [29] (Fig. 6).

Water Quality Management: At least four billion people worldwide lack access to safe drinking water or have no faith in the water to consume.

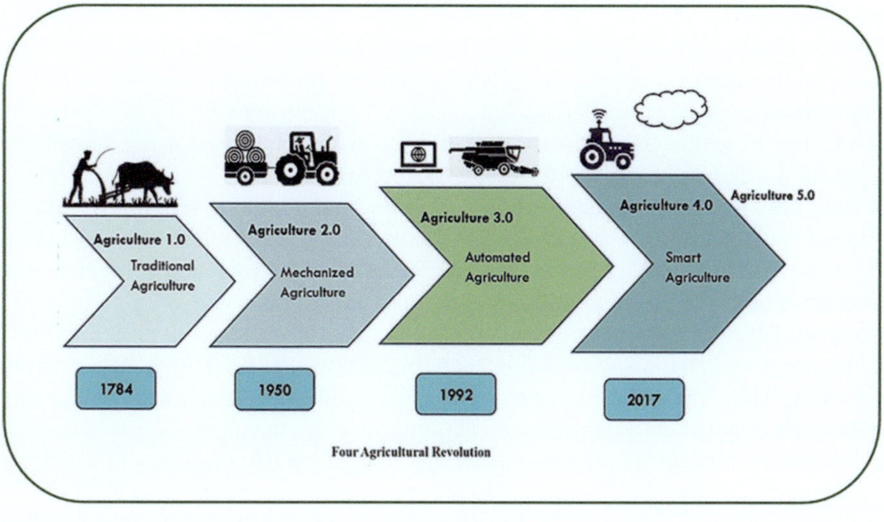

Fig. 6 Four agricultural revolution

It is difficult to locate a single city, town, or village in the more than 1.7-billion-person South Asian region where most residents believe the tap water is safe to drink and has no health risks. It should be evoked that water that is deemed "nearly drinkable" is not.

It also consists of wealthy nations like the USA and France, where sociocultural norms and quality alarms are triggering an increasing number of people to avoid drinking water conventionally from the tap.

Water quality has rarely been a primary concern in discussions around the right to access water, despite the recent global attention on water as a human right. This is further demonstrated by the fact that effective treatment and waste water collection are still not receiving enough attention. Therefore, less than 10% of the world's population has access to wastewater collection and proper treatment [7].

Smart water quality monitoring (SWQM) technology helps with the nonstop measurement of water quality using the electric conductivity, temperature, pH, and turbidity qualities as its four physical parameters. To detect the water parameters, an Arduino Uno is discretely coupled to four sensors. The sensor's take out data is sent to a desktop application built on the NET platform, where it is related to WHO (World Health Organization) standard values. The suggested SWQM system may effectively use a rapid forest binary classifier to assess the water parameters and determine whether the test water sample is drinkable based on the measured results [30].

Soil Nutrient Management: Improving soil health and meeting crop nutrient requirements are the goals of soil nutrient management. The "continued capacity of soil to function as a vital living ecosystem that sustains plants, animals, and humans" is the NRC's definition of healthy soil. According to this concept, achieving crop productivity and nutrient requirements is necessary for both short- and long-term production objectives.

For growers, nitrogen, potassium, and phosphorus are frequently the nutrients that are most important to consider. Any of these nutrients applied excessively or in excess might cause environmental damage by releasing nutrients into the surrounding area. In addition to identifying nutritional inadequacies, soil testing can tell a producer about the acidity and capacity of the soil to store water.

For food producers, processors, distributors, and retailers, striking a balance between sustainability aims and food preservation and safety regulations is becoming a more crucial goal. After harvest, a product is handled at every step of the handling process: cleaning, packing, shipping, storage, preparation, and retail. There are numerous ways to lessen losses and enhance food safety, including pasteurization, cold storage, chilled transportation, and handling procedures that stop pathogens from spreading. The same objectives of reducing losses and contamination are pursued by sustainable postharvest management techniques in ways that maximize productivity and minimize the consumption of energy and non-renewable resources [18, 21].

3.1 Sustainable Supply Chain Management in Agriculture

Supply chain management in agribusiness: Supply chain management (SCM) in agribusiness refers to overseeing the connections among the companies in charge of effectively producing and delivering goods from the farm level to the end user, guaranteeing that the latter's needs are satisfied in terms of quantity, quality, and cost.

Values-Based Supply Chains: Data distribution in association is gradually dominant. This study procedures the shareholder value created by E-Collaboration along a supply chain. A values-based supply chain (VBSC) is a manufacture, marketing, and supply channel that upholds the communal, ecological values that are combined into production while also protecting the individuality of the farmers and planters who cultivate or raise the product. The buyer can be informed via the VBSC that their values are upheld across the whole supply chain. Since they effectively connect crops to markets and provide customers confidence that important practices or principles have been kept, VBSCs are vital to sustainable agronomy. This could contain fair compensation for farmers and other food system employees, ecological sustainability, the opportunity to establish communities, the availability of healthful food, or a mix of values motivated by customer demand [31, 32].

There is a growing consciousness regarding the manufacture, supply, and consumption of food throughout the nation. Customers are preferring food producers who take into consideration the external, upstream, or downstream effects of their products more and more, and they are demanding openness in the food chain. In addition to fostering local economic growth, this consumer interest supports small and mid-sized growers in the development and upkeep of their enterprises' sustainable environmental and social components (Fig. 7).

Regulations pertaining to the sector's environmental effects have become increasingly stringent as a result of the institutions in charge of its governance implementing the sustainability paradigm. But the adoption of novel farming methods frequently necessitates institutional assistance in the form of subsidies and other forms of public support, raising doubts about their actual economic viability both now and down the road [33].

- It is necessary to conduct studies that measure the economic profitability of the industry and draw comparisons with other economic sectors.
- The original study documents in the field of sustainable agriculture, mainly from an financial perspective, are the focus of this. We are looking for unique work for this that addresses the current sustainability concerns, including economically analyzing more sustainable agricultural systems and processes, as well as more ecologically friendly technological solutions for livestock and agriculture.
- Research that suggests an economic evaluation of agricultural sustainability-related strategies as well as studies that assess and contrast sustainability from an economic standpoint is both encouraged.
- Specific topics include but are not restricted to the following:

Fig. 7 Cycle of food system

  - Economic valuation of new agricultural and livestock management practices.
  - Policy tools to inspire the acceptance of sustainable practices.
  - Estimation of agronomic systems/techniques based on-field researches.
  - Applications for measuring and comparing sustainability (in time and space).

- The goal of sustainable agriculture is to light society's current food and textile demands while preserving the capacity of future generations of farmers to meet their own needs. It might be predicated on an awareness of ecological services [4]. There are numerous ways to make agriculture more sustainable. Developing adaptable business procedures and farming techniques is crucial for the development of agricultural production within sustainable food systems. Due to its massive environmental impact, agriculture is a major contributor to climate change (Fig. 8).

Before agricultural commodities are produced, they must go through a number of processes like collection, shredding, cutting, bagging, transportation, storage, processing, and exchange. Numerous studies conducted across the nation have shown that there are significant losses in crop output at every stage of these processes [33–36].

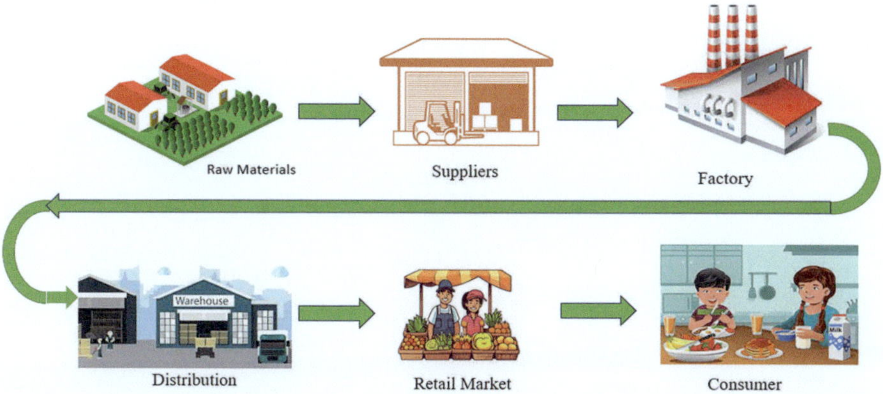

Fig. 8 Food supply chain process

Agri-supply networks are commercial structures in which members share risks and profits. As a result, supply chains create incentives for the entire chain and enforce internal processes to guarantee that manufacturing and distribution commitments are met on time. Due to promises about transaction volume, reciprocal scheduling, shared information, and product quality guarantees, they are intertwined and interrelated. Process links enhance the value of agricultural goods and necessitate that each participant coordinate their efforts as part of an ongoing process of improvement [36–40].

The elements of a well-organized agri-supply chain are as follows:

(1) Sourcing or procurement, (2) Transportation, material management, logistics, and the assumption that most supply comes from production rather than stock six: (a) Storage, (b) Modeling of logistics networks. (3) Management of organizations Contracting: (a) Partnerships and strategic alliances; (b) Vertical integration; (i) Long-term storage, (ii) Cold chain management, (iii) Packaging technology, (iv) Energy-efficient transportation, (v) Safety and quality. (4) Using an Efficient Consumer Response (ECR) System: (a) At the point of sale, scan the product and price electronically, (b) Simplify the whole supply chain (Figs. 9, 10 and 11).

Labor market flexibility and domestic value-added trade: Labor market flexibility and domestic value-added trade: Non-SOE firms, foreign-invested firms, processing firms, coastal firms, and businesses in the ambitious sectors of GVCs are more likely to experience this effect. This complete spectrum of action businesses takes to develop a product from inception to ultimate use and elsewhere is referred to as the value chain. Design, manufacturing, advertising, distribution, and customer service are all included in this. A value chain's activities spread among several businesses or kept within the purview of one company. Value chain operations can generate products or services, and they might take place across larger geographic areas or in a single place.

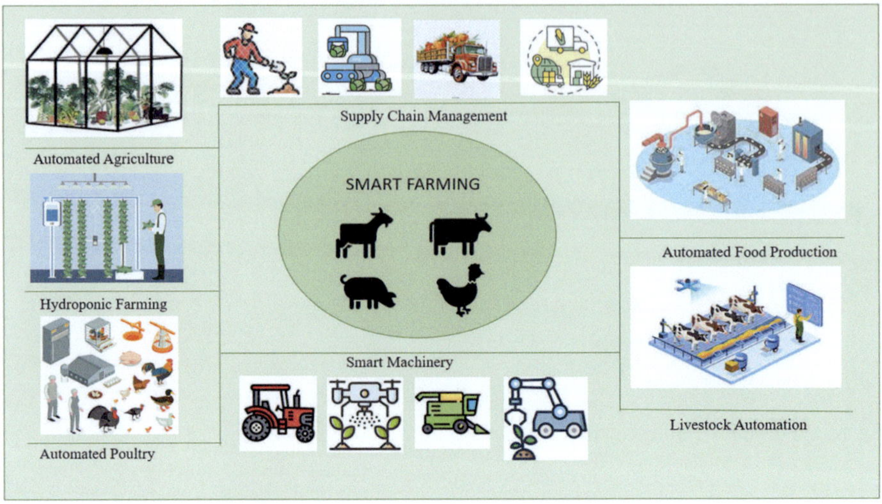

Fig. 9 Different areas of smart farming

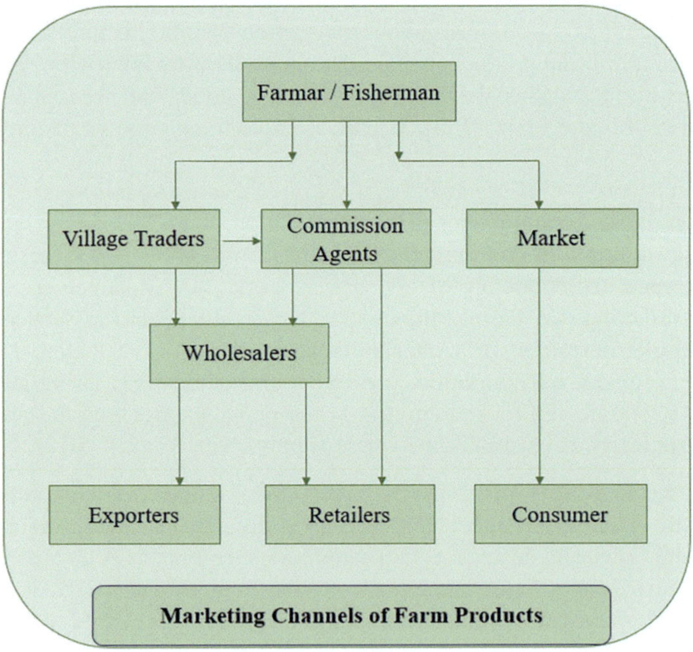

Fig. 10 Different marketing aspects of farm products

Production	Supply Chain	Processing	Marketing
• Poor extension • Quality inputs • Low productivity • Deficient and inefficient production management • Non demand linked production • Improper post harvest management resulting in poor quality	• Lack of storage • Poor transportation • High wastages • Multiple intermediaries • Fresh produce transported to mandis in open baskets or gunny bags stacked one on top of the other • Cold chain absent or broken, produce deteriorates rapidly • Food safety is major concern: Hygiene and pesticide MRL not monitored	• Low processing • Lack of quality • Poor returns • Low capacity utilization	• Poor infrastructure • Lack of grading • No linkages • Non-transparency in prices • Long delays from producer to retailer
Each segment working in an isolated manner resulting in multiple losses across the value chain			

Fig. 11 Broken links in agri supply chain in India

Urban agriculture: Urban agriculture is defined by agriculture and natural resources as the production, marketing, and distribution of food and other goods both inside and outside of metropolitan regions. This can include backyard and rooftop gardening, community and school gardens, and creative production techniques that optimize yield in a limited physical space.

The first is to growing food inside city bounds, and the second is peri-urban agriculture, which is produced by regional farmers who sell their products directly to cities. A range of social, economic, and environmental advantages can be developed for the areas where these two types of UA initiatives are located. These initiatives affect a wide range of stakeholders in addition to specific community members, such as local, regional, or state governments and nonprofit or private organizations.

Water management technology: A key technique for supplying drylands—which typically receive insufficient rainfall to support agriculture—with water is irrigation. Even though this is a vital component of modern farming, a lot of farmers still irrigate their crops with excessive amounts of water.

More environmentally friendly methods of giving plants enough water are available to farmers because to agricultural innovation and technology. For example, the N-Drip micro drip irrigation system creates the ideal conditions for crop growth by allowing water to slowly trickle to the roots of plants. With this method, crop quality is improved and water usage can be reduced by up to 50%.

Since agriculture uses 70% of the world's water resources, we need to adapt how we utilize water in agriculture to help address the global water shortage. The world is rapidly running out of water. 85% of all irrigated land is still used for flood irrigation, which has been doing so for the last 5000 years using the same approach. Nevertheless, flooding results in 70% water loss, reduced yield, mineral and soil waste, depletion of the land, and contaminated water. With forty years of experience in water and irrigation, Professor Uri Shani, a former Water Commissioner for Israel and a leading expert worldwide, has created an innovative Gravity Micro Irrigation System.

The system makes efficient irrigation possible at no further cost by using gravitational force and already-existing infrastructure. This discovery has the potential to drastically alter the world's water scarcity [7, 41].

3.2 Agri-waste

Biofuels: Investigating the development, properties, and uses of several types of biofuels, including algal biofuels, biodiesel, bio-alcohols (bioethanol, biobutanol), biogas (biomethane, biohydrogen), and other newly discovered biofuel sources. Moreover, BRJ promotes bioproduct research, encompassing bio-composites, bio-based innovative substances, and inventive uses in the food, pharmaceutical, and other industries [42].

Agro-tourism: One option to increase the earnings and potential financial sustainability of small farms and rural areas is agritourism, often known as agricultural tourism. Any agriculturally oriented business or activity that attracts tourists to a farm or ranch is referred to as agritourism or agro-tourism. The term "agritourism" denotes to a type of business venture that associates the cultivation and/or processing of crops with tourism to draw guests to a farm, ranch, or other farming enterprise with the intention of entertaining or educating them while making money for the owner of the property.

3.3 Future Directions for Sustainable Agronomic Technologies and Economic Integration

Looking ahead, the integration of AI, edge computing, and sustainable agronomic practices holds tremendous promise for revolutionizing agriculture economics and environmental stewardship. Emerging technologies such as precision agriculture, hydroponics, and Integrated Pest Management (IPM) are reshaping farming landscapes by optimizing input usage, improving crop yields, and reducing environmental externalities. The proliferation of smart sensors, IoT, and data-driven platforms like FarmFox can facilitate real-time monitoring and decision-making at the farm level. Further research is needed to evaluate the long-term economic viability and environmental trade-offs of these technologies, especially in developing regions. Additionally, institutional support and policy frameworks must evolve to ensure equitable access to these innovations for smallholder farmers. Studies assessing value-based supply chains and their impact on transparency, fair trade, and sustainability metrics will be essential. The convergence of technology, economics, and ecology in future agri-systems will pave the way for resilient, inclusive, and environmentally aligned food systems.

4 Conclusion

In this chapter, we present a detailed study on agricultural economics. We have discussed about the system that studies mechanisms of agriculture trade that has impact on agricultural economy. We have stated how agricultural economics impacts food production along with crop yield. Becoming a high-income nation is every developing nation's one of the main objectives. Due to the supply of many of the commodities used to create other goods and services for other economic sectors, agriculture in this aspect contributes the lion's share to the whole economy. Agriculture yield and water usage can be sustained with the application of federated learning where there is a paradigm shift of IoT and dew-edge-cloud [43]. The yield of the crop increases with the involvement of IoT and edge computing where we can determine the size of the crop yield and growth of the stem and branches of the crop or tree [44, 45]. With the involvement of Agriculture 5.0, intelligent crop recommendation system can be developed for the betterment of application of machine learning in agriculture [46]. The future of the agriculture with the involvement of IoT, edge computing, and federated learning shall reach a new height to develop country's economy.

References

1. Li Z, Jin M, Cheng J (2021) Economic growth of green agriculture and its influencing factors in china: based on energy theory and spatial econometric model. Environ Dev Sustain 23:15494–15512. https://doi.org/10.1007/s10668-021-01307-1
2. Lichtfouse E, Navarrete M, Debaeke P, Souchère V, Alberola C, Ménassieu J (2009) Agronomy for sustainable agriculture: a review. In: Lichtfouse E, Navarrete M, Debaeke P, Véronique S, Alberola C (eds) Sustainable agriculture. Springer, Dordrecht. https://doi.org/10.1007/978-90-481-2666-8_1
3. Gutiérrez J, Villa-Medina JF, Nieto-Garibay A, Porta-Gándara MÁ (2013) Automated irrigation system using a wireless sensor network and GPRS module. IEEE Trans Instrum Meas 63(1):166–176
4. https://www.ifpri.org/blog/agriculture-key-economic-transformation-food-security-and-nutrition
5. https://sarep.ucdavis.edu/sustainable-ag/employment
6. https://www.sciencedirect.com/topics/agricultural-and-biological-sciences/aeroponics
7. Pouget L et al (2023) PRIMA MAGO project: open-source applications based on Copernicus data for agricultural water management. In: 2023 IEEE international workshop on metrology for agriculture and forestry (MetroAgriFor), Pisa, Italy, pp 531–536. https://doi.org/10.1109/MetroAgriFor58484.2023.10424068
8. Khouryieh HA (2021) Novel and emerging technologies used by the US food processing industry. Innov Food Sci Emerg Technol 67:102559
9. Shangguan S, Afshin A, Shulkin M, Ma W, Marsden D, Smith J, Saheb-Kashaf M, Shi P, Micha R, Imamura F et al (2019) A meta-analysis of food labeling effects on consumer diet behaviors and industry practices. Am J Prev Med 56:300–314. https://doi.org/10.1016/j.amepre.2018.09.024

10. Cecchini M, Warin L (2016) Impact of food labelling systems on food choices and eating behaviours: a systematic review and meta-analysis of randomized studies. Obes Rev 17:201–210. https://doi.org/10.1111/obr.12364
11. Mora-García C, Tobar L, Young J (2019) The effect of randomly providing nutri-score information on actual purchases in Colombia. Nutrients 11:491. https://doi.org/10.3390/nu11030491
12. de Morais Sato P, Mais LA, Khandpur N, Ulian MD, Bortoletto Martins AP, Garcia MT, Spinillo CG, Urquizar Rojas CF, Jaime PC, Scagliusi FB (2019) Consumers' opinions on warning labels on food packages: a qualitative study in Brazil. PLoS ONE 14:e0218813. https://doi.org/10.1371/journal.pone.0218813
13. Miller L, Cassady D, Applegate E, Beckett L, Wilson M, Gibson T, Ellwood K (2015) Relationships among food label use, motivation, and dietary quality. Nutrients 7:1068–1080. https://doi.org/10.3390/nu7021068
14. Sogari G, Li J, Lefebvre M, Menozzi D, Pellegrini N, Cirelli M, Gómez MI, Mora C (2019) The influence of health messages in nudging consumption of whole grain pasta. Nutrients 11:2993. https://doi.org/10.3390/nu11122993
15. Oostenbach LH, Slits E, Robinson E, Sacks G (2019) Systematic review of the impact of nutrition claims related to fat, sugar and energy content on food choices and energy intake. BMC Public Health 19:1296. https://doi.org/10.1186/s12889-019-7622-3
16. Benson T, Lavelle F, McCloat A, Mooney T, Bucher T, Egan B, Dean M (2019) Are the claims to blame? A qualitative study to understand the effects of nutrition and health claims on perceptions and consumption of food. Nutrients 11:2058. https://doi.org/10.3390/nu11092058
17. Bongiovanni R, Lowenberg-Deboer J (2004) Precision agriculture and sustainability. Precision Agric 5:359–387. https://doi.org/10.1023/B:PRAG.0000040806.39604.aa
18. Sengupta A, Debnath B, Das A, De D (2021) FarmFox: a quad-sensor-based IoT box for precision agriculture. IEEE Consum Electron Mag 10(4):63–68. https://doi.org/10.1109/MCE.2021.3064818
19. Lin Y-B, Chen W-E, Chang TC-Y (2023) Moving from cloud to fog/edge: the smart agriculture experience. IEEE Commun Mag
20. Fizza K, Jayaraman PP, Banerjee A, Georgakopoulos D, Ranjan R (2021) Evaluating sensor data quality in internet of things smart agriculture applications. IEEE Micro 42(1):51–60
21. Xiao B, Wang Q, Yu D (2011) Effects of conservation tillage on soil, water loss and crop yield on sloping land in Beijing, China. In: 2011 international conference on new technology of agricultural, Zibo, China, pp 382–385. https://doi.org/10.1109/ICAE.2011.5943824
22. https://byjus.com/free-ias-prep/genetically-modified-crops-upsc-notes/
23. Yuan Z, Li S, Peng R, Leybourne D, Yang P, Li Y (2023) PestDSS: an integrated decision support system for sustainable pest management in agriculture. In: 2023 IEEE 32nd international symposium on industrial electronics (ISIE), Helsinki, Finland, pp. 1–6. https://doi.org/10.1109/ISIE51358.2023.10228070
24. Misra NN, Dixit Y, Al-Mallahi A, Bhullar MS, Upadhyay R, Martynenko A (2020) IoT, big data, and artificial intelligence in agriculture and food industry. IEEE Internet Things J 9(9):6305–6324
25. Li X, Hou B, Zhang R, Liu Y (2023) A review of RGB image-based internet of things in smart agriculture. IEEE Sens J
26. Singh SP, Dhiman G, Juneja S, Viriyasitavat W, Singal G, Kumar N, Johri P (2023) A new QoS optimization in IoT-smart agriculture using rapid adaption based nature-inspired approach. IEEE Internet Things J
27. Farooq MS, Sohail OO, Abid A, Rasheed S (2022) A survey on the role of IoT in agriculture for the implementation of smart livestock environment. IEEE Access 10:9483–9505
28. Saxena S, Suman S, Naveen, Yadav N (2021) Laser scarecrow technology for prevention of bird damage. Just in Agriculture Magazine, September
29. Biswas A, Prakash S (2015) Farming technology for India agriculture based sensorics and indicative systems. In: 2015 IEEE technological innovation in ICT for agriculture and rural development (TIAR), Chennai, India, pp 72–78. https://doi.org/10.1109/TIAR.2015.7358534

30. Mukta M, Islam S, Barman SD, Reza AW, Hossain Khan MS (2019) IoT based smart water quality monitoring system. In: 2019 IEEE 4th international conference on computer and communication systems (ICCCS), Singapore, pp 669–673. https://doi.org/10.1109/CCOMS.2019.8821742
31. Zhang X, Lam JSL (2021) Measuring the impact of E-collaboration on supply chain parties: a value-based management approach. IEEE Access 9:118181–118193. https://doi.org/10.1109/ACCESS.2021.3103738
32. Sai DGSM, Venkatraman K, Chellammal P, Natarajan B, Sridevi R (2023) A novel reinforcement learning based optimization approach for supply chain management. In: 2023 international conference on computing, communication, and intelligent systems (ICCCIS), Greater Noida, India, pp 655–661. https://doi.org/10.1109/ICCCIS60361.2023.10425106
33. Zhang P, Liu S (2021) Application of big data technology in water conservancy project construction. In: 2021 7th international conference on hydraulic and civil engineering & smart water conservancy and intelligent disaster reduction forum (ICHCE & SWIDR), Nanjing, China, pp 1642–1645. https://doi.org/10.1109/ICHCESWIDR54323.2021.9656372
34. Somashekhar IC, Agriculture supply chain management: a scenario in India
35. Nikookar E, Stevenson M, Varsei M (2024) Building an antifragile supply chain: a capability blueprint for resilience and post-disruption growth. https://doi.org/10.1111/jscm.1231
36. Mowla MN, Mowla N, Shahen Shah AFM, Rabie K, Shongwe T (2023) Internet of things and wireless sensor networks for smart agriculture applications—a survey. IEEE Access
37. Kumar P, Gupta GP, Tripathi R (2021) PEFL: deep privacy-encoding-based federated learning framework for smart agriculture. IEEE Micro 42(1):33
38. Akbari M, Syed A, Sean Kennedy W, Erol-Kantarci M (2023) Constrained federated learning for AoI-limited SFC in UAV-aided MEC for smart agriculture. IEEE Trans Mach Learn Commun Netw
39. Vangala A, Das AK, Kumar N, Alazab M (2020) Smart secure sensing for IoT-based agriculture: blockchain perspective. IEEE Sens J 21(16):17591–17607
40. Ram SK, Das BB, Mahapatra K, Mohanty SP, Choppali U (2021) Energy perspectives in IoT driven smart villages and smart cities. IEEE Consum Electron Mag 10(3):19–28. https://doi.org/10.1109/MCE.2020.3023293
41. Kadar HH, Sameon SS, Rafee PA (2019) Sustainable water resource management using IOT solution for agriculture. In: 2019 9th IEEE international conference on control system, computing and engineering (ICCSCE), Penang, Malaysia, pp 121–125. https://doi.org/10.1109/ICCSCE47578.2019.9068592
42. Chakrabarty R et al (2023) Weed segmentation: a novel dataset, benchmark and YOLO baseline. In: 2023 7th international conference on electronics, materials engineering & nano-technology (IEMENTech), Kolkata, India, pp 1–6. https://doi.org/10.1109/IEMENTech60402.2023.10423448
43. Bera S, Dey T, Mukherjee A, De D (2024) FLAG: federated learning for sustainable irrigation in Agriculture 5.0. IEEE Trans Consum Electron 70(1):2303–2310. https://doi.org/10.1109/TCE.2024.3370373
44. Sengupta A, Mukherjee A, Das A, De D (2021) GrowFruit: an IoT-based radial growth rate monitoring device for fruit. IEEE Consum Electron Mag 11(3):38–43. https://doi.org/10.1109/MCE.2021.3119276
45. Sengupta A, Mukherjee A, Das A, De D (2022) AgriStick: an IoT-enabled agricultural appliance to measure growth of jackfruit using 2-axis JoyStick. IEEE Instrum Meas Mag 25(3):58–62. https://doi.org/10.1109/MIM.2022.9759351
46. Dey T, Bera S, Latua LP, Parua M, Mukherjee A, De D (2024) iCrop: an intelligent crop recommendation system for Agriculture 5.0. IEEE Trans AgriFood Electron 2(2):587–595. https://doi.org/10.1109/TAFE.2024.3454109

Energy Economics: Unraveling Sustainable Convergence of Oil Markets and Macroeconomic Signals

Shrey Arora, Rishabh Kaushik, and Palash Bairagi

Abstract This study investigates the relationship between the stock performance of the oil and gas industry and crude oil prices, alongside key macroeconomic indicators such as the Index of Industrial Production (IIP), inflation rate, and exchange rate. We focus on two Indian stock market indices—BSE Sensex (a broad market indicator) and the BSE Oil and Gas Index (a sector-specific index). The analysis is based on monthly data from January 2014 to December 2024. To explore these relationships, we employed correlation analysis, the Granger causality test, and Generalized Least Squares (GLS) regression. The findings reveal that crude oil prices significantly influence the BSE Oil and Gas Index but do not have a direct causal effect on the broader BSE Sensex. Among the macroeconomic indicators, only the IIP was found to Granger cause the BSE Sensex, indicating a predictive relationship between industrial output and overall market performance. Interestingly, all the selected macroeconomic indicators—along with crude oil prices—demonstrated predictive power over the BSE Oil and Gas Index. This study highlights the sensitivity of sector-specific indices to commodity prices and macroeconomic trends, offering valuable insights for investors, policymakers, and analysts in understanding the dynamics of the oil and gas sector within the broader economic context.

Keywords Commodity · Crude oil prices · Sensex · BSE oil and gas · Index of industrial production · Inflation · GLS regression · Real effective exchange

S. Arora
Symbiosis Centre for Information Technology, Symbiosis International (Deemed University), Pune, India

R. Kaushik
J. C. Bose University of Science and Technology, YMCA, Faridabad, Haryana, India

P. Bairagi (✉)
Department of Business Management, Central University of Odisha, Koraput, Odisha, India
e-mail: pbairagi@cuo.ac.in

S. Roy et al. (eds.), *Humanized Intelligence in Sustainable Finance and Economic Behaviours*, Studies in Computational Intelligence 1226,
https://doi.org/10.1007/978-981-95-0599-9_11

1 Introduction

The coronavirus pandemic had an intense effect on the oil economies. Recently, we witnessed a plummet in the oil prices [1]. The global oil demand has fallen largely due to low industrial activity, reduction in global travel, etc. Oil prices have significant impact on some industries like the Airline Industry [8]. India ranks third among all countries when it comes to the import of crude oil. Whereas, stocks are an important component of the Indian economy [17]. Oil has been considered as an important indicator of the exchange rate between countries [18]. Some studies have proven that macroeconomic indicators such as inflation rate affect the stock market indices [19].

As an importer of oil, India is observed to be affected by the variations in the international oil prices, as it determines the position of Indian rupee against the U.S. dollar. When there is escalation in the prices of crude oils, we see a depreciation in the currency. A decline in the Indian Rupee occurs as large amount of oil trades are done in U.S. dollars [7]. The depreciation of the currency also accounts for a surge in the inflation rate of the country [19, 20]. Various studies have been conducted for investigating the linkage between prices of crude oil and the rate of exchange observed by the economy of a country that imports oil. These studies provide support to the inference that any pattern followed by oil prices impact the rate of exchange in the country [7].

Investors and portfolio manager have been invested in commodities more than before. These traders are constantly watching the movements in the indices of stock market and the prices concerned with commodities. When a country's economy is not doing well, investors consider commodities as a safer investment option [7].

It is essential for an investor or a portfolio management company to understand the dynamic relationship the stock market share with commodities like oil price, gold price, and exchange rates and their dynamic relation [11, 21, 23]. The relation is also important to regulators and policymakers, as they can appropriately initiate actions for the guarding the economy and stock markets against the fluctuations of oil prices [22].

In above, we discussed about the people who have vested interest in the stock market. It is essential for them to be calculative and cautious. In a time of crisis, the stock performance greatly depends on the performance of the macroeconomic factors than the company's own performance [21, 22]. Overall, crude oil price plays a vital role in an oil importing economy India [16].

The answers to the following questions need to be ascertained:

RQ1. Do the crude oil prices have an impact on the Indian Stock Market?
RQ2. Do the crude oil prices specifically impact the oil and gas stock market index?
RQ3. Is there any causal relationship between the stock market indexes and macroeconomic factors?
RQ4. Is there any causal relationship between oil prices and macroeconomic factors?

In the context discussed above, we need to anticipate the influence a commodity and macroeconomic factors can have on the stocks and sector-specific stocks. In order to do that, we need to understand the linkages between them by empirically investigating the causality and correlation the variables may have. We employed various statistical tests to ascertain the relationship the variables might have. The study analyzes India which is an oil importing economy. We selected India as there are less studies conducted attributing to the backdrop of the countries which import oil.

Further, remaining paper is organized into sections which present the literature review, paper's objective, research methodology, analysis, discussions, conclusion, and limitations pertaining to the study.

2 Review of Literature

There are significant studies and researches conducted to study the relationship and the dynamicity of indices in stock markets along with the prices of crude oils. Along with it, there are extensive literature that tells us about the relationship among stock markets, crude oil prices, and macroeconomic factors. Macroeconomic factors are aggregated indicators or factors that reflect the performance of the economy as a whole. Any changes in them may influence the economy [14]. An investigation done by Gkillas et al. [6] observed that an incorporation of the financial situation enhances forecasting mechanisms for oil prices. The literature review would discuss about studies related to the association of prices of crude oils with the stocks and indicators of macroeconomic fundamentals, followed by the relationship of stocks with macroeconomic indicators. The discussion on past studies conducted is divided into sections, one dedicated to the crude oil and another to macroeconomic factors.

2.1 Crude Oil Prices

Crude oil may have a negative or positive influence depending upon the nation, supply demand situation, and exchange rate [16]. Study specific to Turkey conducted by Caner Demir observed that an upward rise of the prices of crude oil creates a downturn in the performance of the stock market [4]. Performance of stock market relies on how the company performs, a company that relies on oil directly or indirectly for their industrial activities would get impacted by oil prices soaring [22].

However, in accordance with the observations drawn for multiple researches, it was found that the demand and supply of a country also influences the fluctuation in the prices of crude oil [23, 24]. In reference to the countries who are oil importers, a surge in the prices of oil is observed due to supply might impact stocks in a negative aspect on whereas a price rise due to demand might have a positive impact [24].

Therefore, the effect that the oil price variation may also be determined by the factor that is causing the variation [26].

Study conducted by Kumar [27] indicated that the stock market of particular country can react to a surge in the prices of oil in an aspect either negative or positive. It largely depends on the type of economy the country is, i.e., either oil importing or exporting country [28].

Study conducted by Lui and all [29] analyzes the influential impact the variations in prices of crude oil have on the returns received from the stock market. A study provided support to the argument that an oil importing economy like India observes a negative influence or a downfall in the stock prices after witnessing a rise in oil prices.

However, it is not entirely true, a study said that in a situation where an economy is healing after a recession or is aggressively expanding as an economy; therefore, a surge in the price of crude oil may create a positive effect on the stock market [30].

For an economy, an experience to the surge in the prices of oils may result in an escalation of the inflation rate and decline in exchange rates [23]. However, for oil exporting country like Iran, a rising inflation leads to an escalation the value of real effective exchange rate [25].

A study conducted by Maghyereh and Abdoh [10] investigated the co-movements of the shocks in the prices of crude oil and the returns obtained based on the composite index, i.e., BSE oil and gas, using a novel quantile cross-spectral dependence approach and observed that the relationship the oil price and returns of oil and gas index have directly. Jain & Biswal [7] observed a short-term causality between fall in the prices involving the crude oil and the stock index along with the exchange rate. Several studies have indicated that a dip in the prices of crude oil can be considered as a positive note for the economy and the stock market performance of India [16].

2.2 Macroeconomic Factors

India uses the IIP as a ready indicator relying on the production data of sample data for the manufacturing units. IIP is an index that reflects the rate of growth in various industry groups associated with the economy for a specific time frame. It is computed monthly by the Central Statistical Organization (CSO) [31]. It used to track the growth of the industrial sector in India.

In a study conducted specific to Russia, it was observed that oil prices Granger cause the industrial production and manufacturing.

Studies like [9, 32, 33] investigated the correlation between Index of Industrial Production and stock returns for UK. It was found that IIP indicated higher industrial activity therefore higher earnings for company. However, post the year of 2010s, it was observed that IIP didn't serve as independent variable in order to predict the stock. The reason was the transition of the economy of UK from manufacturing to services.

The external competitiveness of any country is indicated by the REER (Real Effective Exchange Rate) of the country. It is attributed to a specific country, in order to evaluate its currency against other major currencies. According to Reserve Bank of India (RBI), REER is defined "as a weighted average of nominal exchange rate adjusted for relative price differential between the domestic and foreign countries" (Reserve Bank of India, NEER report, 2022, [2]). It is connected to the purchasing power parity (PPP) hypothesis which states that rates of exchange and prices are related [34].

Studies have indicated that the prices of oil and the rate of exchanges are correlated on long-term basis [35]. Macroeconomic indicator like exchange rate has been linked to rate of inflation for an oil importing country like India [7]. A study specific to Turkey using ARDL Bounds Test suggested that a rise in REER increases stock indices [4].

An investigation conducted by Jain and Biswal [7] found out that a dip in the prices of crude oil can lead to a decline in the Indian Rupee exchange rate and Sensex Index. They indicated that a dip in the prices of crude oil leads to a downturn in the currency which in turn leads to a drop in the Sensex.

A study conducted on Southeast Asian countries examined that the oil prices affected the REER of these countries, namely Indonesia, Malaysia, Singapore, Thailand, and Philippines [5]. In an investigation into the rate of exchange and stock price interactions specific to Vietnam, it was observed that exchange rate Granger causes the stock prices [36]. Therefore, it can be used in forecasting of stock prices [37].

Furthermore, we referred to the study done by Fama in the year 1981. The analysis in the paper said the rate of inflation and the activities within the economy of a country influences its stock market. Some studies like indicated that index of industrial production can be used to measure the activities within the economy [25, 26].

3 Chapter Objective

This chapter aims to test how the oil and gas sector index are affected by the prices of a specific commodity, i.e., crude oil, macroeconomic indicators such as IIP, REER, rate of inflation, etc.

4 Research Methodology

4.1 Data

The study purposed the monthly observations made for the following India specific macroeconomic indicators namely between January 2014 to December 2024. A period of COVID-19 poses series threat to economies than any other crisis [3].

We used the data service available on the webpage of PPAC, i.e., Petroleum Planning and Analysis Cell (www.ppac.gov.in) to collect the prices of single barrel of crude oil in U.S. dollars for aforementioned period.

The data for IIP and Inflation were collected from the website of Ministry of Statistics and Program Implementation (www.mospi.nic.in). Whereas, the data for real effective exchange rate were collected form RBI's website (www.rbi.org). We used BSE Sensex which is a composite stock index recorded by the Bombay Stock Exchange (BSE) and BSE Oil and Gas Index which is a market specific index for Oil and Gas industries. The selection of the observation period was based on availability and subjected to the data points along with the other features in consideration. For the period mentioned, there are 84 observations.

4.2 Methodologies

4.2.1 Test and Analysis

We carry out a univariate analysis with the time-series data. To ascertain the stationarity of all the variables in the model, the chapter employed unit root test. It is essential that we examine the stationarity otherwise the results would be spurious [3]. Further, for Augmented Dickey and Fuller test formulated by Dickey and Fuller in 1979, to examine the presence of unit root in the variables separately. It is a commonly used approach in verifying the stationarity in a time-series data [6]. The test follows "augmenting" and adds lagged values of the dependent variable [16]. Further, we will be denoting the series as integrated. A series is said to be integrated in order of d, if we apply differencing$d$ times to render stationarity in the series. $I(d)$ is the denotation used to represent it, *as said by* Engle and Granger in the year 1987.

To comprehend the relation between the variables, we execute the Granger casualty test as developed by Granger in the year 1969s. It is again a commonly used method to test causality [7, 8]. The attempt is to predict if a time series x Granger causes another time series y. For this, both x and y have to be stationary series. For running Granger casualty test, the time-series data needs to be stationary [16, 22]. The null hypothesis is that x is not Granger causing Y. We reject the null hypothesis if we find the value of p to be less than 0.05.

4.2.2 Model

We built a basic regression model using Generalized Least Squares (GLS) estimator. It is more suitable than Ordinary Least Squares (OLS) under autocorrelation and heteroscedasticity [3, 5, 6, 10, 11]. Heteroscedasticity refers to the condition when the variations of regression analysis differ [12]. It uses data and earlier information to create model with best-fitting parameters.

5 Analysis of Results and Interpretation

5.1 Descriptive Statistics

It tells us the measured values of the central tendencies, i.e., mean and median along with the variation using standard deviation, skewness, and kurtosis. To check for normal distribution, we performed two tests: namely D'Agostino–Pearson test and Jarque–Bera test. In D'Agostino–Pearson test, the null hypothesis is that the population has distributed values. It uses a statistic with a chi-square distribution with 2 degrees of freedom.

Jarque–Bera (Jarque & Bera, 1980) uses chi-square (CSQ) distribution, the null hypothesis is that data points are normal distributed. D'Agostino–Pearson test for normality works well for short-tailed distribution whereas for long-tailed Jarque–Bera works best [17, 18].

Figure 1 shows the price movements of all the six variables during the time period. There is a notable dip in the prices of crude oils in the year 2016. The World Bank attributes this plunge to the increase in the efficiency of U.S. in oil production, and ability to sell at cheaper prices in the international market [13].

Table 1 displays the descriptive analysis on the variables selected which are crude oil prices, BSE Sensex Index, BSE Oil and Gas Index, IIP, REER, and inflation.

Based on the results from Table 1, the data on the indexes of BSE Sensex and BSE Oil and Gas have higher standard deviation. Therefore, we can say that the data points are span over a wide range of values. Whereas, for macroeconomic variable, the data points are close to each other. Similarly, the crude oil price doesn't show a huge standard deviation.

On comparing the results from Table 1 and Fig. 2, we can confirm the skewness of the data points. The crude oil price, BSE Sensex, IIP, and inflation are positively skewed or right skewed distribution. The distribution for dimensions BSE Oil and Gas Index and REER are negatively skewed or left skewed. The distribution for BSE Sensex, BSE Oil and Gas, and IIP are approximately symmetric, as their skewness lies closer to 0. Only REER is a heavily skewed distribution. The kurtosis for BSE Sensex, BSE Oil, and Gas and IIP is less than 3. They are platykurtic with shorter tails and broader and lower peak, whereas as rest of the distribution are leptokurtic with longer and thicker tailers with higher peaks.

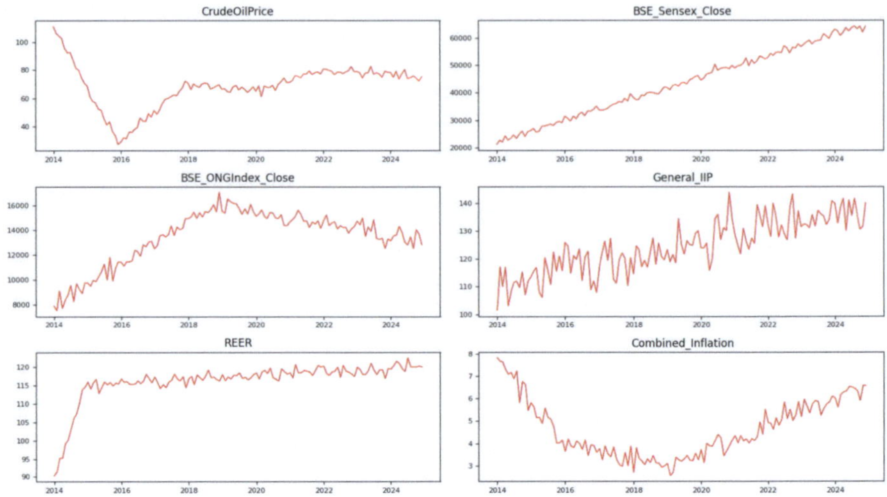

Fig. 1 Movements of the dimensions in the considered time frame. *Source* Author own computations

After performing the D'Agostino–Pearson test, we obtained the p-value for all the dimensions. We reject the null hypothesis which states the population has distributed values for crude oil price, BSE Sensex, BSE Oil and Gas, and real effective exchange rate. Whereas, there is failure while rejecting the null hypothesis for IIP and rate of inflation dimensions. To support the test results for normality of D'Agostino–Pearson test, we perform Jarque–Bera test and the null hypothesis that the population has distributed values for crude oil prices, BSE Oil and Gas and REER is rejected.

Null hypothesis for the rest of the variables is accepted other than the two discussed.

From both the tests, we can confirm that the IIP and Inflation rate are normally distributed. Therefore, they follow a Gaussian distribution. It is important for other variables to follow normal distribution for obtaining an optimal statistical model for predicting them. It allows us to obtain good results and it increases the speed of calculations as quoted by Sola and Sevilla in the year 1997.

5.2 *Correlation Test*

Pearson's correlation coefficient is employed to compute the linear association that the two variables have between each other. The correlation coefficient values might take values between positive and negative one. A positive coefficient indicates the one variable can either increase or decrease with another variable. A negative coefficient indicates that an opposite movement happens in one variable when another variable increases or decreases [20] (Fig. 3).

Table 1 Descriptive statistics of the dimensions selected

	Price of crude oil	BSE sensex	BSE oil and gas index	IIP	REER	Inflation
Mean	63.070007	30,732.112500	12,286.287778	120.970833	112.849583	4.597639
Standard deviation	19.641791	5355.665885	2516.116419	8.813457	6.334370	1.622106
Gaussian/ non-Gaussian (D'Agostino–pearson test)	Non-Gaussian (0.007)	Non-Gaussian (0.007)	Non-Gaussian (0.000)	Gaussian (0.336)	Non-Gaussian (0.000)	Gaussian (0.079)
Skewness	0.90401	0.27626	− 0.00917	0.25897	− 2.27602	0.61574
Kurtosis	0.28848	− 1.00045	− 1.44695	-0.58907	6.76672	0.11745
Jarque–Bera test	10.05645 (0.00655)	3.91854 (0.14096)	6.28199 (0.04323)	1.84579 (0.397)	199.52827 (0.0)	4.59101 (0.101)
No. of observations	72	72	72	72	72	72
Normally distributed?	No	No	No	Yes	No	Yes

Note Figures in parenthesis indicates *p*-value
Source Author own computations

Table 2 Augmented Dickey–Fuller test results

	No. of lags chosen	ADF test statistic	Critical value 1%	Critical value 5%	Critical value 10%	*p*-value
Crude oil price	0	− 2.2065	− 3.526	− 2.903	− 2.589	0.2039
BSE sensex	2	− 0.4166	− 3.529	− 2.904	− 2.59	0.9073
BSE oil and gas index	0	− 1.4889	− 3.526	− 2.903	− 2.589	0.539
IIP	12	− 0.7707	− 3.546	− 2.912	− 2.594	0.8276
REER	0	− 4.4854	− 3.526	− 2.903	− 2.589	0.0002
Inflation	12	− 2.1918	− 3.546	− 2.912	-2.594	0.1092

Source Author own computations

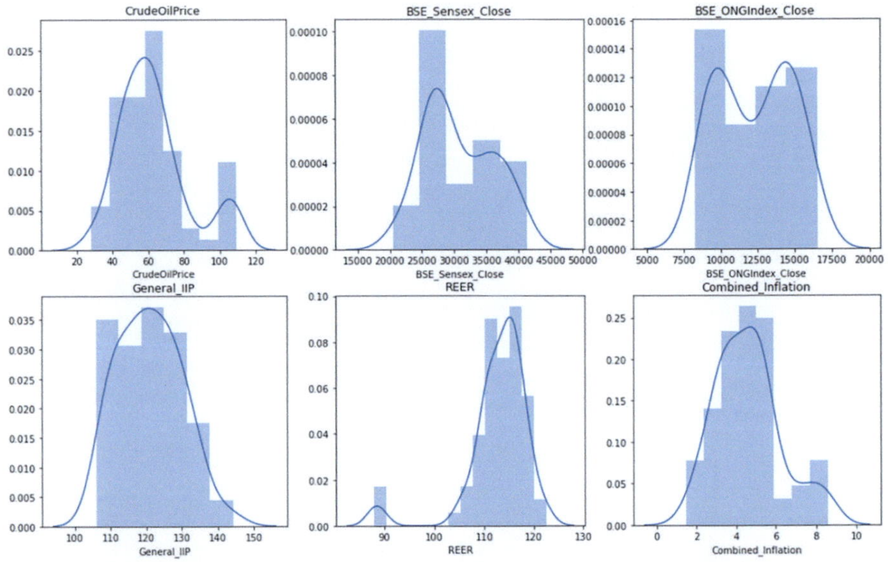

Fig. 2 Frequency and normal distributions of the dimensions in the considered time frame. *Source* Author own computations

From the above results, we can observe that BSE Sensex and Oil and Gas Index are positively correlated to each other, IIP, and REER. However, they are negatively correlated to inflation. That means a decrease in inflation would increase the stock prices and the other way around. We can see that there isn't a notable correlation between the prices of oil and the data for stocks according to Pearson's Correlation Test. Although, there is a negative correlation of oil prices and Sensex, whereas a positive correlation exists for BSE Oil and Gas Index.

Although, Pearson's correlation test doesn't take into account causation, it only analyzes the nature of dependence that the statistical relationship of the variables has

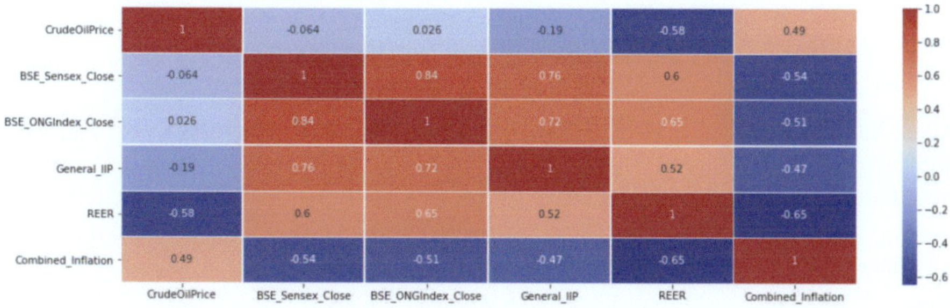

Fig. 3 Correlation test. *Source* Author own computations

among the data points in observation [6]. The pair-wise correlation coefficients for variables under study are static. They do not reflect the changes that had happened over a course of time [7]. We will have to perform other tests to further verify whether the crude oil prices influence the stock market performance.

5.3 *Results of ADF Test*

We obtained the following results on performing a univariate analysis on the ADF test conducted for both the response and predictors.

The null hypothesis for ADF test states that a time-series data in nature has the presence of a unit root and it has displays non-stationarity. If the value of p is lower than the level of significance, i.e., 0.05, we cannot reject the null hypothesis. From the results obtained in Table 1, we can say that the variable REER is a stationary data.

Differencing is the general method to transform a non-stationary time-series data into stationary, and we performed differencing till we obtain all the variables stationary individually. Table 3 displays the results obtained after first differencing for the dimensions of crude oil price, BSE Sensex, BSE Oil and Gas Index, IIP and Inflation.

After performing the first differencing, all the variables are now stationary. Therefore, as discussed earlier as per Engle and Granger (1987), we said that a series is integrated in the order of d if it was differenced d times to transform it into a stationary series. The denotation $I(d)$. Therefore, REER is a $I(0)$ series, whereas the dimensions Crude oil price, BSE Sensex Closing Price, BSE Oil and Gas Index Closing price, Index of Industrial Production, Inflation are $I(1)$ series. Therefore, we need to ensure stationarity in order to perform any sort of forecasting on the results.

Table 3 ADF test results after performing the 1st differencing

	No. of lags chosen	ADF test statistic	Critical value 1%	Critical value 5%	Critical value 10%	p-value
Crude oil price	0	– 5.8308	– 3.527	– 2.904	– 2.589	0.00
BSE sensex	1	– 7.2552	– 3.529	– 2.904	– 2.59	0.00
BSE oil and gas index	1	– 6.4261	– 3.529	– 2.904	– 2.59	0.00
Index of industrial production	11	– 3.3713	– 3.546	– 2.912	– 2.594	0.012
Inflation	11	– 2.9145	– 3.546	– 2.912	– 2.594	0.1092

Source Author own computations

5.4 *Results of Granger Causality Test (GCT)*

We execute the test to investigate whether there is causation between the pair of variables in discussion. We checked the Granger causality of all possible combinations of the variables we have. The null hypothesis that one variable doesn't Granger causes another.

In the table, the rows are the response variables, whereas columns are predictor variables. We want to see whether predictor Granger causes the response variables. The table shows the p-value obtained after the test. If the value of p is less than 0.05, then the null hypothesis is rejected (Table 4).

Table 5 is the null hypothesis, and we test them on the basis of the obtained p-value (Table 6).

We can observe the following results from the analysis as:

(a) BSE Oil and Gas Granger causes BSE Sensex
(b) Index of industrial production Granger causes BSE Sensex
(c) Crude oil prices Granger causes BSE Oil and Gas Index
(d) Index of industrial production Granger causes BSE Oil and Gas Index

Table 4 GCT

x y	Price of crude oil	BSE sensex	BSE oil and gas	IIP	REER	Inflation
Price of crude oil	1.0000	0.0000	0.0000	0.0101	0.0262	0.0403
BSE sensex	0.1039	1.0000	0.0000	0.0000	0.0813	0.1152
BSE oil and gas	0.0125	0.3067	1.0000	0.0045	0.0122	0.0135
IIP	0.0344	0.0000	0.0000	1.0000	0.0088	0.0085
REER	0.1410	0.5185	0.0004	0.0029	1.0000	0.0007
Inflation	0.0000	0.3485	0.0062	0.0066	0.0000	1.0000

Source Author own computations

Table 5 Null Hypothesis testing with the Granger causality for BSE Sensex

	Null hypothesis	p-value	Accept → $p > 0.05$ Reject → $p < 0.05$
1	Crude oil prices do not Granger cause BSE Sensex	0.1039	Accept
2	BSE oil and gas index prices do not Granger cause BSE Sensex	0.0000	Reject
3	IIP do not Granger cause BSE Sensex	0.0000	Reject
4	REER do not Granger cause BSE Sensex	0.0813	Accept
5	Inflation do not Granger cause BSE Sensex	0.1152	Accept

Table 6 Null hypothesis testing with the Granger causality for BSE oil and gas

	Null hypothesis	p-value	Accept → $p > 0.05$ Reject → $p < 0.05$
1	Crude oil prices do not Granger cause BSE oil and gas index	0.0125	Reject
2	BSE Sensex prices do not Granger cause BSE oil and gas index	0.3067	Accept
3	IIP do not Granger cause BSE oil and gas index	0.0045	Reject
4	REER do not Granger cause BSE oil and gas index	0.0122	Reject
5	Inflation do not Granger cause BSE oil and gas index	0.0135	Reject

(e) Real effective exchange rate Granger causes BSE Oil and Gas Index
(f) Inflation Granger causes BSE Oil and Gas Index

5.5 *Regression Model and Analysis*

We investigated the influence the crude oil prices and macroeconomic variables had on the two respective indices one is the BSE Sensex and another is BSE Oil and Gas Index. It was examined using Generalized Least Squares approach for regression.

Model is:

$$y_{\text{bse\_oil\_}n\text{\_gas}} = \alpha_0 + \alpha_1 \text{Crude Oil Price} + \alpha_2 \text{IIP} + \alpha_3 \text{REER} + \alpha_4 \text{Inflation} + e_t,$$

where e_t is the error term and all the $\boldsymbol{\alpha_n}$ are coefficients of the variables.

From the table below, we can see that t-value is high for all the variables that means each of them is significant. It implies that a unit rise in the price of oil will lead to a growth of BSE Oil and Gas Index by 72.7645, assuming all other variables and factors are constant. Similar interpretation can be extended to all the variables based on their coefficient (Table 7).

The p-value of all the variables is close to 0.0000 implies that the model is significantly high. We obtain the R-squared (R^2) and adjusted R-squared (Adj. R^2) of the model as 0.825 and 0.814, respectively. This measure tells us how well our model fits

Table 7 Regression output

Component	Coefficient	*t*-value	*p*-value
Intercept (α_0)	– 3.481e + 04	– 8.966	0.0000
Crude oil price (α_1)	72.7645	8.702	0.0000
IIP (α_2)	119.1897	6.686	0.0000
REER (α_3)	259.4085	8.293	0.0000
Inflation (α_4)	– 258.3957	– 2.359	0.021

Source Author own computations

the actual data. It lies between 0 and 1 [3]. It implies that the 82.5% of the variance found in the response variable is explainable by our predictor variables.

6 Discussion

Based on the correlation test, we found out that inflation and the stock indices are negatively correlated. Therefore, if the inflation increases, it implies a negative impact on the stock indices. Which is sensible, as the inflation increases the purchasing power decreases. It makes the stock investment less attractive; therefore, there might be a decrease in the high-dividend stocks like BSE Sensex and Oil and Gas Index. BSE Sensex and BSE Oil and Gas Index are correlated to each other as per correlation test. The reason is that some of the companies like ONGC and Reliance are part of both these indexes. The crude oil prices were positively correlated to Oil and Gas Index, and negatively correlated to Sensex. Therefore, it justifies the study conducted by Islam e al. [22] and Yadav et al. [16]. They stated a negative effect on stock prices on price rise in oil, and a positive effect on stock performance if the oil prices fall.

According to the outcome obtained from the Granger causality test, we see that the crude oil only Granger causes the BSE Oil and Gas Index. Similarly, macroeconomic factors namely IIP, REER, and inflation Granger causes the BSE Oil and Gas Index. Therefore, we can say that these variables can allow us to predict the prices of BSE Oil and Gas Index. In a crisis period like COVID-19, macroeconomic variables can be helpful in forecasting the stocks as per the study conducted by Gkillas et al. [6, 15]. The causality tells us that we can incorporate these macroeconomic factors to predict stock prices.

Based on the regression output, we see that there is a negative co-movement between inflation and BSE Oil and Gas Index, which supports the result obtained in Pearson's correlation coefficient test. However, based on the t-value which is significant and higher than any other variables, we see a co-movement between the crude oil prices and BSE Oil and Gas Index in a positive manner.

7 Conclusion

We can conclude that a commodity like crude oil price has significant causal effect on its sector-specific index which is oil and gas sector. However, as crude oil doesn't have any causal effect on BSE Sensex, we cannot say that it can be used to predict BSE Sensex. Therefore, we observed a specific impact on the BSE Oil and Gas Index. BSE Oil and Gas Index causes the BSE Sensex, there is a unidirectional causality relationship between them. It implies that we can predict BSE Sensex using Oil and Gas, but not the other around.

Out of all the macroeconomic variables, only IIP Granger caused BSE Sensex, whereas all the macroeconomic variables Granger caused BSE Oil and Gas Index. We observed a bidirectionality in the causation observed between the prices of crude oil and all other macroeconomic variables. Therefore, the prices of crude oil and the variables can used to predict each other. However, based on the correlation coefficient, crude oil price negatively impacts REER and positively impacts the inflation. It is aligned with past studies like the one conducted by Jain and Biswal [7].

The study conducted in the paper will be of relevance to investors, regulators, and policymakers. As there is a causation between the BSE Oil and Gas Index and crude oil prices which is bidirectional in nature, we can predict the fall or rise in crude oil prices based on global composite indexes on oil and gas. It's important to anticipate the fall in oil prices as it might be beneficial for the regulatory authorities and policy-making bodies in India as India imports 85% of its crude oil requirement and can add to its macroeconomy [16, 21].

8 Limitations

We haven't considered the COVID-19 pandemic period, i.e., January 2020 onwards. We have only considered composite indexes of BSE Sensex and Oil and Gas. Crude oil prices may have influence over other indexes on automobiles sector and energy sector. We didn't look at the long-term causality and we didn't test for multicollinearity which may support the results obtained. Granger causality test faces problem of spurious regression and specification bias.

The study doesn't take into account the events that could lead to structural breaks which can lead to fluctuations in the relationship for a specific time frame.

9 Future Scope

The prediction model is limited to regression, we can incorporate vector autoregression based on long-term causality to individually predict all the macroeconomic variables. Forecasting plays a vital role for accurate profit maximization and risk minimization for the scenario concerned with oil trading and trading done for stocks. A crisis like coronavirus has led to a volatility in these markets, and pose huge challenge in forecasting. Taking into account the information gathered from social media, we can incorporate influential factors like conflicts, political stability or instability [15] along with the macroeconomic fundamentals can help us to forecast in a much better way by use deep learning algorithms.

References

1. Albulescu CT (2020) Coronavirus and oil price crash. ArXiv, April. https://doi.org/10.2139/ssrn.3553452
2. Bank DR (2020) Reserve bank of india bulletin revision of nominal effective exchange rate (NEER) and real effective exchange rate (REER) indices, pp 1061–1066
3. Chaudhary R, Bakhshi P, Gupta H (2020) The performance of the Indian stock market during COVID-19. Invest Manag Financ Innov 17(3):133–147. https://doi.org/10.21511/imfi.17(3).2020.11
4. Demir C (2019) Macroeconomic determinants of stock market fluctuations: the case of BIST-100. Economies 7(1). https://doi.org/10.3390/economies7010008
5. Forhad MAR, Alam MR (2021) Impact of oil demand and supply shocks on the exchange rates of selected Southeast Asian countries. Glob Fin J 100637. https://doi.org/10.1016/j.gfj.2021.100637
6. Gkillas K, Gupta R, Pierdzioch C (2020) Forecasting realized oil-price volatility: The role of financial stress and asymmetric loss. J Int Money Financ 104:102137. https://doi.org/10.1016/j.jimonfin.2020.102137
7. Jain A, Biswal PC (2016) Dynamic linkages among oil price, gold price, exchange rate, and stock market in India. Resour Policy 49:179–185. https://doi.org/10.1016/j.resourpol.2016.06.001
8. Kang W, Perez de Gracia F, Ratti RA (2021) Economic uncertainty, oil prices, hedging and U.S. stock returns of the airline industry. N Amn J Econ Financ 57. https://doi.org/10.1016/j.najef.2021.101388
9. Le TH, Chang Y (2015) Effects of oil price shocks on the stock market performance: do nature of shocks and economies matter? Energy Economics 51:261–274. https://doi.org/10.1016/j.eneco.2015.06.019
10. Maghyereh A, Abdoh H (2021) The impact of extreme structural oil-price shocks on clean energy and oil stocks. Energy 225:120209. https://doi.org/10.1016/j.energy.2021.120209
11. Singh NP, Sharma S (2018) Phase-wise analysis of dynamic relationship among gold, crude oil, US dollar and stock market. J Adv Manag Res 15(4):480–499. https://doi.org/10.1108/JAMR-12-2017-0124
12. Singhal S, Ghosh S (2016) Returns and volatility linkages between international crude oil price, metal and other stock indices in India: evidence from VAR-DCC-GARCH models. Resources Policy 50(C):276–288. https://doi.org/10.1016/j.resourpol.2016.10.001
13. Stocker M, Baffes J, Vorisek D (2018) Global economic prospects, January 2018: broad-based upturn, but for how long? World Bank, Washington. https://doi.org/10.1596/978-1-4648-1163-0
14. Times TE (2020) Macroeconomics definition | macroeconomics meaning—the economic times. The Economic Times. https://economictimes.indiatimes.com/definition/macroeconomics
15. Wu B, Wang L, Wang S, Zeng Y-R (2021) Forecasting the U.S. oil markets based on social media information during the COVID-19 pandemic. Energy 226:120403. https://doi.org/10.1016/j.energy.2021.120403

16. Yadav N, Tandon P, Tripathi R, Shastri RK (2020) A dynamic relationship between crude oil price and Indian equity market: an empirical study with special reference to Indian benchmark index Sensex. Benchmarking 28(2):582–599. https://doi.org/10.1108/BIJ-06-2020-0306
17. Ph H, Rishad A (2020) An empirical examination of investor sentiment and stock market volatility: evidence from India. Financ Innov 6(1):34. https://doi.org/10.1186/s40854-020-00198-x
18. Beckmann J, Czudaj RL, Arora V (2020) The relationship between oil prices and exchange rates: revisiting theory and evidence. Energy Economics 88:104772. https://doi.org/10.1016/j.eneco.2020.104772
19. Hilmola OP (2021) Inflation and hyperinflation countries in 2018–2020: Risks of different assets and foreign trade. J Risk Financ Manag 14(12):618. https://doi.org/10.3390/jrfm14120618
20. Egilsson JH (2020) How raising interest rates can cause inflation and currency depreciation. J Appl Econ 23(1):450–468. https://doi.org/10.1080/15140326.2020.1795526
21. Sheikh UA, Asad M, Ahmed Z, Mukhtar U (2020) Asymmetrical relationship between oil prices, gold prices, exchange rate, and stock prices during global financial crisis 2008: evidence from Pakistan. Cogent Econ Financ 8(1):1757802. https://doi.org/10.1080/23322039.2020.1757802
22. Islam U, Awwad EM, Sarhan NM, Fattah Sharaf MA, Ali I, Khan I, Ahmad S, Khan F (2024) Enhancing economic stability with innovative crude oil price prediction and policy uncertainty mitigation in USD energy stock markets. Fluct Noise Lett 23(02):2440021. https://doi.org/10.1142/S0219477524400212
23. Wang G, Sharma P, Jain V, Shukla A, Shabbir MS, Tabash MI, Chawla C (2022) The relationship among oil prices volatility, inflation rate, and sustainable economic growth: evidence from top oil importer and exporter countries. Resour Policy 77:102674. https://doi.org/10.1016/j.resourpol.2022.102674
24. Peng J, Li Z, Drakeford BM (2020) Dynamic characteristics of crude oil price fluctuation—from the perspective of crude oil price influence mechanism. Energies 13(17):4465. https://doi.org/10.3390/en13174465
25. Yang Y, Liu Z, Saydaliev HB, Iqbal S (2022) Economic impact of crude oil supply disruption on social welfare losses and strategic petroleum reserves. Resour Policy 77:102689. https://doi.org/10.1016/j.resourpol.2021.102304
26. Xiuzhen X, Zheng W, Umair M (2022) Testing the fluctuations of oil resource price volatility: a hurdle for economic recovery. Resour Policy 79:102982. https://doi.org/10.1016/j.resourpol.2022.102982
27. Kumar S (2019) Asymmetric impact of oil prices on exchange rate and stock prices. Q Rev Econ Finance 72:41–51. https://doi.org/10.1016/j.qref.2018.12.009
28. Khalfaoui R, Sarwar S, Tiwari AK (2019) Analysing volatility spillover between the oil market and the stock market in oil-importing and oil-exporting countries: Implications on portfolio management. Resour Policy 62:22–32. https://doi.org/10.1016/j.resourpol.2019.03.004
29. Liu Z, Ding Z, Lv T, Wu JS, Qiang W (2019) Financial factors affecting oil price change and oil-stock interactions: a review and future perspectives. Nat Hazards 95:207–225. https://doi.org/10.1007/s11069-018-3473-y
30. Yang Y, Liu Z, Saydaliev HB, Iqbal S (2022) Economic impact of crude oil supply disruption on social welfare losses and strategic petroleum reserves. Resour Policy 77:102689. https://doi.org/10.1016/j.resourpol.2022.102689
31. Kumar A, Mehta A, Gupta R, Mehta S (2021) Tracing the contours of official economic statistics in India. Dr Arjun Kumar, IMPRI Impact and Policy Research Institute. Book
32. Altig D, Baker S, Barrero JM, Bloom N, Bunn P, Chen S, Davis SJ, Leather J, Meyer B, Mihaylov E, Mizen P (2020) Economic uncertainty before and during the COVID-19 pandemic. J Publ Econ 191:104274. https://doi.org/10.1016/j.jpubeco.2020.104274
33. Chowdhury EK, Khan II, Dhar BK (2022) Catastrophic impact of Covid-19 on the global stock markets and economic activities. Bus Soc Rev 127(2):437–460. https://doi.org/10.1111/basr.12219

34. Raut DK (2021) Behavioural equilibrium exchange rates in emerging market economies. Reserv Bank India Occas Pap 42(2)
35. Churchill SA, Inekwe J, Ivanovski K, Smyth R (2019) Dynamics of oil price, precious metal prices and the exchange rate in the long-run. Energy Economics 84:104508. https://doi.org/10.1016/j.eneco.2019.104508
36. Dang VC, Le TL, Nguyen QK, Tran DQ (2020) Linkage between exchange rate and stock prices: evidence from Vietnam. J Asian Financ Econ Bus 7(12):95–107
37. Ning Y, Wah LC, Erdan L (2019) Stock price prediction based on error correction model and Granger causality test. Clust Comput 22:4849–4858. https://doi.org/10.1007/s10586-018-2406-6

Economic Sustainability and Humanized Intelligence in Online Buying Behavior: A Human-Centric Perspective

Palash Bairagi, Sheetal Bansude, and Samarjit Roy

Abstract The chapter explores the integration of human-centered intelligence reshaping online consumer behavior and redefining the modern e-commerce practices. It focuses the growing importance of user-focused innovations, such as personalized website interfaces (PWI), responsive digital support system (RDSS), and referral programs that strengthen customer loyalty. The discussions extended to dynamic pricing models and psychographic analysis. By combining the theoretical perspectives with the real-world applications, the chapter also addresses the challenges and ethical concerns that deploying intelligence digital tools. In doing so, it contributes to ongoing discourse in interaction between technology and human behavior, which offering valuable implication for industry practices and academic research. The narrative is enriched with forward-thinking strategies and real-world applications, presenting a compelling case for positioning humanized intelligence as the cornerstone of future innovations in the online retail domain. Concluding with a call for continued exploration, the chapter advocates for a holistic approach that harmonizes technological sophistication with human-centric principles, paving the way for a more inclusive and engaging digital commerce ecosystem.

P. Bairagi (✉)
Department of Business Management, Central University of Odisha, Koraput, Odisha, India
e-mail: pbairagi@cuo.ac.in

S. Bansude
School of Commerce and Management, D. Y. Patil International University, Pune, Maharashtra, India
e-mail: sheetal.bura@dypiu.ac.in

S. Roy
Department of Software Engineering, School of Computing and Information Technology, Eastern International University, Becamex IDC Corp, Ho Chi Minh City, Binh Duong Province, Vietnam
e-mail: roy.samarjit@eiu.edu.vn

S. Roy et al. (eds.), *Humanized Intelligence in Sustainable Finance and Economic Behaviours*, Studies in Computational Intelligence 1226,
https://doi.org/10.1007/978-981-95-0599-9_12

Keywords Economic sustainability · Humanized intelligence · Online buying behavior · Human-centric AI approach · e-commerce platforms · Sports brand

1 Introduction

Analyzing the sustainable economic aspects of online buying behavior is crucial in understanding how consumers interact with e-commerce platforms and how their choices impact the environment, society, and economy [1]. Sustainability in this context refers to the responsible consumption of resources, reduction of waste, and promotion of ethical business practices. But while recommending the products from a sustainable economic perspective, the key examination is required to drive the factors impacting buying behavior and the consumers' purchasing decisions. These factors are directly or indirectly supporting the economic aspects of online buying decisions in e-commerce platforms. Recommending one to make purchase decision in e-commerce platforms become a big challenging task for the online venders as well as for the clients by looking on the sustainable economic guidelines such as economic aspects like brand preferences, cost- benefit analysis, supply chain efficiency, local economic burdens, circular economic initiatives, transparency and ethical sourcing, value creation for long-term usage, marketing dynamics and competitive advantages, and alien with the policy and regulation environment and social and ethical consideration.

Looking at the challenges in economic behavior, the human-centric AI approach ensures that AI systems are designed and implemented in alignment with human values, ethical standards, and societal needs. This alignment fosters trust, inclusivity, and effective collaboration, ultimately leading to more sustainable and equitable economic growth in buying decisions. Moreover, the human-centric AI accounts for the contextual understanding of human behavior, recognizes and mitigates the biases in economic decision making, provides transparency in improving user trust and ethical consideration, leading to increased efficiency and creativity in buying behavior of customers.

In this chapter, first, we encompass a model on the driven factors of online buying decisions which are integrating ethical principles, user-centric design, and responsible data governance into the development and users' application. Second, focused on human-centric intelligence in the analysis of sustainable economic aspects of online buying behavior involves leveraging AI models in a way that prioritizes human values, ethical considerations, and empathy. Third, we explore case-based analysis on the online buying behavior of consumers about the purchase of sports goods with an aim to identify the factors influencing consumer decision-making, including online behavior, attributes of online shopping, influence of referrals and word-of-mouth, and the psychographic profile of consumers.

2 Conceptual Background

2.1 Economic Sustainability

Economic sustainability is defined as the way the company conserves or economizes the valuable financial and natural (environmental, economic, and social) resources to strengthen long-term stability [2, 3]. Lovarelli et al. [3] and Rai et al. [4] means economic sustainability adheres to economic feasibility in accessing the sustainable initiatives in any of the economic and social dimensions that balance between the interests of the consumer and sustainable responsibilities. The feasibility of economic sustainability in buying behavior can be driven by human-centric intelligence through AI interventions, which link consumers' attitude toward products and its association with their purchase intention, which we call balancing the demand–supply side of the business cycle in a sustainable decision.

2.2 Economic Sustainability in Online Buying Behavior (ESoBB)

Economic sustainability in online buying behavior (ESoBB) can be defined as "the practice of making purchasing decisions through digital platforms in a manner that supports long-term economic health and stability." This involves choosing products and services that promote environmental stewardship, ethical practices, and social equity. Sustainable online buying behavior aims to minimize negative impacts on the environment and society while fostering economic growth and resilience [5]. It includes considerations such as buying durable and high-quality products, supporting local and fair-trade businesses, reducing waste, and encouraging mindful and responsible consumption [6, 7]. The breakdown of key economic aspects to consider the sustainability in online buying behavior is shown in Fig. 1.

The diagram represents the various economic factors influencing online buying behavior in the sports industry and how these factors interact through e-commerce platforms to promote sustainability in economic buying behavior. Here is an elaboration on each component with examples from the sports industry perspectives.

2.2.1 Economic Aspects of Online Buying Behavior of Consumers Includes

A. **Brand Preferences**: A driving force that affects how consumers interact with e-commerce platforms, their considerations of supply chain efficiency, and their evaluation of the long-term value of products, ultimately empowering

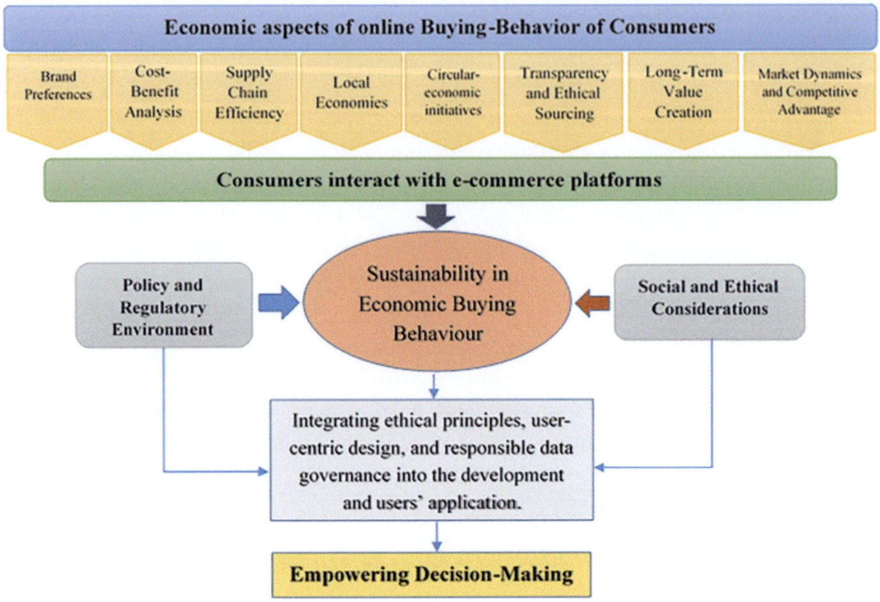

Fig. 1 Key economic aspects to consider the sustainable online buying behavior

them to make decisions that align with their economic ethical priorities and sustainable practices [8, 9]. It includes:

- Customer interaction with e-commerce platforms and influence consumers visit and trust. For instance, loyal Nike customers are more likely to shop on Nike's official website or app.
- Emphasize ethical sourcing and transparency, such as Patagonia, appeal to consumers who prioritize sustainability, influencing their buying decisions.
- It maintains a competitive advantage, continuously attracting and retaining customers.
- Ultimately empowering them to make decisions that align with their economic and ethical priorities.

Take for an example, consumers might prefer brands like Nike or Adidas, which are known for their reputed brand and high-quality sports apparel and innovative designs [9]. These brands often emphasize sustainability by incorporating recycled materials in their products.

B. **Cost–Benefit Analysis**: It is integral to know how consumers evaluate their purchases, influencing their interactions with e-commerce platforms, considerations of supply chain efficiency, and assessments of long-term value, thereby empowering them to make economically and ethically sound decisions [10].

Let say, when purchasing sports equipment, such as a high-end bicycle, consumers may consider the long-term benefits of durability and performance versus the initial higher cost. A brand like Trek offers bikes with a higher upfront cost but promises long-term performance and durability.

C. **Supply Chain Efficiency**: An efficient supply chain plays crucial role in minimizing operational costs, boosting reliability, and ensuring timely product delivery. Companies such as Decathlon streamline their logistics to keep prices competitive, providing greater value to consumers. Beyond the cost saving, optimal supply chains contribute to local economic development and promote circular economic model, which exemplified by initiatives like Patagonia's Worn Wear [10, 11]. They foster greater transparency and ethical sourcing practices, as demonstrating like puma. Such improvements not only help brands meet regulatory standers and sustain market competitiveness, but also reinforce ethical commitments and purchasing choices.

D. **Local Economies**: This approach encourages the sustainable economic purchasing by cutting down the transportation-related emissions and cost. It simultaneously generating employment within the local communities. It helps in motivating consumers to choose nearby suppliers by reducing the burden of long-distant shipping, results economic resilience and fosters long-term sustainability [12]. In e-commerce platform, it not only offers sporting goods but also bolsters local economy through partnerships of regional suppliers and by providing in-store pickup options [13]. In essence, this strategy cultivates both environmentally conscious and economically responsible consumer habits.

E. **Circular Economic Initiatives**: This initiative in sustainable online buying behavior can be promoted by minimizing the packaging waste and encouraging the adoption of reusable and recyclable materials. This approach also aligns with the product-as-a-service models and shaping economy, where goods are rented or leased instead of being purchased outright. Circular economy practices such as repair and refurbishment programs help extend product life cycles, thereby reducing the environmental footprint of e-commerce activities [14]. These efforts motivate consumers to make more thoughtful and eco-friendly purchasing decision [15].

F. **Transparency and Ethical Sourcing**: Providing consumers with transparent information about the origins and environmental impact of their purchases promotes sustainable buying behavior. When company openly share their supply chain practices, it fosters consumer trust and encourages choices that reflect personal values. Ethical sourcing ensures fair labor conditions and eco-friendly production methods, reducing negative impacts on people and the environment. Transparency empowers consumers to make thoughtful, responsible decisions and supports business committed to sustainability [16]. Such practices help shape a marketplace that values ethical conduct and sustainability, driving the industry wide improvements. For instance, Puma and Nivia Sports (https://www.niviasports.com/), a leading Indian manufacturer of sports gear and apparel, exemplify these values. Nivia emphasizes sustainable materials, ethical labor standards, and domestic sourcing, reinforcing its deduction to both environmental and social responsibility. By choosing such brads, consumers actively contribute to a more sustainable and ethically grounded sports industry.

G. **Long-Term Value Creation**: This prioritizes product durability, quality, and the full lifecycle of goods. This approach extends product longevity, reduces the frequency of replacements, and minimizes waste. By adopting eco-conscious materials and responsible production methods, business can attract and retain customers who vale sustainability. Such long-term outlook encourages innovations in both design and manufacturing, supported by ESG considerations [17]. This model enhances operational efficiencies and promote economic benefits. A notable example is Decathlon India, a leading sports retailer that champions long-term value creation. The company focuses on crafting durable, high-quality products that lesson the need for repeated purchases. Its sustainability efforts include the use of eco-friendly materials and a strong product innovation to reduce environmental harm. Through this strategy, Decathlon fosters customer loyalty and encourages sustainable consumer behavior in online shopping.

H. **Market Dynamics and Competitive Advantage (MDCA)**: The MDCA can be driven by sustainable practices by encouraging companies to differentiate themselves through eco-friendly practices. As consumers increasingly prioritize sustainability, businesses that adopt green initiatives can gain a competitive edge [18]. These companies attract environmentally conscious customers, leading to increased market share and brand loyalty. By integrating sustainability into their business models, companies not only meet consumer demand but also reduce their environmental impact. This competitive advantage fosters a market environment where sustainability becomes a key driver of success, promoting widespread adoption of sustainable practices. Nivia sports focus on sustainable practices not only meets growing consumer demand for eco-friendly products but also sets a standard in the sports industry. This strategic approach enhances the competitive advantage while encouraging sustainable online buying behavior.

2.2.2 Interaction with e-commerce Platforms

Consumers use online platforms to purchase goods, where their buying decisions are influenced by the aforementioned economic aspects. The e-platforms such as Amazon, Nike's, Puma, Nivia Sports, Decathlon's official website, or specialized sports retailers like Sports Direct to purchase sports goods. These platforms facilitate the buying process by offering detailed product information, reviews, and competitive pricing.

1. **Policy and Regulatory Environment**: Policies and regulations ensure fair practices, consumer protection, and sustainable ecosystem and provide tax benefits or subsidies for company adopting green practices. It promotes sustainable online buying behavior by setting standards and incentives for eco-friendly

practices [19, 20]. Governments can implement regulations that require companies to adhere to environmental guidelines, ensuring products are sustainably sourced and produced.

2. **Social and Ethical Considerations**: Social habits are when companies emphasize values like fare labor practices, community engagement, and environmental care [21]. By placing importance on ethical sourcing and ensuring the fair treatment of workers, business appeal to consumers who prioritize social responsibility in their buying decisions. This shift toward ethical considerations drives the overall market toward more sustainable practices. The rise of ethical sports brands like Sneaker veja and Nivia India, which uses sustainable materials and fair labor practices, appeals to socially conscious consumers.
   Adhering to all these key aspects, sustainable online buying behavior integrates ethical principles, user-centric design, and responsible data governance and e-commerce platforms enhances consumer experience and trust [16, 21]. These combined factors empower consumers to make informed and sustainable buying decisions.

2.3 Humanized Intelligence in Online Buying Behavior

Humanized intelligence involves the incorporation of AI systems that are crafted with an in-depth comprehension of human psychology, behavior, and socio-economic needs. These systems go beyond the mere automation of tasks, focusing instead on improving human experiences and aiding in decision-making processes [22]. The goal of humanized AI is to foster a synergistic relationship between technology and its users, wherein the AI systems evolve and adjust based on human actions and preferences [22, 23].

Adhering to the following we can define the term "humanized intelligence in online buying behavior" as:

> The intelligent systems that can interpret and adapt to human emotions, preferences, and social cues within online shopping platforms.by drawing on psychological understanding and behavioral patterns, these systems tailor the shopping journey to individual users, leading to higher level of satisfaction and deeper consumer engagement.

For example, recommendation tools that rely on advanced analytics can assess a user's previous purchase and browsing activities to suggest products aligned with their preferences, thereby enhancing personalization in the shopping experience [24]. Furthermore, systems designed to emulate human understanding that can help identify and anticipate consumer behavior trends, enabling business to refine their marketing efforts more precisely [25].

The use of such human-like digital interfaces, such as chatbots and virtual assistants, supports the development of emotional connections between consumers and brands. These interactive features offer a more engaging and supportive shopping environment, which contributes to greater consumer loyalty [26].

In examining the role of humanized intelligence in shaping online purchase patterns, what often comes into force are digital tools rooted in artificial intelligence (AI) or more precisely, human-centered approach. These technologies are transferring how consumers engage with e-commerce platforms by reshaping economic behaviors. However, alongside their growing adoption, concerns surrounding user privacy, algorithmic bias, and the ethical use of data have become increasingly significant. To fully realize the potential of such systems, it is essential to focus on the designed AI tools that reflect human values, incorporate ethical safeguards, and demonstrate empathy. These human-centric systems aim to align technological progress with the broader goals of sustainability and social well-being [27].

2.3.1 The Concept of Human-Centric AI

Definition The Human-centric artificial intelligence focuses on designing and implementing system that reflect core human value, ethical standards, and the broader needs of society. These systems aim not to perform tasks effectively but also to uphold principles such as dignity, privacy, fairness, and overall human well-being [28, 29].

Principles of Human-Centric AI:

Transparency: These allowing users to understand how decisions are made in clear and transparent manner.

Fairness: Intelligent system should be developed to avoid discriminatory outcomes and ensure that all individuals are treated justly and impartially.

Accountability: A clear structures must exist to ensure responsibility for the actions and decisions made by these systems, including the roles of developers and organizations.

Empathy: These technologies are sensitive to users' emotional and psychological experience's, helping to build trust and strengthen user relationships.

2.3.2 Applications of Human-Centric AI in Online Buying Behavior

Embedding human-centered intelligent technology into online purchasing practices is reshaping digital commerce by making the process more user-oriented, responsive, and secure. This technology not only improve buyer satisfaction but also sustainable business performance and growth. The core applications are the following:

Personalized shopping experiences: By interpreting consumer behavior and preferences, intelligent tools offer tailored suggestions, recommend related items, and personalize promotional content. This individualized approach enhances the shopping experience and encourages repeat purchases.

Customer Service and Support: Automated digital assistants and virtual support systems engage customers with prompt answers and guidance during their shopping process. These systems replicate empathetic human responses, which helps in building trust and reducing incomplete transactions.

Dynamic Pricing strategy: Based on real-time changes in market conditions, consumer interest, and demand trends, pricing can be flexibly adjusted. This allows companies to remain competitive while optimizing earnings.

Smarter Inventory Coordination: With insights drawn from purchase patterns and seasonal shifts, inventory levels are managed efficiently. This ensures better availability, cuts down on surplus or shortages, and contributes to operational savings.

Security and Risk Monitoring: Transaction monitoring tools detect irregular activities by recognizing unusual patterns, reducing potential risks and helping maintain consumer trust through secure shopping environments.

Improved Product discovery: Search features become more effective when systems understand everyday language and intent, making it easier for customers to find what they're looking for quickly and accurately.

Customer Feedback insights: User reviews and comments are examined to highlight areas where improvements are needed. This helps brands align their offerings more closely with customer expectations.

Smart Promotional Strategy: Repetitive tasks like targeted emails, promotional posts, and advertising can be scheduled and customized based on consumer profiles. This results in more relevant outreach and higher levels of engagement.

Augmented Reality (AR) Shopping: AI-driven AR applications allow customers to visualize products in their real environment before making a purchase. For instance, customers can see how furniture would look in their home or how clothes would fit. This immersive experience enhances decision-making and reduces return rates.

Customer Journey Mapping: Human-centric AI maps the entire customer journey, identifying key touchpoints and optimizing the path to purchase. By understanding the steps customers take, businesses can improve the overall shopping experience and address pain points effectively.

2.4 AI Architecture Intelligence in Sustainable Online Buying Behavior

A multi-layered framework for an AI architecture and intelligence system integrates various data sources and analysis techniques to provide personalized and sustainable recommendations for buying decisions [30]. At the top, the "Users and Customer Layer" captures user profiles, which are informed by social, economic, and environmental data collected in the "Data Collection Layer." This raw data undergoes "Feature Engineering" and "Normalization" in the "Data Preprocessing and Analysis Layer," preparing it for further analysis. The Behavioral Modeling Layer incorporates fundamental principles from utility theory and behavioral economics to represents user preferences and decision-making patterns effectively. Following this, the recommended system layer applies a combination of contents-based filtering (CB), collaborative filtering (CF), and reinforcement learning (RL) to deliver personalized suggestions that align with individual user profiles [31].

To enhance clarity and user trust, the Explainability and XAI layer is integrated, that leveraging interpretability techniques such as LIME and SHAP to build a repository of explanatory data [32]. This promotes transparency and allows users to understand the rationale behind each recommendation.

Sustainability considerations are embedded within the sustainability analysis layer, where ESG indicators are analyzed using optimization models. This ensures that recommendations support environmentally and socially responsible decision-making. The entire framework is designed with adaptability in mind, that each component undergoes periodic feedback-driven adjustments to refine performance. Over time, this learning-oriented structure helps the system maintain relevance and improve its effectiveness [27]. Each stage of this architecture has clearly defined purposes, methodology, and output, offering a structural explanation of the system as illustrated in Fig. 2.

The stepwise description of Fig. 2 are as follows:

(a) **Users and Customer Layer**: The main objective of this layer is to capture comprehensive user profiles. It is processed through gathering detailed information about users, including their preferences, behaviors, and demographics. The output of the user profiles serves as the foundation for personalized recommendations.

(b) **Data Collection Layer**: The prime role is to collect diverse data inputs for a holistic view of the user. The source data collected from the social data includes data from social media, user interactions, and online behavior, Economic data encompasses financial transactions, purchasing history, and economic indicators, and environmental data covers geographical information, weather patterns, and other environmental factors.

(c) **Data Preprocessing and Analysis Layer**: In this layer, the raw data is used for analysis and processes through feature engineering which identify and create relevant features from the collected data to enhance model performance and through normalization process of standardized data to ensure consistency and comparability across different sources and formats.

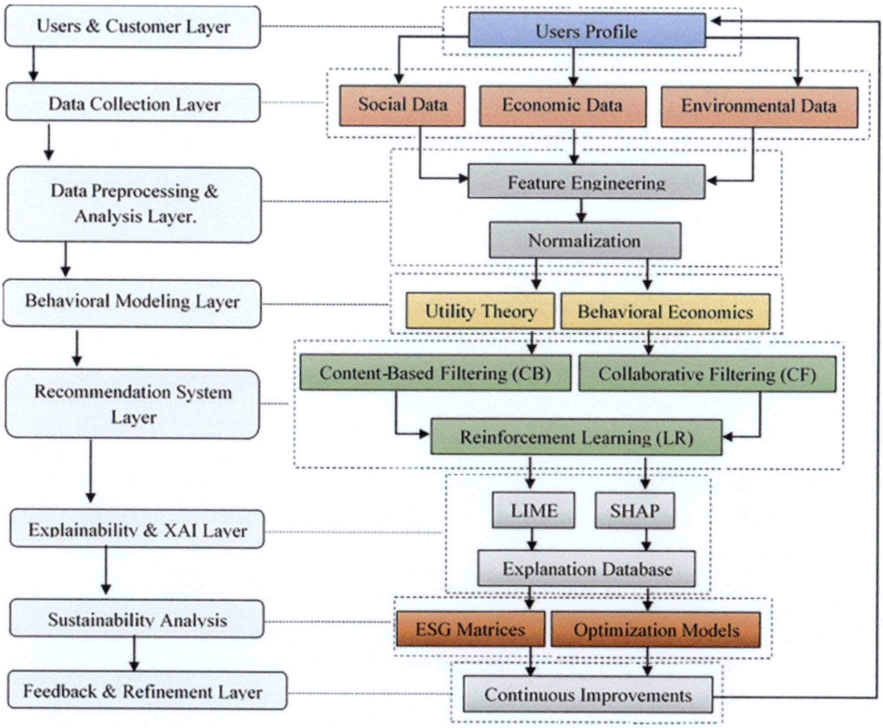

Fig. 2 Intelligence in online buying behavior enabled by AI architecture

(d) **Behavioral Modeling Layer**: The layer will generate a model for user behavior and preferences to inform recommendations. These are generated through the economic theories such as utility theory, which measures how users make choices based on the perceived utility or value of options and through behavioral theory which studies the effects of psychological, social, cognitive, and emotional factors on user pre and post-purchase decisions.

(e) **Recommendation System Layer**: It generates personalized recommendations using advanced algorithms like CB, CF and RL. Content-based filtering (CB) recommends items similar to those the user has liked in the past, based on item features. Collaborative filtering (CF) suggests items based on the preferences of users with similar tastes. Reinforcement learning (RL) helps in continuous learning and improves recommendations through user interactions and feedback.

(f) **Explainability and XAI (Explainable AI) Layer**: It provides transparent and interpretable recommendations through LIME and SHAP. Local Interpretable Model-agnostic Explanations (LIME) is to explain individual predictions by approximating the model locally. SHAP (SHapley additive explanations) provides consistent and accurate feature importance values. The output of the explanation database can store and manage explanations for the recommendations buying model.

(g) **Sustainability Analysis**: It ensures recommendations align with sustainability goals and principles. The components include ESG Matrices which evaluate the environmental, social, and governance impact of recommendations and optimization models which optimize recommendations to balance user satisfaction with sustainability considerations.

(h) **Feedback and Refinement Layer**: This layer continuously improves the recommendation system through continuous improvement and feedback loop. Continuous improvement incorporates user feedback and performance metrics to refine algorithms and models and feedback loop regularly updates the system based on user interactions and new data, ensuring the recommendations remain relevant and effective over time.

3 Analyzing Factors of Online Buying Behavior of Sports Goods

Observing the above context and paradigm, an analysis is employed on the driving factors of online buying behavior and satisfaction of consumers on sports category goods. The analysis is based on the primary pilot survey method conducted in the PCMC region of Pune City. A well-designed questionnaire was employed for the survey. In which 300 questionnaires were returned from the respondent. After scrutiny and verification of 300 filled samples, 216 (72%) complete filled questionnaires were taken for the final analysis, to check the internal consistency. The result of the reliability score of 0.82 Cronbach's Alpha, which is above 0.7 are considered as good for analysis by Lee Cronbach in 1951 [33].

3.1 Demographic Profile of Consumer Purchasing Sports Products

Table 1 illustrates a comprehensive demographic profile of consumers purchasing sports products online, including information on gender, age, monthly income, and education levels. Significantly, 66.7% of the shoppers are male, with a mean of 1.33 and a standard deviation (S.D.) of 0.475, while females account for 33.3%. The largest age group is 18–24 years, making up 56.9% of the consumers, with a mean of 1.72 and an S.D. of 1.038. Regarding income, 66.7% of the shoppers earn less than 20,000 per month, with a mean income category of 1.76 and an S.D. of 1.327. Most shoppers, 56.9%, hold a Master's degree, with the mean education

Table 1 Demographic information of consumers using online shopping

Variables	Groups	Count	Percentage	Mean	S.D
Gender	Male	144	66.7	1.33	0.475
	Female	72	33.3		
Age	18–24 years	123	56.9	1.72	1.038
	25–32 years	54	25.0		
	33–40 years	21	9.7		
	41–48 years	12	5.6		
	49 years and above	6	2.8		
Monthly income	<20,000	144	66.7	1.76	1.327
	20,000–40,000	27	12.5		
	40,000–60,000	18	8.3		
	60,000–80,000	12	5.6		
	80,000–100,000	9	4.2		
	Above 100,000	6	2.8		
Education	Graduates	72	33.3	3.76	0.617
	Masters	123	56.9		
	Ph.D	21	9.7		

Source Author computation from survey data

level being 3.76 and an S.D. of 0.617. These demographic insights are crucial for understanding the online purchasing behavior of consumers in the sports products market in Pune City.

3.2 *Online Purchase Frequency of Different Brand and AI Network*

The provided data outlines in Table 2. shows the frequency and percentage of time spent or purchases of sports products from various brands on online websites. Amazon leads with the highest percentage of frequent purchases at 40.3%, indicating strong consumer preference and trust. PUMA and Flipkart both have a significant share of frequent buyers at 34.7%, while Nike follows closely with 31.4%. Adidas shows a fair share with 26.4%. For weekly purchases, Flipkart stands out with 26.4%, followed by Amazon at 22.2%. Several brands, including Khelmart, Gucci, Paragon, and Runner, each have a moderate weekly engagement of 19.4%. In terms of fortnightly purchases, Homeshop18 has the highest percentage at 16.7%, followed by Action and Khelmart at 15.3%. Monthly purchases are lower overall, with PUMA at 8.3% and Adidas at 6.9%. Occasional purchases are significant for Khelmart at 27.8% and Gucci at 26.4%, while festive season purchases are notable for Homeshop18 and Action at 23.6% each. This analysis reveals varied buying behaviors, with Amazon being the go-to platform for regular purchases, while brands like Khelmart and Gucci are more popular for occasional and festive season purchases.

Table 2 Percentage of time spent/purchase of sports products on online websites

Different sports brand	Frequency of time spent or purchase in online website (In %)					
	Frequently	Weekly	Fortnightly	Monthly	Occasionally	Festive seasons
PUMA	34.7	18.2	6.9	8.3	19.4	12.5
Amazon	40.3	22.2	9.7	2.8	15.3	9.7
Flipkart	34.7	26.4	2.8	5.6	19.4	11.1
Nike	31.4	15.3	6.9	5.6	22.2	18.1
Adidas	26.4	15.3	8.3	6.9	25	18.1
Decathlon	30.6	16.7	11.1	4.2	22.2	15.3
Khelmart	16.7	19.4	15.3	1.4	27.8	19.4
Sports 365	22.2	16.7	13.9	4.2	23.6	19.4
Gucci	18.1	19.4	13.9	1.4	26.4	19.4
Action	22.2	13.9	15.3	1.4	23.6	23.6
Runner	19.4	19.4	13.9	0	25	22.2
Homeshop18	19.4	16.7	16.7	0	23.6	23.6
Paragon	18.1	19.4	12.5	4.2	23.6	22.2

Source Author computation from survey data

Although the interconnections of AI intelligence and the computed data shown in above Table 2, encompass more importance in this chapter to show the usage of humanized intelligence in buying behavior through graph which is shown in Fig. 3. The graph insights into the purchasing habits of consumers for different sports brands. Amazon is the most frequently purchased, followed by PUMA and Flipkart. Many brands have a mix of regular and occasional customers. Some brands, like Homeshop18 and Paragon, are popular during festive seasons. Brands with frequent purchases likely have strong customer loyalty. Understanding these patterns can help brands tailor marketing strategies and product offerings to meet customer needs and preferences.

3.3 *Factors of Buying Behavior and Its Impact on Satisfaction in Decision*

Here, we intended to study the reason behind buying attempts made by respondents and even want to understand the most important reasons that affect consumers' satisfaction in buying decisions of sports category goods. Hence, a direct extracted components sports goods buying pattern such as availability of product, website rating, economy-in-pricing, risk-in-purchase, experience and referrals, return policy and after-sales-service are taken for consideration to evaluate the level of correlation and their impact with the satisfaction level in buying decision in further points.

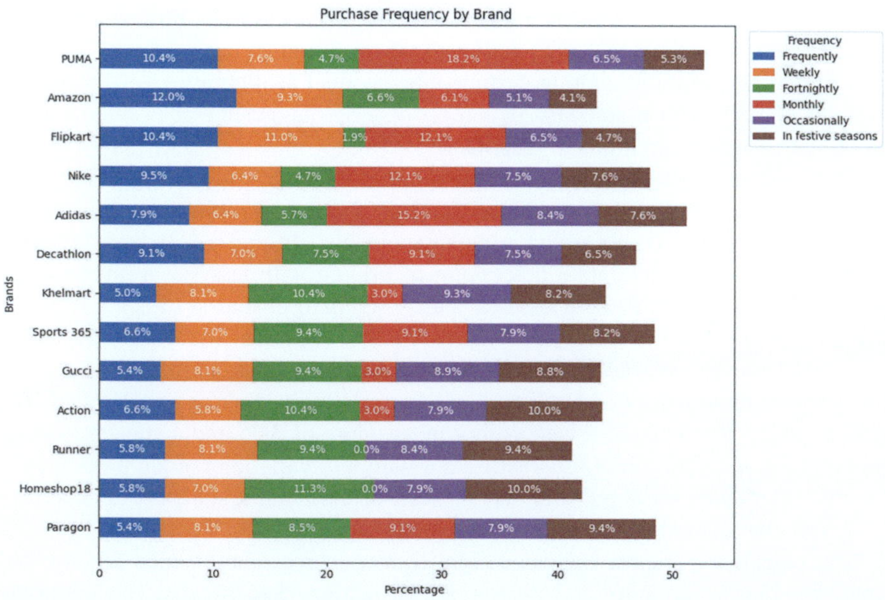

Fig. 3 Purchase frequency graph of sports brand

3.3.1 Descriptive Statistics and Correlation of the Buying Factors

The descriptive statistics shown in Table 3 offer a comprehensive view of customer perceptions based on various factors as availability of product, website rating, economy-in-pricing, risk-in-purchase, experience and referrals, return policy,

Table 3 Description of key factors and its correlations

Descriptive statistics			Pearson's correlations (N-216)							
Variables	Mean	S.D	SBD	AoP	WR	EoP	RIP	EXP and Ref	Rp	ASS
SBD	16.79	5.44	1.00	–	–	–	–	–	–	–
AoP	9.19	2.85	**0.81**	1.00	–	–	–	–	–	–
WR	7.38	2.56	0.66	**0.79**	1.00	–	–	–	–	–
EoP	5.46	1.77	**0.87**	**0.79**	**0.71**	1.00	–	–	–	–
RIP	6.15	1.99	*0.67*	*0.53*	0.48	*0.63*	1.00	–	–	–
EXP and Ref	7.86	2.47	*0.67*	*0.65*	*0.68*	*0.68*	*0.69*	1.00	–	–
Rp	3.75	1.38	*0.65*	**0.75**	**0.78**	**0.70**	0.45	*0.67*	1.00	–
ASS	5.85	2.07	*0.68*	*0.67*	*0.64*	*0.67*	*0.68*	**0.79**	**0.70**	1.00

Note 1. SBD—satisfaction in buying-decision, *AOP*—availability of product, *WR*—website rating, *EOP*—economy-in-pricing, *RIP*—risk-in-purchase, *EXP and Ref*—experience and referrals, *RP*—return policy and *ASS*—after-sales-service. 2. All the predictor variables and criterion variables are significant at 1% level as $p<0.001$
Source Author computation from survey data

and after-sales-service. The average score for satisfaction in buying decisions (SBD) is 16.79, with a standard deviation of 5.44, indicating moderate satisfaction and some variability among participants. Product availability (AoP) scores an average of 9.19 with a standard deviation of 2.85, suggesting it is rated highly, albeit with notable variations in responses.

The website rating (WR) has a mean of 7.38 and a standard deviation of 2.56, reflecting generally favorable views but with some differences in opinion. Economy-in-pricing (EoP) shows a mean score of 5.46 and a standard deviation of 1.77, indicating a moderate perception of pricing with less variability. Risk in purchase (RIP) scores a mean of 6.15 with a standard deviation of 1.99, suggesting a moderate level of perceived risk. Similarly, experience and referrals (EXP and Ref) have an average score of 7.86 and a standard deviation of 2.47, showing positive experiences and reliance on referrals with moderate variability. The return policy (Rp) is rated lower, with a mean score of 3.75 and a standard deviation of 1.38, indicating less satisfaction but with less variation among participants. Finally, after-sales service (ASS) has a mean of 5.85 and a standard deviation of 2.07, showing moderate satisfaction with some variability.

The correlation matrix indicates significant positive relationships among various predictor's variables and customer satisfaction. Specifically, the economy shows a strong positive correlation with satisfaction ($r=0.872$, $p<0.001$), indicating that as economic conditions improve, customer satisfaction tends to increase. Similarly, the availability of products also has a strong positive correlation with satisfaction ($r=0.812$, $p<0.001$), suggesting that ensuring products are readily available significantly enhances customer satisfaction in buying decisions.

Moderate positive correlations were found between satisfaction and the website ($r=0.655$, $p<0.001$), risk ($r=0.665$, $p<0.001$), experience/referrals ($r=0.672$, $p<0.001$), return policy ($r=0.649$, $p<0.001$), and after-sales service ($r=0.684$, $p<0.001$). These results indicate that these factors are also important for customer satisfaction but to a slightly lesser extent compared to the economy and product availability.

The matrix also shows strong inter-correlations among predictors. For instance, the availability of products correlates strongly with the website ($r=0.790$, $p<0.001$) and the return policy ($r=0.754$, $p<0.001$). Similarly, the return policy has strong correlations with the website ($r=0.780$, $p<0.001$) and experience/referrals ($r=0.671$, $p<0.001$). These inter-correlations suggest that these factors are interconnected and collectively influence customer satisfaction in buying decisions. Overall, it indicates a strong and moderate correlations with significant at 1% level, which give importance the factors influencing customer satisfaction in buying decisions.

3.3.2 Regression Model

The regression model resulted that the factors influencing customer satisfaction in buying decisions and their levels of significance. It highlighted that the constant and the regression coefficient (Beta = 1.564) is highly significant at the 1% level ($p = 0.007$), indicating a baseline level of satisfaction in buying decision even when all other variables are zero. Availability of product emerges as the most critical factor, also showing a highly significant with positive effect (Beta = 0.923, $p < 0.001$). This recommended that greater product availability is crucial for enhancing customer satisfaction, increases convenience for customers, and reduces frustration on product availability.

In terms of economic-in-pricing, the coefficient beta of 0.424 satisfies the influence of consumer satisfaction during purchase decisions at 1% ($p < 0.001$) level of significance. These findings suggest that when products are reasonably priced or align well with prevailing economic conditions, customers tend to feel more satisfied with their purchase. Essentially, consumers value getting good quality for the amount they spend, reinforcing the importance of price-value alignment in shaping overall satisfaction. Similarly, Experience/Referrals has a coefficient of 0.421 and is significant at the 1% level ($p < 0.001$). Positive customer experiences and referrals play a substantial role in improving satisfaction in buying sports goods, as they build trust and reinforce the perceived quality of the product or service. Conversely, risk-in-purchase has a significant negative effect on satisfaction ($B = -0.271$, $p = 0.014$). Higher perceived risks associated with a purchase decrease satisfaction, as customers experience increased anxiety and uncertainty about their decisions. Return policy also negatively affects satisfaction with a coefficient of − 0.352 and is significant at the 1% level ($p < 0.001$). Restrictive return policies can significantly lower satisfaction by making customers feel less secure about their purchase choices. In contrast, website rating ($B = 0.219$, $p = 0.105$) and after-sales-service ($B = 0.262$, $p = 0.148$) show positive coefficients but are not statistically significant at the 5% level. This suggests that while these factors may influence satisfaction, their effects are not strong enough to be considered significant in this model (Table 4).

The regression equation for customer satisfaction in buying decision incorporating all the significant variables is as follows:

$$\begin{aligned}\mathbf{SBD} = {} & 1.564 + 0.923(\text{AoP}) + 0.219(\text{WR}) + 0.424(\text{EoP}) + 0.421(\text{Exp.\&Ref}) \\ & - 0.271(\text{RIP}) - 0.352(\text{Rp}) + 0.262(\text{ASS})\end{aligned}$$

Overall, the regression model explains a substantial portion of the variance in customer satisfaction, with an R Square of 0.906 and an adjusted R Square of 0.903, indicating that approximately 90.6% of the variance is accounted for by the model. The Durbin–Watson statistic of 1.985 suggests that there is no significant autocorrelation in the residuals, reinforcing the reliability of the results. Availability of product, economy-in-pricing, experience/referrals, risk-in-purchase, and return policy are the key significant factors affecting customer satisfaction, while website rating and after-sales-service show positive impacts but are not statistically significant.

Table 4 Estimated coefficients of the regression model

Independent variable	Coefficients		Std. coefficients	t	Significance level
	β	Std. error	Beta		
Constant	1.564	0.576		2.714	0.007**
Availability of product	0.923	0.101	0.578	9.832	0.000**
Website rating	0.219	0.135	0.072	1.627	0.105
Economy-in-pricing	0.424	0.115	0.386	3.689	0.000**
Risk-in-purchase	− 0.271	0.110	− 0.091	− 2.472	0.014**
Experience and referrals	0.421	0.111	0.144	3.787	0.000**
Return policy	− 0.352	0.055	− 0.392	− 6.455	0.000**
After-sales-service	0.262	0.181	0.072	1.450	0.148
R square		0.906			
Adjusted R square		0.903			
Durbin–Watson		1.985			

** Significant at 1% level ($p<0.001$)
Source Author computation from survey data

4 Conclusion

In view of digital transformation, this chapter ends with a concluding remark that explores the intersection of economic sustainability and human-centric AI in online buying behavior, in which AI-powered technologies are revolutionizing the online shopping experience. Personalized recommendations, voice assistants, and chatbots enhance customer satisfaction and drive sales. The ethical considerations, such as data privacy and algorithmic bias, must be carefully addressed to ensure responsible AI implementation in online buying. It highlights the crucial role of economic factors like brand preferences, cost–benefit analysis, and supply chain efficiency in fostering sustainable online buying practices, additionally, the significance of human-centric AI in understanding and influencing consumer behavior toward sustainable choices. However, AI intelligence in online buying behavior indicates that the consumers are more focused on their personalized shopping experience, by finding easy ways to explore the website which are user-friendly and having dynamic contents. Second, focused on the improved customer support or real-time support through AI-chatbot, providing virtual assistants and addressing customers concerns. Third, referrals and word-of-mouth point which encourage customer referrals, monitoring social media feedback, and automating referral programs. Fourth, creating optimized sales points with dynamic pricing and personalized discounts. Fifth, making a psychographic profiling based on the past purchase interests and values for tailored marketing strategies (refer Fig. 4.).

Moreover, the analysis results also delve into the factors influencing online buying behavior and satisfaction among sports goods consumers in the PCMC

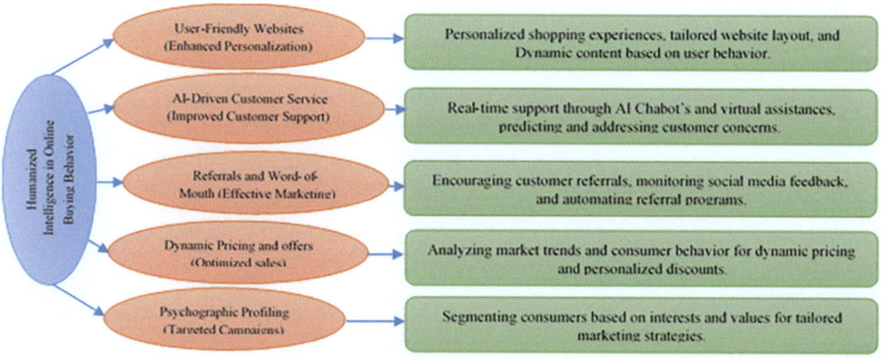

Fig. 4 Recommended AI implementation in buying behavior

region of Pune City. The analysis reveals a preference for Amazon as the most frequently used platform, followed by PUMA and Flipkart. Notably, brands like Homeshop18 and Paragon experience heightened demand during festive seasons. Key factors impacting satisfaction include product availability, pricing, customer experience, perceived risk, return policies, and after-sales service. Understanding these factors is crucial for brands to optimize marketing strategies and product offerings, thereby aligning with consumer needs and preferences. By gaining insights of these elements and utilizing advance intelligent technologies, business can refine their strategies, foster deeper connections with consumers, and promote long-term growth within the competitive market conditions.

Furthermore, the chapter emphasizes integration of human-centered intelligence systems has reshaped online retail experiences like:

1. Examines the extensive user data such as browsing habits, preferences, and past purchase through advance algorithms.
2. Use of voice-controlled tools for simplifying purchases without needing to touch a device.
3. Virtual assistant and chatbots for support customer instantly.
4. Image-based search features for local product visibility, streamlining the process of finding desire goods.
5. Intelligent system forecast future trends and behaviors.

Moreover, it also holds significant implications for both consumers and business.

First, for consumers:

1. It acts as a smarter choice, as it equipped to choose products that reflects values around sustainable users-intensive.
2. The user experiences offer recommendations and transparency, that allowing consumers to feel more in control of their decisions.
3. It will encourage sustainability as awareness grows, consumers are more environmentally friendly options, encourages greener practices and adopt in terms of brand selections.

Second, for Business:

1. To make sustainable business model by improving brand image, appeal to eco-conscious customers, and increase efficiency.
2. Building trust about the AI tools that strengthens consumers' loyalty.
3. Business can use ethical data and behavior analysis that help firms to develop focused marketing efforts that highlights sustainable offerings.

5 Future Scope

While this chapter provides a solid foundation for understanding the factors influencing online buying behavior and the potential of human-centric AI in promoting economic sustainability, it argues that consumers can make more sustainable choices with the help of AI-powered platforms that consider economic, social, and environmental factors. Several avenues, which are uncovered in this research which could be scope of discussion with applicability of advanced AI techniques for creating innovative and personalized shopping experiences, ethical AI, and privacy to protect user data while enabling personalized recommendations, are could also be focus on optimizing supply chains to reduce environmental impact and promote ethical sourcing AI practices, AI-driven social media, and influencer marketing on online buying behavior. By delving into these areas, future research can further advance our understanding of online buying behavior and contribute to the development of more sustainable and ethical e-commerce practices.

References

1. Kennedyd SI, Marjerison RK, Yu Y, Zi Q, Tang X, Yang Z (2022) E-commerce engagement: a prerequisite for economic sustainability—an empirical examination of influencing factors. Sustainability 14(8):45–54
2. Robinson J, Brase G, Griswold W, Jackson C, Erickson L (2014) Business models for solar powered charging stations to develop infrastructure for electric vehicles. Sustainability 6(10):7358–7387. https://doi.org/10.3390/su6107358
3. Oláh J, Kitukutha N, Haddad H, Pakurár M, Máté D, Popp J (2018) Achieving sustainable e-commerce in environmental, social and economic dimensions by taking possible trade-offs. Sustainability 11(1):89
4. Lovarelli D, Bacenetti J, Guarino M (2020) A review on dairy cattle farming: Is precision livestock farming the compromise for an environmental, economic and social sustainable production? J Clean Prod 262:121409. https://doi.org/10.1016/j.jclepro.2020.121409
5. Rai SS, Rai S, Singh NK (2021) Organizational resilience and social-economic sustainability: COVID-19 perspective. Environ Dev Sustain 23:12006–12023. https://doi.org/10.1007/s10668-020-01154-6
6. Baratian S, Safaei AS, Goodarzian F (2023) The effect of sustainability on customers' intention to buy from an online group-buying website. نشریه بین المللی مهندسی صنایع و تحقیقات تولید 34(4):1–18

7. Khan NA, Hassan S, Pravdina N, Akhtar M (2023) Drivers of sustainability: technological and relational factors influencing young consumers' green buying intentions and green actual consumption behavior. Young Consumers 24(6):686–703. https://doi.org/10.1108/YC-09-2022-1610
8. Huynh T (2021) Online consumer behavior toward sports apparel in Finland. https://www.theseus.fi/bitstream/handle/10024/506379/Thuy_Huynh_Thesis_Supplement_20_09.pdf?sequence=2
9. Srivastava PR, Eachempati P, Panigrahi R, Behl A, Pereira V (2022) RETRACTED ARTICLE: analyzing online consumer purchase psychology through hybrid machine learning. Ann Oper Res 1–2. https://doi.org/10.1007/s10479-022-05023-5
10. Reddy VM, Nalla LN (2024) Optimizing e-commerce supply chains through predictive big data analytics: a path to agility and efficiency. Int J Mach Learn Res Cybersecur Artif Intell 15(1):555–585
11. Lahkani MJ, Wang S, Urbański M, Egorova M (2020) Sustainable B2B E-commerce and blockchain-based supply chain finance. Sustainability 12(10):3968
12. Burch S, DiBella J, Rao-Williams J, Forrest N, Hermelingmeier V, Ninomiya SDM, Chisholm K (2020) Sustainable business practices can build resilient local economies for a post-COVID-19 recovery. TRANSFORM project paper (1)
13. Tang L, Zhang B (2024) Linking prosperity and sustainability: the economic impact of energy transition and financial stability. Econ Chang Restruct 57(3):107. https://doi.org/10.1007/s10644-024-09678-x
14. Ghisellini P, Cialani C, Ulgiati S (2016) A review on circular economy: the expected transition to a balanced interplay of environmental and economic systems. J Clean Prod 114:11–32
15. Sudusinghe JI, Seuring S (2022) Supply chain collaboration and sustainability performance in circular economy: a systematic literature review. Int J Prod Econ 245:108402
16. Prasanna S, Verma P, Bodh S (2024) The role of food industries in sustainability transition: a review. In: Environment, development and sustainability, pp 1–21. https://doi.org/10.1007/s10668-024-04642-1
17. Schoenmaker D, Schramade W (2019) Investing for long-term value creation. J Sustain Financ Invest 9(4):356–377
18. Shehata GM, Montash MA (2020) Driving the internet and e-business technologies to generate a competitive advantage in emerging markets: evidence from Egypt. Inf Technol People 33(2):389–423
19. Berger V (2019) Social norm-based gamification to promote eco-friendly food choice. J Consum Mark 36(5):666–676. https://doi.org/10.1108/JCM-01-2018-2547
20. Sheoran M, Kumar D (2022) Benchmarking the barriers of sustainable consumer behaviour. Soc Responsib J 18(1):19–42. https://doi.org/10.1108/SRJ-05-2020-0203
21. Gardner TA, Benzie M, Börner J, Dawkins E, Fick S, Garrett R, Godar J, Grimard A, Lake S, Larsen RK, Mardas N (2019) Transparency and sustainability in global commodity supply chains. World Dev 121:163–177
22. Hermann E (2022) Anthropomorphized artificial intelligence, attachment, and consumer behavior. Mark Lett 33:157–162. https://doi.org/10.1007/s11002-021-09587-3
23. Fenwick A, Molnar G (2022) The importance of humanizing AI: using a behavioral lens to bridge the gaps between humans and machines. Discov Artif Intell 2:14. https://doi.org/10.1007/s44163-022-00030-8
24. Anderson J, Johnson D (2024) The role of artificial intelligence in enhancing E-commerce customer experience (No. 13355). EasyChair
25. Govindaraj M, Gnanasekaran C, Khan P, Gnanamanickam SV, Hoang SD (2024) Humanizing the digital touch: strategies for balancing automation in modern marketing. In: Balancing automation and human interaction in modern marketing. IGI Global, pp 175–202

26. Sarin AB, Sharma S (2024) Harmonizing AI and human interaction: enhancing modern marketing strategies. In: Balancing automation and human interaction in modern marketing. IGI Global, pp 151–174
27. Sharma M, Shail H, Painuly PK, Kumar AS (2023) AI-powered technologies used in online fashion retail for sustainable business: AI-powered technologies impacting consumer buying behavior. In: Sustainable marketing, branding, and reputation management: strategies for a greener future. IGI Global, pp 538–561. https://doi.org/10.4018/979-8-3693-0019-0.ch028
28. Zhou J, Chen F (2023) AI ethics: from principles to practice. AI & Soc 38:2693–2703. https://doi.org/10.1007/s00146-022-01602-z
29. Hagendorff T (2020) The ethics of AI ethics: an evaluation of guidelines. Mind Mach 30:99–120. https://doi.org/10.1007/s11023-020-09517-8
30. Nishant R, Kennedy M, Corbett J (2020) Artificial intelligence for sustainability: challenges, opportunities, and a research agenda. Int J Inf Manage 53:102104. https://doi.org/10.1016/j.ijinfomgt.2020.102104
31. Dhelim S, Aung N, Bouras MA, Ning H, Cambria E (2022) A survey on personality-aware recommendation systems. Artif Intell Rev 1–46. https://doi.org/10.1007/s10462-021-10063-7
32. Chen TCT (2023) Sustainable smart healthcare: lessons learned from the COVID-19 pandemic. Springer, p 11
33. Cronbach LJ (1951) Coefficient alpha and the internal structure of tests. Psychometrika 16(3):297–334

Human-Centered Intelligence in Financial Decision-Making

Anchal Saini, Nitin Kulshrestha, and Shubhanker Yadav

Abstract In the rapidly evolving landscape of twentyfirst-century finance, digitized financial services have become ubiquitous, with Artificial Intelligence (AI) emerging as a transformative force. This chapter investigates consumers' contemporary dilemma in personal finance, navigating the intersection between traditional reliance on social networks for financial wisdom and the allure of AI-driven FinTech solutions. The study, encompassing 163 individuals, explores their financial decision-making preferences to discern whether they gravitate toward human connections or AI. The literature review categorizes findings into Digital Financial Literacy, FinTech, and Financial Socialization, offering insights into aspects such as autonomy, confidence, subjective well-being, and the digital divide. The chapter employs regression analysis, revealing the substantial impact of human connection and artificial intelligence on financial well-being. The significant relationships are the predictive power of these variables in shaping financial outcomes. This comprehensive exploration contributes valuable insights into the dynamic interplay between human connections and AI in personal finance. It provides a structured overview of key themes and emphasizes the need for a balanced approach that leverages the strengths of both human connections and AI to enhance overall financial well-being. The findings are helpful for individuals navigating the contemporary financial landscape, emphasizing the importance of informed decision-making that integrates the strengths of interpersonal relationships and cutting-edge AI technologies. The nexus of personal relationships, social networks, and AI has wide-ranging socioeconomic implications for various domains, individuals, households, and financial institutions. Financial well-being and artificial intelligence are emerging subjects of paramount importance in the current phase of the economy. Closer collaboration between these

A. Saini (✉) · N. Kulshrestha
CHRIST (Deemed to be University), Delhi, NCR, India
e-mail: anchal.saini@res.christuniversity.in

N. Kulshrestha
e-mail: nitin.kulshrestha@christuniversity.in

S. Yadav
Faculty of Finance, Jaipuria Institute of Management, Lucknow, India
e-mail: shubhanker.yadav@jaipuria.ac.in

S. Roy et al. (eds.), *Humanized Intelligence in Sustainable Finance and Economic Behaviours*, Studies in Computational Intelligence 1226,
https://doi.org/10.1007/978-981-95-0599-9_13

will likely generate new and useful insights into the financial decision-making of individuals and households.

Keywords Human connections · Artificial intelligence · Personal relationships · Interpersonal relationships · Social relationships · Social connections · Financial decision-making · Financial well-being · Households · Household financial management

1 Introduction

The twenty-first century has witnessed a seismic shift in the financial landscape as digitized financial services have become ubiquitous. Artificial intelligence has risen as an influential tool with the potential to revolutionize the financial industry. Grounded in various streams, AI seeks to replicate human intelligence using intelligent machines. Within finance, AI is increasingly being harnessed by firms to analyze vast data sets, make predictions, and provide valuable insights to individuals. Researchers continually study the nexus of artificial intelligence and finance, and various terminologies have developed. Researchers have deployed terms like FinBrain [1] and FinTech over time. FinBrain stands for "Financial Brain," and FinTech, which stands for Financial Technology, has become a household name. AI is the central technology of the Industrial Revolution, also called Industrial Revolution 4.0, and has a significant potential impact on people across all fields, and finance is one of these fields. With the emergence of artificial intelligence, there has been a major change in the financial industry following the series of financial innovations that have transformed the way banks, financial institutions, and individuals perform operations to some extent. Artificial intelligence has plunged into the finance industry and, in some way, molded individuals to strive their rigid mindsets to deviate from the fixed patterns of investments practiced till now. The deployment of technologies in the financial sector is beneficial to advancing the financial sector's growth [2]. Intending to build intelligent machines, there is a constant attempt to emulate, augment, enhance, and reproduce human intellect utilizing artificial technologies. Financial well-being and artificial intelligence have prevailed for a long time, though they only gathered attention/gained prominence when mankind stood at the most vulnerable stage. The Global Financial Stability Report by IMF states the increasing financial stability risk worldwide. Inflation hikes and the series of extended lockdowns during the Covid pandemic made humans realize the importance of being financially sound. Simultaneously, the awareness of the rising applications of artificial intelligence instilled fear in humans of replacing them at their jobs and created an urge to be equipped with several relevant applications. Both are important for human sustenance and the economy's growth and development. Individuals and households are the key players in the growth of financial markets and, ultimately, in developing the country's economy.

1.1 Motivation

In the ever-evolving landscape of personal finance, consumers stand at a crossroads, juggling two seemingly contradictory sources of information and advice. On one side the age-old tradition of relying on social networks and personal relationships for financial wisdom. For generations, we have turned to trusted friends, family members, and colleagues for guidance when making significant financial decisions. The human touch, personal anecdotes, and real-life experiences have often been our compass in financial decisions. In contrast, we live in an era marked by the rapid progress of AI, FinTechproducts, and algorithm-driven advice. The digital revolution has ushered in a new wave of financial services and tools that promise efficiency, convenience, and data-driven insights. Algorithms, machine learning, and robo-advisors now offer us the allure of making financial decisions based on numbers and calculations.

This chapter delves into the heart of this contemporary dilemma i.e., the pros and cons of traditional personal relationships versus the allure of FinTech and artificial intelligence. On one side, the warmth and empathy of human interaction provide comfort and a sense of trust. On the other, the cold precision and data-driven recommendations of technology promise efficiency and objectivity. How do we reconcile these seemingly disparate approaches? How do we strike the right balance between the personal and the digital when managing our financial futures? This book chapter addresses these questions and the intersection of human connections and artificial intelligence (AI) in financial decision-making. It explores how personal relationships and social networks influence financial decisions and how AI technologies can be employed in household financial management.

1.2 Chapter Contributions

In today's dynamic landscape, the choices of individuals stand at crossroads. Are they embracing artificial intelligence as their financial guidance companion, or are they still placing trust in personal and social connections? Do they yearn for the familiar support of financial advisors, family, and friends, or are they curious about AI's possibilities in managing their finances? We first conducted a comprehensive literature review search to study the existing literature and work around our research questions. Then, we conducted a comprehensive study to illuminate this evolving paradigm, gathering data from 163 individuals to gain insights into their financial decision-making preferences and tendencies. The following research questions are addressed in the chapter.

RQ1: What is the current state of literature around human connections, AI, and financial well-being?
RQ2: What is the role of human connections in financial well-being? RQ3: What is the role of artificial intelligence in financial well-being?

1.3 Chapter Organization

The remaining sections of the chapter highlight the Literature Survey in Sect. 2, Research Methodology in Sect. 3, Data Analysis in Sect. 4, Discussion in Sect. 5, and conclusions are drawn in Sect. 6.

2 Literature Survey

2.1 Personal Relations and Financial Decision-Making

The financial decision-making process is influenced by a wide range of elements, from personal qualities to external economic conditions. The function of personal relationships within a household is an often overlooked but crucial part of this process. Their role is strong and unavoidable in any field of work. However, it has coiled interplay in relation to financial decision-making within households. Over the years, it has gained attention from different domains like economics, psychology, and sociology.

Individuals' financial choices are formed based on multiple factors, including personal relationships and social ties, discussed in the further section. Personal relationships encompass familial bonds, friendships, and romantic partnerships that are formed by emotional bonds and shared values.

Within households, family members often serve as informal educators, transmitting their financial experiences, beliefs, and skills. Familial relationships like parents and extended family impact financial behaviors and play a significant role in shaping the financial attitude and outcomes of individuals and households. Financial socialization has been identified as a remedy for fostering financial well-being as it is directly associated with financial attitude, knowledge, and capabilities. It is instrumental in shaping the future, and parents are considered the primary socialization agents in a household. Financial socialization refers to the "process of acquiring and developing values, attitudes, standards, norms, knowledge, and behaviors that contribute to the financial viability and well-being of the individual, Danes (1994)." Recent research has observed that parents have the most influence over positive financial behavior and financial well-being compared to peers, neighbors, and self-learning [3]. They influence the financial decisions of their children both directly and indirectly. Seeking financial advice in a direct nature [4]. Indirectly, a person's familial environment and parents' financial literacy impact their behaviors and financial literacy from a young age [5]. Parents contribute to setting financial expectations and modeling financial behavior through financial conversations and financial support.

According to an explicated process, financial outcomes are improved by personal connections and financial literacy [6]. It has been analyzed that long-lasting relationships influence individuals' financial decisions, which is greater when compared with the influence of various social contacts a person possesses [7]. Romantic relationships have a greater influence on how individuals make their financial decisions. Several studies have observed that the greater the partner's support, communication, and relationship satisfaction, the greater the chances of better financial outcomes. Romantic partners may also impact spending patterns by encouraging or discouraging each other to exhibit similar behavior to their partner [8].

2.2 *Social Connections and Financial Decision-Making*

Beyond the boundaries of immediate familial relationships, social networks contribute substantially in the change in the financial behavior of households. Social networks encompass communication, networks with neighbors, peers, co-workers, friendships, and social media connections. The central facet of social networks is the social comparison. Individuals inherently partake in comparative evaluations of financial choices and status against their peers. This social comparison process is instrumental in shaping individual financial choices and behavior. For instance, observing a friend or acquaintance's investments or extravagant spending might lead to a desire to emulate the same lifestyle [9]. Alternatively, witnessing prudent financial choices within one's social circle might motivate an individual to adopt a more frugal approach, focusing on savings and long-term investments. When it comes to the informal financial markets, especially in developing countries, social networks shape individual financial decision-making to a larger extent. The literature underscores these connections' role in forming financial identities, saving styles, consumption patterns, and investment strategies. It is not limited to the impact on the purchasing and saving decisions but also to the skewed lending patterns of individuals between formal and informal markets. The network structure has substantially affected the financial decision-making process, positively impacting the household's participation in the financial market [10]. The network structure determines the network's size and composition. According to studies, the broader the social network, the greater the participation in formal and informal financial markets. The broader the social network, the greater the risky asset holdings in the portfolios. This finding of a broader network structure and increased financial market activity is also pertinent to the various evolving features of social networks. Social connections are more pronounced in investment decisions. Peer pressure can strongly influence investor behavior, which frequently causes people to follow the herd. The phenomenon of herding behavior, where individuals make decisions based on the actions of others, has been observed for ages in the financial market. Professional investors are not the only ones who exhibit this behavior; when choosing investments, households, too, are susceptible to herd mentality. Information diffusion strongly influences social [11]. In this digital era, access to information and advice is readily available through various social media platforms and varied online communities. When in dire need of information, individuals turn to their social connections for information and suggestions about different financial products, services, and opportunities. Peer recommendations significantly affect financial choices, from choosing a bank to picking stocks. These are based on trust and shared experiences, leading to the adoption of certain financial choices that align with the preferences and values of the social network. Moreover, social relationships can also contribute to financial awareness and education. While social connections offer numerous benefits in financial decision-making, they can also present challenges. The pressure to live up to social norms and expectations can cause people to make financial decisions that are out of step with their actual financial needs. Long-term financial security may be jeopardized by conspicuous purchases or risky investments encouraged by the desire to impress others or join a particular social group.

The phenomenon of social comparison is highlighted as a driving force behind individual financial decisions. Collectively, these findings underscore the significance of human connections, both personal and social, in enhancing financial well-being, as they affect financial choices and financial identities, saving styles, and investment strategies, ultimately contributing to the financial well-being of individuals. Based on the discussion in the above two sub-sections, the following hypothesis is put forward:

H1: Human connections have a significantly positive impact on financial well-being.

2.3 *Artificial Intelligence and Financial Behavior*

The financial sector will undoubtedly become more intelligent and efficient as a result of the implementation of artificial intelligence technology. A plethora of studies have addressed how AI impacts various business models and operations of organizations, but in comparison, AI around individuals and households is not largely covered. How is AI shaping human decision-making, and what are its contributions in making the individual's decision process more effective? What are the advancements in finance from the individual's point of view? How can the technology be leveraged for our betterment? These questions largely remain unanswered in an academic context. Using artificial intelligence to manage household finances can be a game changer, giving several benefits to consumers looking to optimize their financial decisions and methods.

By utilizing AI technologies, people can have better financial planning, wiser investment decisions, effective budgeting, and a fairly basic knowledge of taxation. AI-driven technologies provide individualized insights, monitor spending patterns, and spot possible cost-cutting opportunities [12]. AI can also assist people in improving their credit scores, protecting themselves from fraud, and better managing their debt. These developments allow people to manage their money, fostering increased financial security and long-term prosperity. Not only is AI entering into the finance industry, making all financial institutions ponder upon their traditional financial products [13], but it has also entered into Personal Financial Management (PFM). Some applications enable you to transfer your money internationally, dealing in several currencies simultaneously make many parts of our lives easier.

Multiple tools utilize robo-advisors where we can take advantage of personalized, tailored information based on their current financial situation, which is always accessible [14]. These automated financial advisors also have fewer conflicts of interest that could arise when compared to a traditional financial advisor. According to studies, people are afraid of getting back, recording their expenses, and checking their bank balances while feeling guilty for their extravagant spending. Many people went through the trouble of manually keeping track of these costs. We are entering a time when budgeting is simple and more comprehensive for everyone. Investments could be evaluated through tools that assess our needs and capacity to invest, enhancing our investment decisions. Most of these advancements are cost-effective and inexpensive, making them reach more people. These technologies have made "financial management for all," where many people can access financial services and not hesitate to knock on the doors of traditional financial service providers [15].

2.4 *The Role of AI in Analyzing Social Network Data for Financial Insights*

The rapid growth of social networks and the spread of digital communication have generated vast amounts of interconnected information, making it possible to draw financial insights. Social networks have become a treasure trove of information, provide a real-time window into consumer behaviors, scenario shifts, and changing market trends [16]. To mine the massive amounts of data created on websites like Twitter, LinkedIn, and Facebook, financial institutions and investment businesses are increasingly using systems driven by artificial intelligence. These insights can help investors make better investment choices, improve risk management, and provide them an advantage in the highly competitive financial market. AI contributes to finding the behavioral aspects of individuals' financial decisions through data processing, analysis, and interpretation.

Several AI techniques facilitate effective analysis of social network data, like natural language processing, network analysis, and machine learning algorithms [17]. Natural language processing (NLP) extracts meaningful information from various unstructured data in social media posts, content, and articles. As a subset of NLP, sentiment analysis does a similar task by discovering the prevailing sentiments around particular stocks, sectors, and other events. Sentiments are generally categorized as positive, negative, or neutral. Sentiment analysis assists analysts in gauging market sentiments in real time.

Network analysis contributes to identifying the influential individuals, communities, and information dissemination patterns, making it easier to understand the complex interactions among the users [18]. Network analysis enables professionals to identify the potential source of information asymmetry, market influencers, and sentiment leaders. Market movements can be predicted through machine learning algorithms using patterns and past data. These algorithms can find connections between online activities and market performance by combining social network data, benefiting analysts in making more informed investing choices. These AI models can make effective investment strategies and reduce exposure to volatile market conditions. Risk management, consumer behavior analysis, and competitor analysis are other applications of AI-driven techniques. In social network data, AI algorithms can detect odd or questionable. Financial activity, such as unexpected money transfers or irregular spending patterns. This helps with risk evaluation and fraud detection. The combined impact of AI on individual financial behavior and market insights identified through the literature underscores its significantly positive influence on financial well-being, which is proposed by the following hypothesis:

H2: Artificial intelligence have a significantly positive impact on financial well-being.

3 Methodologies

3.1 Literature Review Analysis

An extensive search was conducted using Google Scholar and the Scopus database to conduct a thorough literature review encompassing the impact of social networks and artificial intelligence on financial well-being. Additionally, Google Scholar was employed to broaden the scope of studies related to financial decision-making. The search terms comprised "financial decision-making," "financial management," "financial decision-making strategies," and "financial decision-making factors." The search was delimited to articles and research studies published in English between 2019 and 2023. The search syntax utilized for this exploration included a combination of terms related to human connections, personal relationships, social networks, artificial intelligence, and technology, coupled with terms pertaining to financial well-being and consumer finance. The search was focused on the subject areas of business, sociology, and economics. Only articles published in English were considered, specifically targeting keywords related to finance, financial literacy, decision-making, information technology, humans, adults, financial knowledge, investments, FinTech, financial capability, financial behavior, and interpersonal relations. The inclusion criteria extended beyond peer-reviewed research papers to ensure a comprehensive coverage of relevant literature. Furthermore, additional studies were incorporated to enrich the review by examining the reference lists of already included studies. Notably, 4 studies were identified through the references of the most pertinent study, expanding the breadth of the literature review.

The search strategy was intentionally designed to be inclusive, aiming to provide an exhaustive mapping of the pertinent literature without restricting the scope to peer-reviewed research papers. The following keyword search was run on Scopus specifically: ("Human Connections" OR "Personal Relation*" OR "Social Network*" OR "Network*" OR "Artificial Intelligence" OR "Technology") AND ("Financial

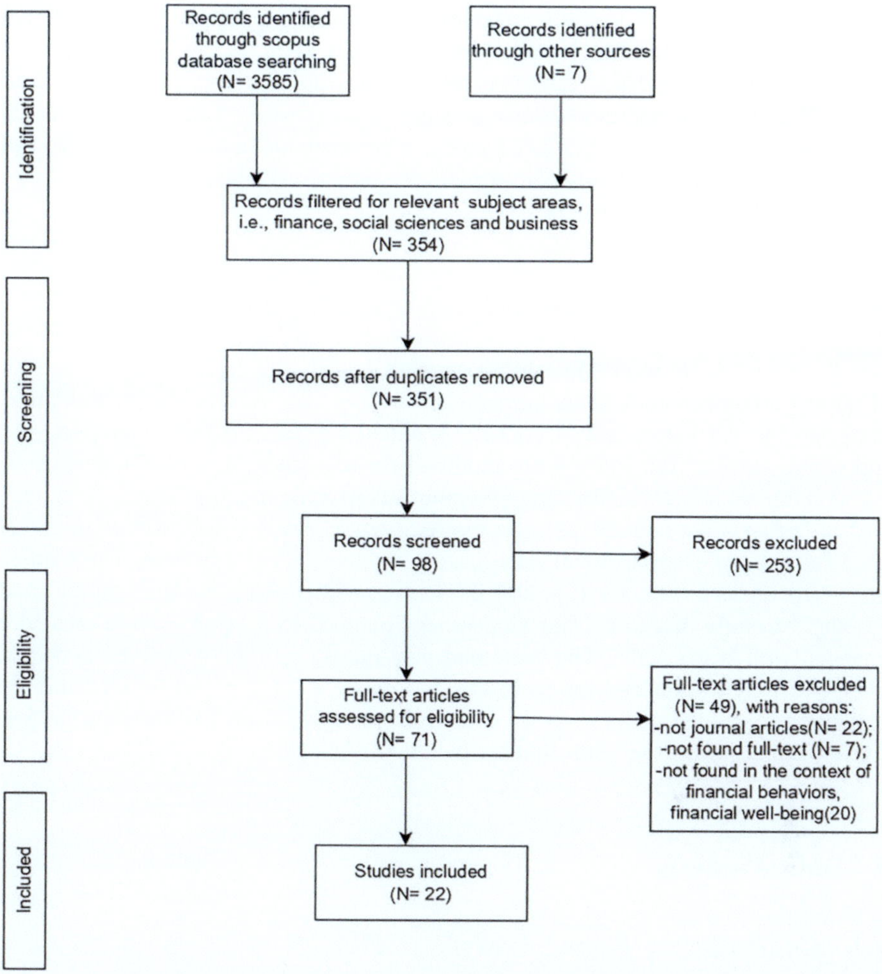

Fig. 1 Selection process of studies included in the review

well-being" OR "Consumer Finance" OR "Personal Finance"). Figure 1 shows a diagrammatic representation of how the selected papers have been identified.

3.2 *Data Collection*

We collected data from 163 respondents for this chapter in order to investigate individuals' financial choices, preferences, and patterns in the context of artificial intelligence and human connections. The survey instrument was sent to 275 people using various online platforms. We received 163 completed responses from the initial respondents. Because of its ability to handle non-normally distributed data and small sample sizes, PLS-SEM has been selected over CB-SEM. According to [19], the sample size should be at least ten times larger than a model's largest number of indicators. We have ten indicators in our conceptual framework. The data are deemed adequate for conducting a PLS-SEM analysis because our sample size crosses 110 (ten times the total number of indicators for the latent variables).

3.3 Measures

The Brief Social Network Questionnaire (SNQ) [20] was used to assess the role of human connections in financial decision-making. The dimensions of this questionnaire are as follows: social contact, practical support, emotional support, and supportive partner. Ten items were modified for this study, and respondents were asked to rate their experiences on a five-point Likert scale ranging from 1 = "never" to 5 = "often." We created self-constructed items to assess respondents' comfort level and trust in using artificial intelligence for financial management. These questions were answered on a five-point Likert scale, with 1 being "strongly agree" and 5 being "strongly disagree." The Consumer Financial Protection Bureau created a financial well-being scale. The scale had nine items, with the first five rated on a five-point Likert scale ranging from 1 = "completely" to 5 = "not at all," and the remaining four rated on a five-point Likert scale ranging from 1 = "always" to 5 = "never." Table 1 lists the items that are included.

4 Data Analysis

4.1 Confirmatory Factor Analysis (CFA)

A total of 24 items across the constructs of Human Connections, Artificial Intelligence, and Financial Well-being were subjected to Confirmatory Factor Analysis (CFA). CFA is a statistical technique to examine the relationships between observed variables (measurement items) and their underlying latent constructs. Within the context of CFA, we considered standardized CFA loadings to assess the strength of the relationships between the observed variables and their respective latent constructs. A factor loading above the standard threshold of 0.70 [21] indicated a strong relationship, substantiating the construct's validity. A stringent approach was undertaken to streamline the measurement model as part of the item reduction process. Items displaying factor loadings below the standard threshold of 0.70 were removed from the analysis, ensuring that the measurement model only included the most reliable and relevant items. Table 2 displays the factor loadings for the selected items. Figure 2 depicts the model to be tested.

Table 1 Indicators for each latent variable

Variable	Indicators
Human connection	I call up or meet friends or relatives not living with me in case of any financial advice. (HC1) Friends and relatives not living with me talked with me about their own financial problems. (HC2) I felt that I was a member of a group, and this was important in my life. (HC3) Had I been in some financial difficulties, there would have been somebody to take care of me. (HC4) When I needed help for my financial planning, I easily found somebody who gave me a hand. (HC5)
Artificial Intelligence	When I felt anxious or worried about my finances, I had somebody who reassured me. (HC6) In case of a financial gain or profit, there was somebody I could share them with. (HC7) I had a trustworthy friend or relative to talk to about my own financial problems. (HC8) When things are going particularly badly around my finances, I can ask other people (relatives or friends) to help or support me. (HC9) In an emergency concerning finances, I am confident that help will be provided by my relatives or friends. (HC10) How likely are you to engage in financial discussions with a chatbot? (AI1) How comfortable are you in shifting your trust to technologies in helping you with your finances? (AI2) The presence of Artificial Intelligence and technology has simplified my financial planning. (AI3) Due to advancements in financial technologies, I have found assistance in accessing credit when needed. (AI4) I experience fear or apprehension when using financial technologies to manage my finances more effectively. (AI5)
Financial well-being	I could handle a major unexpected expense. (FWB1) I am securing my financial future. (FWB2) Because of my money situation, I feel like I will never have the things I want in life. (FWB3) I can enjoy life because of the way I'm managing my money. (FWB4) I am just getting by financially. (FWB5) I am concerned that the money I have or will save won't last. (FWB6) Giving a gift for a wedding, birthday or other occasion would put a strain on my finances for the month. (FWB7) I have money left over at the end of the month. (FWB8) I am behind with my finances. (FWB9) My finances control my life. (FWB10)

4.2 Testing the Reliability and Validity of Measures

The research utilized Composite Reliability (CR) to gauge the constructs' reliability. This method was employed to examine the constancy and internal reliability of the measurement items within each construct. Furthermore, we examined discriminant validity to confirm that the considered constructs remained distinct. In this context,

Table 2 CFA results and measurement quality assessment

Construct	Item	Std. factor loadings	CR	VIF	Discriminant reliability AI	FWB	HC
Artificial intelligence	AI1	0.731	0.889	1.934			
	AI2	0.749		2.229			
	AI3	0.779		1.919			
	AI4	0.734		1.689			
Financial well-being	FWB2	0.939	0.795	1.487	0.524		
	FWB4	0.93		1.555			
	FWB8	0.806		1.220			
	FWB9	0.756		1.201			
Human connections	HC5	0.779	0.751	1.211	0.219	0.348	
	HC6	0.762		1.211			

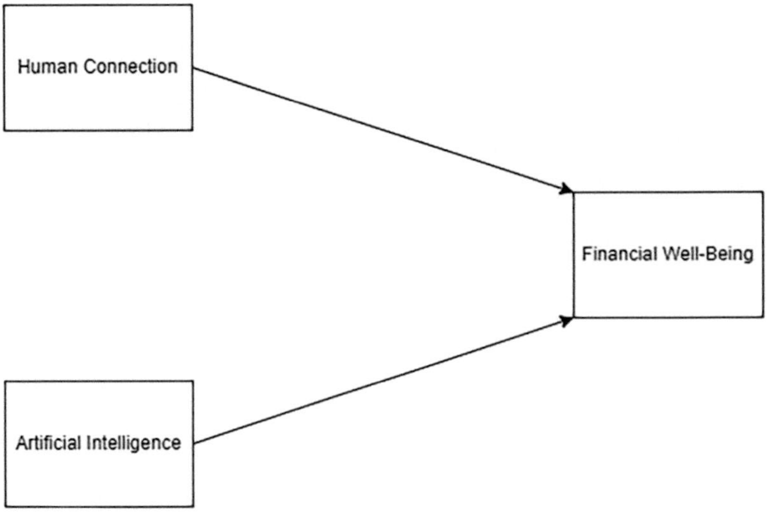

Fig. 2 Theoretical model

it's worth noting that some researchers prefer CR over Cronbach's Alpha due to the latter's susceptibility to criticisms regarding its tendency to provide lower-bound reliability estimates, which could potentially underestimate the true reliability of a scale [22]. However, it is essential to highlight that CR serves as a suitable alternative, offering a marginally higher CR value than Cronbach's Alpha, with any discernible difference between the two typically being of minor importance [22]. The CR values were evaluated to ensure that they exceeded the recommended threshold of 0.7, indicating good reliability for the latent constructs.

Discriminant validity was evaluated to ensure the distinctiveness of the constructs, employing the Heterotrait–Monotrait (HTMT) ratio of correlations. The examination of HTMT values confirmed their adequacy, which was below 1, signifying acceptable discriminant validity. This implies that each construct exhibits a stronger correlation with its own measures than with measures of other constructs. The corresponding values for composite reliability and discriminant validity are detailed in Table 2. Additionally, this investigation employed the variance inflation factor (VIF) to identify multi-collinearity. Results indicated that the VIF for all constructs (ranging from 1.201 to 2.229) is below 3, adhering to the recommended threshold of 5 [21]. The specific values are presented in Table 2.

5 Discussion

The analysis of all the selected papers exhibits a wide array of findings. However, they can be grouped into three distinct categories: Digital Financial Literacy, FinTech, and Financial Socialization. These are classified and presented in category, focus, and authors. These categories encapsulate various authors' focal points and contributions, providing a structured overview of the diverse insights garnered from the literature. The grouping is shown in Table 3. This category comprises four papers contributing to the evolving discourse on digital financial literacy. Each paper within this category explores facets of knowledge, skills, and awareness in the context of digital financial tools and technologies. Authors within this category explain the dimensions of how individuals comprehend and navigate the digital landscape of financial services. The second category encompasses findings from nine papers that converge on the overarching theme of FinTech. Scholars in this category explore the transformative impact of financial technology on various aspects of the financial ecosystem. By examining individual FinTech applications and engaging in broader conversations about the industry's consequences, these papers add to the expanding knowledge base regarding the convergence of finance and technology. The third category focuses on the social dimensions of financial behavior, encapsulated in nine papers. These works investigate the role of social networks, familial influences, and broader socialization processes in shaping individuals' financial decisions. Authors within this category unravel the intricate interplay between social connections and financial outcomes, offering valuable insights into the socio-cultural dynamics of financial behavior. This structured classification provides a roadmap for understanding the varied dimensions explored in the literature, offering researchers and practitioners a clear delineation of key themes and insights within the realms of Digital Financial Literacy, FinTech, and Financial Socialization.

Table 3 Literature synthesis

Category	Focus	Authors
Digital financial literacy	Digital financial literacy, FinTech products and services enhancing the role of digital financial literacy, financial autonomy	Kumar P., Pillai R., Kumar N., Tabash M. I. [24]
	Digital financial literacy, financial confidence	Respati D. K., Widyastuti U., Nuryati T., Musyaffi A. M., Handayani B. D., Ali N. R. [25]
	Subjective financial well-being, financial behavior, financial stress, digital financial services	Mahdzan N. S., Sabri M. F., Husniyah A. R., Magli A. S., Chowdhury N. T. [26]
	Digital divide, the disparity in the use of digital financial services, FinTech	Nam Y., Lee S. T. [27]
FinTech	Financial well-being, Information technology, financial capability, resilience	Carton F. L., McCarthy J., Xiong H. [28]
	Robo-advisor resistance, S–O-R framework, Artificial Intelligence	Aw E. C.-X., Leong L.-Y., Hew J.-J., Rana N. P., Tan T. M., Jee T.-W. (2023)
	Quality of life, financial literacy, FinTech adoption	Kakinuma Y. [29]
	FinTech adoption, financial socialization, UTAUT, self-control, financial literacy, financial behavior	Sabri M. F., Anthony M., Law S. H., Rahim H. A., Burhan N. A. S., Ithnin M. [34]
	FinTech, InsureTech	Zarifis A., Cheng X. [30]
	FinTech services, digital financial technology, actual financial knowledge, and perceived fin knowledge	Nguyen T. A. N. (2022)
	Access to FinTech services, financial knowledge	Hasan M., Noor T., Gao J., Usman M., Abedin M. Z. (2022)
	Artificial Intelligence, FinTech, robo-advisors	Daniel Belanche., Luis V. Casalo, Carlos Flavián. (2019)
	Artificial Intelligence, Personal Financial Planning	Krzysztof Waliszewski 1, Anna Warchlewska. (2020)
Financial socialization	Social connections, ratio of gift expenses to household income (RGI), social networks, household welfare	Hu M., Xiang G., Zhong S. (2021)
	Financial advisors, actor-network theory, financial decisions, financial well-being	Moolman J. [31]
	Social networking sites (SNS), financial attitude, financial behavior, and perceived behavioral control	She L., Ma L., Voon M. L., Lim A. S. S. [32]
	Online social network (OSNs), TAM and valence framework	Haque M. Z., Qian A., Lucky S. A. [33]

(continued)

Table 3 (continued)

Category	Focus	Authors
	Financial socialization, FinTech adoption, UTAUT, self-control, financial literacy, financial behavior	Sabri M. F., Anthony M., Law S. H., Rahim H. A., Burhan N. A. S., Ithnin M. [34]
	Social support, socio-emotional selectivity theory	Weissberger G., Bergman Y. S. [35]
	Social–financial connectedness	Snow S., Vyas D., Brereton M. (2019)
	Role of peers and social media in financial literacy, financial well-being	Yanto H., Ismail N., Kiswanto K., Rahim N. M., Baroroh N. [36]
	Family Financial Socialization, Financial Well-Being	Shirley Y. S., Wei Feng, David Spohn. (2022)

5.1 *Digital Financial Literacy*

Digital Financial Literacy (DFL) is a comprehensive concept that encompasses various dimensions [23]. These dimensions involve a deep understanding of digital financial products and services, awareness of potential risks related to digital finance, knowledge of digital financial risk control, and understanding consumer rights and procedures for redress. By drawing on the goal framing theory, the importance of DFL becomes apparent as a vital resource essential for achieving financial well-being [24]. Research has revealed that DFL significantly influences various aspects, including direct and financial decision-making, skills [25], financial behavior, and subjective financial well-beings [26]. These findings emphasize the extensive impact of DFL on individuals' financial lives and well-being. Additionally, the identification of the digital divide as a research outcome sheds light on the adverse effects linked to uneven access to and utilization of information and communication technology (ICT). This digital divide underscores the necessity of addressing technology access disparities to ensure fair financial outcomes for all [27].

5.2 *FinTech*

Researchers have coined terms such as FinBrain [1] and FinTech, where FinBrain stands for "Financial Brain," FinTech represents Financial Technology, now widely recognized in various fields. Positioned as a central technology in the Industrial Revolution, also known as Industrial Revolution 4.0 [15], Artificial Intelligence (AI) holds substantial potential to impact multiple domains, including finance. Notably, a mixed-method research approach in lower-income households revealed a correlation between mobile phone usage for checking bank balances and meeting financial commitments and resilience [28].

The adoption of FinTech is linked to the quality of life, serving as a catalyst for altering financial behavior [29]. Recognizing AI as a transformative force in enhancing global financial outcomes, it is emphasized that careful consideration of all potential ramifications is essential. Trust, the cornerstone of human interactions, is crucial in personal, social, and financial realms. The research underscores that collaboration between AI and humans, especially in complex or emotionally charged situations, is more effective than sole dependence on technology. This hybrid approach, evident in financial services, addresses resistance to prolonged interaction with complete AI models. Individuals exhibit greater trust in financial advice derived from a blend of human expertise and AI-driven insights attributed to factors such as emotional value, peace of mind, trustworthiness, and confidence. Researches [30] have endeavored to establish trust models specifically tailored for FinTech and InsureTech.

5.3 *Financial Socialization*

Financial socialization is a multifaceted concept that plays a pivotal role in shaping individuals' financial behavior and well-being. It encompasses various factors, including the influence of financial advisors, the Actor-Network Theory (ANT), and the dynamics of financial decisions [31]. Examining the impact of financial socialization requires considering an array of elements such as financial attitudes, behaviors, and perceived behavioral control. Social Networking Sites (SNS) [32] and Online Social Networks (OSNs) have emerged as significant platforms influencing financial socialization. These platforms, embedded within the Technology Acceptance Model (TAM) and valence framework [33], contribute to how individuals perceive and engage with financial information. The adoption of financial technology (FinTech) is also intertwined with financial socialization, with the Unified Theory of Acceptance and Use of Technology (UTAUT) offering insights into the factors influencing its adoption, including self-control, financial literacy, and financial behavior [34].

The concept of financial socialization extends to the role of social support, aligned with the socio-emotional selectivity theory [35], where individuals seek information and guidance that align with their emotional well-being. Social–financial connectedness becomes crucial, emphasizing the interconnectedness of social and financial aspects in an individual's life. Peers and social media wield significant influence on financial literacy and well-beings [36]. Exploring the role of peers and the impact of social media on financial decisions reveals how individuals navigate and acquire financial knowledge within their social circles. Family financial socialization further contributes to shaping financial well-being, reflecting the intergenerational transfer of financial values and behaviors. In conclusion, understanding financial socialization involves a comprehensive exploration of various factors, theories, and frameworks. It is an intricate interplay of individual choices, technological advancements, social networks, and familial influences, all of which collectively contribute to individuals' financial behavior and well-being.

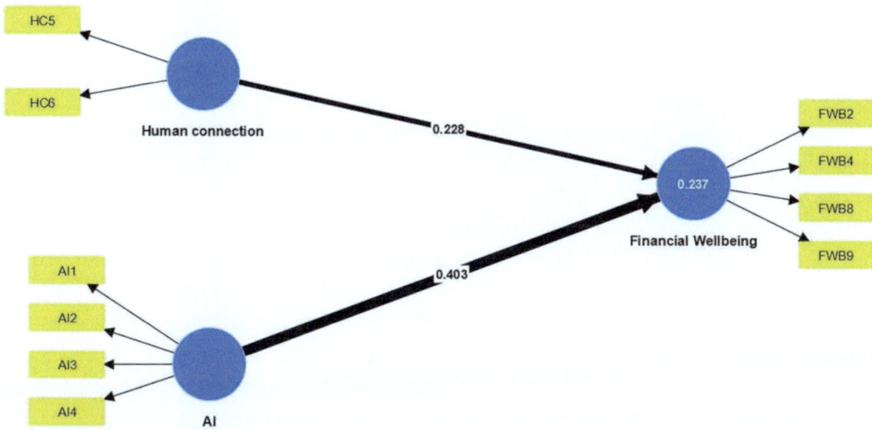

Fig. 3 PLS-SEM results

5.4 *Regression Results*

Upon examining the models, we made initial observations regarding the variance of the endogenous variables (Fig. 3). The quantity of the circle indicates the extent to which another variable elucidates the variance of the latent variable. Additionally, the arrow signifies the path coefficient.

The coefficient of determination, denoted as r^2, stands at 0.237 in defining financial well-being, an endogenous latent variable. The latent variables, Human Connection and Artificial Intelligence, jointly account for 23.7% of the variance in financial well-being. [21] state that the r^2 values of 0.75, 0.50, and 0.25 indicate substantial, moderate, and weak relationships for endogenous variables, respectively. However, the acceptability of an r-square value is context-dependent and varies across disciplines. For instance, an r^2 as low as 0.10 in the financial market is considered satisfactory [21]. The r^2 value of 23.7 holds significance in explaining financial well-being through Human Connection and Artificial Intelligence. The regression results are detailed in Table 4.

The inner path model examines the size and significance of coefficients. Human Connections prove to be a valuable predictor of financial well-being, with a significant relationship (0.228). Similarly, Artificial Intelligence emerges as a robust predictor of financial well-being, with a significant relationship (0.403).

Table 4 Regression results

Construct	Paths	Decision	$_R2$
Hypothesis 1	HC—FWB	Supported	0.228
Hypothesis 2	AI—FWB	Supported	0.403

6 Conclusions

Exploring this intersection of personal relationships, social connections, AI, and financial decision-making, it is conspicuous that these determinants are inextricably linked, collectively shaping the financial choices and the well-being of the households. As we wind up this chapter, we look into the crucial role that advancements in AI and the importance of human connections play in realizing a comprehensive approach to financial well-being. AI techniques like natural language processing, sentiment analysis, and network analysis, along with personal and social relations like parents, romantic relationships, and peers, depict the collaboration leading to financial prosperity through data-driven insights and emotional support. This synergy emphasizes the balance between technology and human connections, unlocking the full potential of individuals and organizations. AI technology in the financial industry has the potential to transform how people make financial decisions and enhance their overall financial well-being. It also delivers considerable advantages to people through enhancing decision-making tools, personalizing financial services, and enhancing access to financial goods. However, to enhance the financial acumen, a hybrid approach of both AI and humans should be employed to leverage both strengths, i.e., the analytical capabilities of AI and wisdom and emotional intelligence of personal connections, to produce beneficial financial outcomes. Tracing back to the roots, personal relationships and social connections serve as foundational pillars that influence financial decisions and provide emotional support and shared experiences, while AI enhances the analytical and informational aspects of decision-making. This balancing fusion of the convenience of AI-driven recommendations and the critical thinking nurtured through personal relationships is of vital importance. The financial ecosystem will become more effective, inclusive, and secure as the financial sector adopts both AI and the human circle. In conclusion, this nexus of personal relationships, social connections, AI, and financial decision-making reflects the intricate amalgamation of emotional, analytical, and technological aspects. The partnership between these determinants has the potential to empower individuals and households with the knowledge, tools, and support needed to navigate the complexities of modern finance and secure a prosperous future.

References

1. Zhou J, Chen C, Li L, Zhang Z, Zheng X (2022) FinBrain 2.0: when finance meets trustworthy AI. Front Inf Technol Electron Eng 23(12):1747–1764. https://doi.org/10.1631/FITEE.2200039
2. Salman S. Artificial Intelligence (AI) and its application on banking and financial services sector in India–A conceptual study. IJFMR-Int J Multidiscip Res 5(2)
3. Jorgenson B, Savla J (2010) Financial literacy of young adults: the importance of parental socialization. Fam Relat 59(10):465–478. https://doi.org/10.1111/j.1741-3729.2010.00616.x

4. Csiszárik-Kocsir Á (2022) Socialisation determining individual financial decisions from the aspect of financial literacy. Public Financ Q 67(4):567–584. https://journals.lib.uni-corvinus.hu/index.php/penzugyiszemle/article/view/1168
5. Katauke T, Fukuda S, Khan MSR, Kadoya Y (2023) Financial literacy and impulsivity: evidence from Japan. Sustainability 15(9). https://doi.org/10.3390/su15097267
6. LeBaron AB, Kelley HH (2021) Financial socialization: a decade in review. J Fam Econ Issues 42(1):195–206. https://doi.org/10.1007/s10834-020-09736-2
7. Patacchini E, Rainone E (2017) Social ties and the demand for financial services. J Financ Serv Res 52(1):35–88. https://doi.org/10.1007/s10693-017-0279-0
8. Olson JG, Rick SI (2022) "You spent how much? Toward an understanding of how romantic partners respond to each other's financial decisions. Curr Opin Psychol 43:70–74. https://doi.org/10.1016/j.copsyc.2021.06.006
9. Kousta S (2017) Social networks: inferring financial status. Nat Publ Group 1(May):1. https://doi.org/10.1038/s41562-017-0127
10. Chai S, Chen Y, Huang B, Ye D (2019) Social networks and informal financial inclusion in China. Asia Pac J Manag 36(2):529–563. https://doi.org/10.1007/s10490-017-9557-5
11. Jiang M, Sun R, Zhang B (2022) Social networks and household financial decisions: evidence from China. J Appl Econ 25(1):58–92. https://doi.org/10.1080/15140326.2021.2007725
12. Lewis DR (2018) The perils of overconfidence: why many consumers fail to seek advice when they really should. J Financ Serv Mark 23(2):104–111. https://doi.org/10.1057/s41264-018-0048-7
13. Al-Nasser Mohammed SAS (2021) Financial crisis, financial system and the impact of artificial intelligence: an attempt towards discussion. In: Studies in computational intelligence, vol 954, pp 185–195. https://doi.org/10.1007/978-3-030-72080-3_11
14. Shanmuganathan M (2020) Behavioural finance in an era of artificial intelligence: longitudinal case study of robo-advisors in investment decisions. J Behav Exp Financ 27:100297
15. Mhlanga D (2020) Industry 4.0 in finance: the impact of artificial intelligence (AI) on digital financial inclusion. Int J Fin Stud 8(3). https://doi.org/10.3390/ijfs8030045
16. Oshima J, Oshima R, Matsuzawa Y (2012) Knowledge building discourse explorer: a social network analysis application for knowledge building discourse. Edu Tech Res Dev 60(5):903–921. https://doi.org/10.1007/s11423-012-9265-2
17. Heo W, Lee JM, Park N, Grable JE (2020) Using Artificial Neural Network techniques to improve the description and prediction of household financial ratios. J Behav Exp Financ 25:100273. https://api.seman-ticscholar.org/CorpusID:212956796
18. Njegovanović A (2018) Artificial intelligence: financial trading and neurology of decision. Financ Mark Inst Risks. https://api.seman-ticscholar.org/CorpusID:169880829
19. F. Hair Jr J, Sarstedt M, Hopkins L, G. Kuppelwieser V (2014) Partial least squares structural equation modeling (PLS-SEM): an emerging tool in business research. Eur Bus Rev 26(2):106–121. https://doi.org/10.1108/EBR-10-2013-0128
20. Magliano L, Fadden G, Madianos M, Caldas de Almeida JM, Held T, Guarneri M, Marasco C, Tosini P, Maj M (n.d) Social network questionnaire (SNQ) [Database record]. APA PsycTests. https://doi.org/10.1037/t39856-000
21. Hair JF, Risher JJ, Sarstedt M, Ringle CM (2019) When to use and how to report the results of PLS-SEM. Eur Bus Rev 31(1):2–24. https://doi.org/10.1108/EBR-11-2018-0203
22. Peterson RA, Kim Y (2013) On the relationship between coefficient alpha and composite reliability. J Appl Psychol 98(1):194–198. https://doi.org/10.1037/a0030767
23. Morgan PJ, Long TQ (2019) FinTech and financial literacy in the Lao PDR (March 18, 2019). ADBI working paper 933, March 2019. https://doi.org/10.2139/ssrn.3398235
24. Kumar P, Pillai R, Kumar N, Tabash MI (2023) The interplay of skills, digital financial literacy, capability, and autonomy in financial decision making and well-being. Borsa Istanbul Rev 23(1):169–183
25. Respati DK, Widyastuti U, Nuryati T, Musyaffi AM, Handayani BD, Ali NR (2023) How do students' digital financial literacy and financial confidence influence their financial behavior and financial well-being? Nurture 17(2):40–50

26. Mahdzan NS, Sabri MF, Husniyah AR, Magli AS, Chowdhury NT (2022) Digital financial services usage and subjective financial well-being: evidence from low-income households in Malaysia. Int J Bank Mark (ahead-of-print)
27. Nam Y, Lee ST (2023) Behind the growth of FinTech in South Korea: digital divide in the use of digital financial services. Telematics Inform 81:101995
28. Carton FL, McCarthy J, Xiong H (2022) Digital factors supporting decision making in the financial well-being of social housing residents. J Decis Syst 31(sup1):202–213
29. Kakinuma Y (2022) Financial literacy and quality of life: a moderated mediation approach of FinTech adoption and leisure. Int J Soc Econ 49(12):1713–1726
30. Zarifis A, Cheng X (2022) A model of trust in FinTech and trust in Insurtech: how Artificial Intelligence and the context influence it. J Behav Exp Financ 36:100739
31. Moolman J (2022) Proposing a network of advisors that could guide a professional athlete's financial decisions in pursuit of sustainable financial well-being. Managing Sport and Leisure 27(6):577–592
32. She L, Ma L, Voon ML, Lim ASS (2023) Excessive use of social networking sites and financial well-being among working millennials: a parallel-serial mediation model. Int J Bank Mark 41(1):158–178
33. Haque MZ, Qian A, Lucky SA (2023) Exploring the factors of online social networks (OSNs) on individual investors' capital market investment decision: an integrated approach. J Inf Knowl Manag 22(02):2350002
34. Sabri MF, Anthony M, Law SH, Rahim HA, Burhan NAS, Ithnin M (2023) Impact of financial behaviour on financial well-being: evidence among young adults in Malaysia. J Financ Serv Mark 1–20
35. Weissberger G, Bergman YS (2022) The association between subjective age and financial exploitation vulnerability among older adults: the moderating role of social support. J Elder Abuse Negl 34(4):314–324
36. Yanto H, Ismail N, Kiswanto K, Rahim NM, Baroroh N (2021) The roles of peers and social media in building financial literacy among the millennial generation: a case of Indonesian economics and business students. Cogent Soc Sci 7(1):1947579

Humanized Intelligent Analytics in Sustainable Finance and Economic Behaviors: Conclusion and Future Research Directions

Sirshendu Arosh, Samarjit Roy, Palash Bairagi, Siddhartha Bhattacharyya, and Debashis De

Abstract This chapter accomplishes the delineation of the humanized intelligent analytics in the contexts of sustainable finance and economic behaviors by incorporating the relationship among technology, ethics, and long-term economic resilience. It underscores the requirement to extemporize the sustainability of sustainable finance via strategies that forestall makeshift exploitation while harmonizing growth with the environmental, social, and governance (ESG) prototypes. The discourse accentuates intelligent analytics that may assuage financial instability, promote AI-induced ethical decision-making, and ensure intergenerational equality by integrating human-centered values into financial paradigms. Intelligent economic practices, bolstered by new information, can mitigate systemic risks while fostering equity and transparency. Our chapter scrutinizes analytics-dependent methodologies for portfolio optimization, risk persistency evaluation, and efficient portfolio formulation that

S. Arosh
Symbiosis Centre for Information Technology, Symbiosis International (Deemed University), Pune, India
e-mail: sirshendu@scit.edu

S. Roy (✉)
Department of Software Engineering, School of Computing and Information Technology, Eastern International University, Becamex IDC Corp, Binh Duong, Vietnam
e-mail: samarjit.roy@ieee.org; roy.samarjit@eiu.edu.vn

Lincoln University College, Selangor, Malaysia

P. Bairagi
Department of Business Management, Central University of Odisha, Koraput, Odisha, India
e-mail: pbairagi@cuo.ac.in

S. Bhattacharyya
Faculty of Electrical Engineering and Computer Science, VSB Technical University of Ostrava, Ostrava, Czech Republic, and Algebra Bernays University, Zagreb, Croatia

D. De
Department of Computer Science and Engineering, Maulana Abul Kalam Azad University of Technology, Kolkata, West Bengal, India
e-mail: debashis.de@makautwb.ac.in

S. Roy et al. (eds.), *Humanized Intelligence in Sustainable Finance and Economic Behaviours*, Studies in Computational Intelligence 1226,
https://doi.org/10.1007/978-981-95-0599-9_14

demonstrate emerging instruments for risk mitigation and sustainable wealth generation. We project on the consensus finance to enhance decision-making, foster in-depth comprehension of economic behaviors, and ensure equitable stakeholder contemplations, and hence underscoring the human aspect of financial ecosystems. The associated future research focuses on amalgamating intelligence, behavioral economics, and sustainability principles for promoting more adaptive, inclusive, and ethically responsible finance. The chapter eventually advocates for the humanized intelligence analytics as a foundation for reconciling profitability with planetary and societal welfare.

1 Introduction

Science mapping is a type of study that identifies existing information in a field and shows how it is related to additional perception. Two bibliometric analysis techniques are used in VOSviewer to carry out the science mapping of sustainable finance research: The main topics underlying the descriptive framework of sustainable finance research during the past 35 years (1986–2020) are shown using a word-network analysis by employing of keyword co-occurrence and word clouds are used in temporal analysis to break down the main themes that define sustainable finance research in specified time period [1].

These approaches have produced important clusters such as climate financing [2], impact investing finance, green financing [3], socially aware investing, carbon financing [4], and energy financing [5], as themes for research on sustainable finance.

Moreover, as we can see more and more evidences that developed and developing nations are requiring SDG accomplishments through sustainable finance [6], including green, carbon, and climate finance. In order to meet economic demands and contribute significantly to sustainability and sustainable development—particularly in relation to accomplishing the sustainable development goals and lowering carbon emissions in accordance with the Paris Agreement [7]—financial markets are constantly searching for novel sustainable finance instruments. In addition to this, investors are becoming more interested in ESG and socially conscious investment funds [8] these days, and they are instructing fund managers to find and seek funds for impact investing. When considered as a whole, this means that a steady flow of fresh ideas is needed to spark and meet the changing needs for sustainable finance.

Future researchers studying the various causes and effects of sustainable finance, particularly with regard to its performance and return, should anticipate being in a much more privileged position than their predecessors due to the expansion of the discipline of knowledge and the availability of data on transactions specific to sustainable finance [1]. Advancement of big data analysis and financial machine learning models [1, 9–11] are going to produce better efficiency in projecting better financial roadmap. Prediction algorithms also can be effective in handling external, far-fetched variables such as green energy and carbon footprints based on the sustainability model. Even if terms like "green financing," "carbon financing," and "energy

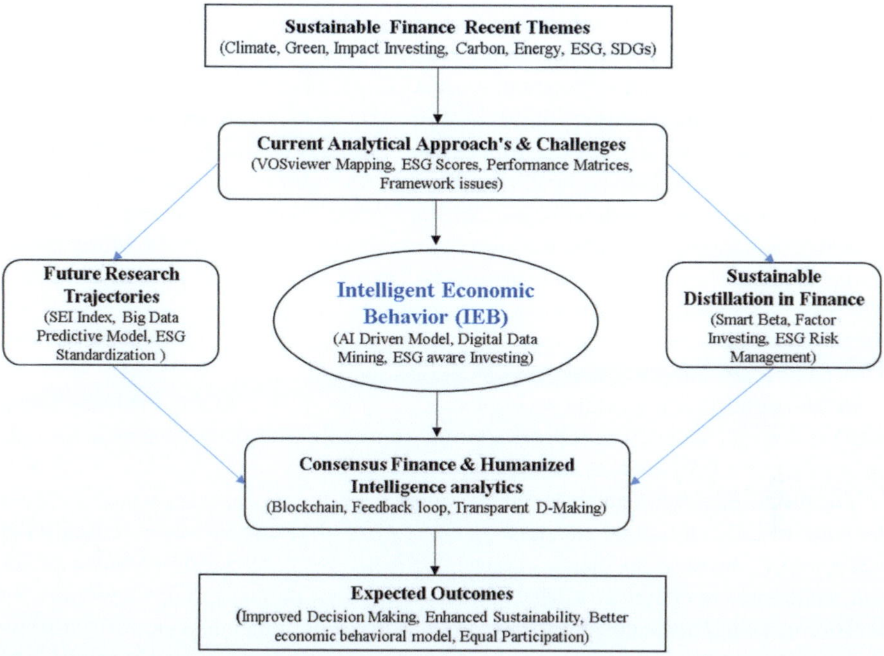

Fig. 1 Framework for humanized intelligent analytics in sustainable finance and economic behaviors

financing" seem identical at first, their study focuses allow for differentiation: While carbon and energy financing is backbone of the same but for sustainable development priorities specific to carbon emission (e.g., greenhouse effects) and energy (e.g., renewable energy), respectively, green financing focuses on expanding the financial flow (e.g., banking, micro-credit, insurance, and investment) across sectors such as public, private, and not-for-profit) [12]. Derived from the scientific mapping and our assessment of the topics within the recognised gaps and prospective research trajectories disclosed in recently published literature.

We have chosen to compile a plan for future composition and trajectory of the field, shown in Fig. 1.

2 Enhancing the Sustainability of Sustainable Finance

It is challenging to evaluate the benefits of impact investing and the sustainability of sustainable finance. Additionally, non-financial performance measurements are frequently demanded by investors for these kinds of investments; ESG ratings, exposure metrics, and carbon footprints [13, 14] are becoming more and more well-liked despite their inherent drawbacks and limits.

Nowadays, the significant amount of funds that venture capital investments make available is crucial for driving innovation in sustainable financing and impact investing. This is evident in financial markets like China [15], where market forces and government interventions have incentivized investments in sustainable and environmentally friendly ways. Furthermore, the introduction of novel sustainable financing tools might help companies achieve stock liquidity (for example, the issue of green bonds has a beneficial impact on stock prices); however, the introduction of such products is not common in developing financial markets like India. More importantly, formal and informational institutions are essential to the growth of supply of innovative sustainable financing instruments as well as the knowledge, comprehension, and demand of consumer and corporate investors regarding the costs and benefits of these financing and investment options in the financial markets.

In this light, it is important to implement a scale of sustainable evidence index (SEI) which can address the following points and score over these aspects to provide more analytical insights over the cases.

The proposed points can be: (i) environmental sustainability index which includes decrease in CO_2 emissions, decrease in energy use, decrease of waste, reduction of water usage, decrease in plastic, material effectiveness, reduction of noise pollution, adherence to chemical safety, (ii) acceleration of the data analysis process for decision-making. Since current society is data-driven, judgments made with a variety of data kinds in mind will be more precise and goal-oriented. (iii) It is possible to manage the performance cost of sustainable finance by making the best possible portfolio modifications. It is necessary to take into account the social impact of the industry's equipment as well as how the items are boosting the economies of nations with poor rates of economic growth. (iv) It is important to investigate how a business model can withstand an ever-changing situation in the event of a tragedy or epidemic, such as the COVID-19 pandemic. It is noteworthy that in the context of a multinational corporate environment, a business model must possess the ability to adapt to global changes. (v) Ultimately, a few key issues need to be addressed, including how investors view sustainable finance, how they profit from it, and how their personalities and behavioral biases affect their decision to choose impact investing-based funds over traditional ones.

3 Intelligent Economic Behavior

Employing child or forced labor, unsafe working conditions, and the environmental impact on protected areas are just a few of the problems that more and more investors are considering when deciding where to put their money in addition to standard financial research. The goal of sustainable finance is to guarantee that future generations can fulfill their requirements as well as to assist society in better meeting its own.

Artificial intelligence (AI) is a valuable tool when it comes to generating a business model [16] that takes into account several social and environmental factors. AI would be the most effective instrument to address such complicated business model-based difficulties in the age of varied data and reports.

These days, most research focuses on creating AI models for different purposes in order to create and evaluate financial distress prediction models for manufacturing and financial services firms. The development of LLMs has made digital information, including digital news and databases, readily available for the creation of sustainable AI economic models [17]. The goal is to provide enduring financial sustainability by serving as a decision-making instrument and a performance management guide for business operations. Advanced AI tools are beneficial for that purpose.

ESG ratings and scores are used by institutional investors and asset managers acting on their behalf to assist them in making allocation decisions that are in line with their long-term performance goals, risk management objectives, and values. In a similar vein, the Intelligent Economic Behavior (IEB) score may be developed to offer information on the number of parameters the models have been taking into account, the emphasis that each element has received, the degree of optimization that has been accepted, etc.

Here we tried to mathematically formulate a score to show how this method can be implemented. Let's define the **IEB score** as a weighted sum of relevant factors influencing sustainable finance and AI-based business models

$$\text{IEB} = \sum_{i=1}^{n} w_i X_i$$

where

IEB Intelligent Economic Behavior score
X_i Economic, social, and environmental factors
w Weights assigned to each factor, determined by optimization techniques
n Number of factors considered

The weights can be assigned as per the long-term sustainability and financial performance. The optimization can be supported by regression techniques, and neural networks as deep learning tools. In this context, we require to consider another aspect as the importance of optimization constraint for long-term sustainability. Long-term sustainability depends on optimization constraints because they make sure that economic choices are made with social, ethical, and environmental obligations in mind rather than only short-term financial advantages. Few points can be mentioned as,

i. **Preventing Short-Term Exploitation**: Without sustainability restrictions, companies could disregard environmental laws, take advantage of workers, or damage local communities in order to maximize short-term profits. Restrictions provide moral limits that promote responsible conduct.
ii. **Balancing Economic Growth with ESG Goals**: While revenue and cost efficiency are frequently given priority in economic models, sustainability limitations make sure that profit-seeking does not come at the expense of

social suffering or environmental deterioration. Maintaining this equilibrium is essential for corporate responsibility.

iii. **Mitigating Risk and Financial Instability**: Companies that disregard sustainability frequently incur long-term operating concerns, regulatory fines, and harm to their brand. Limitations assist businesses in making more informed decisions that lower legal risks and improve stability.

iv. **Encouraging AI-Driven Ethical Decision-Making**: Constraints can direct AI models to emphasize sustainable practices as AI becomes more prevalent in economic activity, ensuring that financial services, production models, and investment strategies all adhere to ESG standards.

v. **Ensuring Intergenerational Equity**: In order for society to flourish over time, optimizing for sustainability entails taking future generations into account, protecting resources, ensuring moral working conditions, and using climate-conscious tactics.

To put it briefly, these limitations compel economic systems to change in a responsible manner rather than prioritizing short-term profits over moral and environmental ramifications. They encourage industries to be more inventive, resilient, and sustainable.

4 Analytics in Sustainable Finance

Financial institutions have been incorporating sustainability ideas into their business strategies for a considerable amount of time. As a result, they are providing environmentally friendly financial products including green mortgages, green savings and bonds, and green auto loans [18]. When it comes to creating a sustainable banking client journey, data analytics and AI/ML can revolutionize the game by adding sustainability elements to current products. This entails providing incentives for using credit cards to buy sustainable products, putting in place systems to offset carbon footprints, and launching green deposit accounts that are only intended for sustainable investments. Furthermore, the "green" component of financial institutions is coming under closer examination, with the percentage of "green loans, advances, and debt securities compared to total assets" emerging as a crucial evaluation criterion.

An efficient grading system that gives preference to companies who follow specified sustainability criteria is necessary for a supply chain to be sustainable. To evaluate and control ESG risk throughout the network, advanced analytics is needed. Future research can focus on minimizing the amount of greenhouse gas produced by transportation, maximizing the usage of automated supply chains, and improving efficient management procedures. The goals of these studies are to reduce waste, promote equitable working conditions, and stimulate the economy. Assessing sustainability, the absence of a standardized industry framework is one of the major obstacles to quantifying sustainability. Not everyone uses the functional frameworks, such as the Task Force on Climate-Related Financial Disclosures (TCFD) and the Business Environment Index (BEI) [19, 20].

The lack of these guidelines leads to variations in methods within the banking industry. Ongoing initiatives by industry consortiums attempt to create standardized protocols for ESG review, supporting a more consistent approach industry-wide.

ESG data is generally acquired by financial institutions via external sources, firm disclosures, or open channels. Gen AI is able to gather this information from several sources and create a trustworthy database for further examination. However, in order to consistently gather, compile, and present a wide range of sustainability data, scaling up necessitates adapting IT systems. Many of the institutions lack a thorough plan for incorporating sustainability data into current reporting and taking action on KPIs and KRIs related to sustainability. Research may focus on two main areas: creating an appropriate pipeline for authenticated data collecting and developing universally accepted indicators.

As the industry moves toward a more sustainable, responsible, and inventive future, artificial intelligence (AI) and data analytics emerge as both a solution and a critical driver.

5 Sustainable Distillation in Finance

Sustainable distillation is a method that uses less energy and produces fewer emissions in an effort to lessen the impact of distillation on the environment. In the food and chemical sectors, distillation is a widely used method, yet it may be energy-intensive and contribute to climate change. Like distillation, the same concepts can be iterated to the finance sector.

(a) **Risk Factors Persistency**: Determine the underlying risk variables that affect asset returns.
(b) **Portfolio Cleaning**: Create portfolios exposed to particular risk variables so that investors may more effectively isolate and manage these risks
(c) **Effective Portfolio Development**: Eliminate highly correlated or illiquid assets to lower the risk of portfolio concentration. Similarly, to build diversified portfolios with specific exposures, use strategies such as smart beta strategies or factor investing.
(d) **Protection and Risk Management**: Use techniques like factor investing or smart beta methods to create diversified portfolios with targeted exposures. It enhances to build ethical and sustainable portfolios, think about integrating ESG considerations into the optimization process.

It is the process of improving economic policies, investment models, and financial strategies in order to maximize long-term sustainability and profitability. Businesses and investors may lower risks, increase long-term profitability, and support a robust global economy by refining financial plans through a sustainability perspective. It guarantees that financial ecosystems are resource-efficient, morally consistent, and flexible enough to adjust to changes in the market.

Case studies of the real-life scenarios include the green bonds and sustainable investments, which are financial instruments that facilitate the funding of environmentally friendly initiatives, such as the carbon reduction programs, sustainable precision agriculture, and renewable energy projects. Decarbonization methodologies of financial organizations to ensure compliance with the net-zero transition aim to have the banks and asset managers associate with climate risk management into several investment portfolios. The ESG-induced lending and sustainable banking: numerous banks are adopting sustainability-centric strategies by offering loans and financial solutions that prioritize social implications and environmental accountability. Sustainable financing is being used by sectors such retail housing, infrastructure, and agriculture to lessen their negative effects on the environment while sustaining economic growth.

In the financial industry, sustainable distilling entails honing investment strategies to strike a balance between long-term environmental, social, and governance (ESG) objectives and profitability. Future developments will concentrate on fusing ethical financial modeling with AI-driven analytics, making sure that sustainability indicators are clear, measurable, and flexible. Financial institutions must use data-driven sustainability evaluations as global markets move toward ESG compliance in order to maximize risk management and promote economic resilience. In order to ensure that financial ecosystems are in line with global sustainability goals without sacrificing economic growth, the development of sustainable distillation will depend on dynamic regulatory frameworks, investor awareness, and technology advancements. In the end, improving sustainable finance will improve financial performance while simultaneously promoting environmental stewardship and wider societal advancement.

6 Consensus Finance

Consensus finance is an agreement based on acceptance-based financing among all parties involved. All parties must agree on all feasible measurements and elements through sustainability, proof-of-work, proof-of-stake, and other such suggested criteria, much like the blockchain idea [22, 23]. Since the framework already inherits such concepts, it may be implemented using blockchain. Even if this result might not be optimal, a voting procedure can help adopt appropriate policies with the greatest amount of consensus, which might serve as the foundation for the economy of the future.

Financial decision-making also has to be more modern, transparent, and feedback-heavy. Continuous input from all stakeholders is necessary for a sustainable blockchain-based system that is safe, shared, and open to all. That makes it possible for the systems to lessen any errors or discrepancies on the model. It can avert a long-term financial and environmental catastrophe and instantly fix the financial model.

Finally, the system will be opaque to all parties, allowing anybody to assess any changes or modifications.

Various research facets and potential unexplored study domains are examined in this section. The convergence of artificial intelligence and human intelligence provides a potent instrument for tackling the intricate problems associated with sustainable finance and economic practices. Humanized intelligent analytics (Industry 5.0) [24] may improve decision-making, spur positive change, and offer insightful information. We can solve urgent global challenges more successfully by combining the talents of humans and machines.

Important **conclusions from this investigation includes**:

(a) **Improved decision-making**: Using humanized intelligent analytics, decision-makers can spot patterns, trends, and dangers that human analysis would miss.
(b) **Increased sustainability**: AI-powered technologies may be used to create financial strategies and goods that are more environmentally friendly, lessen their influence on the environment, and advance social justice.
(c) **Improved comprehension of economic behaviors**: Humanized intelligent analytics can offer more in-depth knowledge of consumer behavior, market patterns, and economic swings by evaluating large datasets.
(d) **Equal considerations**: In order to guarantee that all parties should have equal access to all possible data and take equal participation in taking sustainable decisions.
(e) This methodology can be modeled as an approach to consensus in machine learning. Recently, machine learning techniques have been the foremost tools for automated model generation and precise market analysis throughout time. Hence, we tried to provide a mathematical model for a given situation of the market.

Consensus-based financial model can be defined as

$$\mathrm{CF}_t = \frac{1}{M}\sum_{i=1}^{M} F_i(X, t)$$

where

CF_t	Consensus Finance prediction at time (t)
(M)	Number of financial models participating
$F_i\ (X, t)$	Financial model (i) considering data (X) at time (t)

We may include Federated Learning or Bayesian aggregation to optimize, guaranteeing that models learn from decentralized sources while maintaining data confidentiality.

This methodology can be used in optimal ESG investing, where AI consensus guarantees that investment strategies are shaped by a variety of ethical viewpoints; risk-aware portfolio management, where combining many AI projections improves diversification and lowers uncertainty; and AI-governed financial sustainability, which maintains flexibility in response to market shifts while guaranteeing regulatory compliance.

By replacing top-down management with collaborative intelligence, machine learning in finance has the potential to completely transform international investment and build a more moral, flexible, and open financial system.

The potential uses of humanized intelligent analytics in sustainable finance and economic behavior will only increase with the advancement of technology. Future generations can benefit from a more sustainable and just society if we can harness the potential of AI while upholding human judgment and values. In order to improve moral and open decision-making, future studies in humanized intelligent analytics for sustainable finance and economic behavior should concentrate on fusing cutting-edge AI models with behavioral insights. More flexible and socially conscious investing frameworks will result from the confluence of machine learning, consensus-driven financial strategies, and ESG-focused analytics. Optimizing sustainable economic policies, reducing risks, and enhancing long-term financial stability will all depend on investigating human-AI collaboration. Additionally, AI-powered sentiment analysis and decentralized financial tools such as blockchain could offer predictive insights that align with evolving global economic and environmental challenges. As financial ecosystems become more dynamic and interconnected, the future of sustainable finance will depend on intelligent systems that balance profitability with ethical, social, and environmental responsibilities.

References

1. Kumar S, Sharma D, Rao S, Lim WM, Mangla SK (2025) Past, present, and future of sustainable finance: insights from big data analytics through machine learning of scholarly research. Ann Oper Res 345(2):1061–1104
2. Giglio S, Kelly B, Stroebel J (2021) Climate finance. Annu Rev Financ Econ 13(1):15–36
3. Fu C, Lu L, Pirabi M (2023) Advancing green finance: a review of sustainable development. Digit Econ Sustain Dev 1(1):20
4. Meitern M (2024) Unlocking carbon finance: Empowering energy communities for mutual benefit. Renew Sustain Energy Rev 199:114499
5. Zhang D (2018) Energy finance: background, concept, and recent developments. Emerg Mark Financ Trade 54(8):1687–1692
6. Ziolo M, Bak I, Cheba K (2021) The role of sustainable finance in achieving sustainable development goals: does it work? Technol Econ Dev Econ 27(1):45–70
7. Liu W, McKibbin WJ, Morris AC, Wilcoxen PJ (2020) Global economic and environmental outcomes of the Paris Agreement. Energy Economics 90:104838
8. Gerard B (2019) ESG and socially responsible investment: a critical review. Beta 33(1):61–83
9. Wyrobek J (2020) Application of machine learning models and artificial intelligence to analyze annual financial statements to identify companies with unfair corporate culture. Procedia Comput Sci 176:3037–3046
10. Jiang Y (2022) Application and comparison of multiple machine learning models in finance. Sci Program 2022(1):9613554
11. Kelly B, Xiu D (2023) Financial machine learning. Found Trends® Financ 13(3–4):205–363
12. Aitken R (2013) The financialization of micro-credit. Dev Chang 44(3):473–499

13. Doni F, Johannsdottir L (2020) Environmental social and governance (ESG) ratings. In: Climate action, pp 435–449
14. Pandey D, Agrawal M, Pandey JS (2011) Carbon footprint: current methods of estimation. Environ Monit Assess 178:135–160
15. Khalid F, Su CY, Weiwei K, Voinea CL, Srivastava M (2025) Financial mechanism for sustainability: the case of China's green financial system and corporate green investment. China Financ Rev Int 15(1):93–116
16. Lee J, Suh T, Roy D, Baucus M (2019) Emerging technology and business model innovation: the case of artificial intelligence. J Open Innov Technol Mark Complex 5(3):44
17. Mueller-Saegebrecht S, Lippert I (2024) In tandem with ChatGPT-4: how LLM enhance entrepreneurship education and business model innovation. In: Academy of management proceedings, vol 2024, no 1. Valhalla, NY 10595: Academy of Management, p 15473
18. Devine A, McCollum M (2022) Advancing energy efficiency through green bond policy: multifamily green mortgage backed securities issuance. J Clean Prod 345:131019
19. Ding D, Liu B, Chang M (2023) Carbon emissions and TCFD aligned climate-related information disclosures. J Bus Ethics 182(4):967–1001
20. Liu B, Yang X, Sun Q, Luan F (2024) Does the business environment inhibit or promote corporate greenwashing? Empirical evidence from China. Environment, Development and Sustainability, pp 1–27
21. Meadowcroft J, Rosenbloom D (2023) Governing the net-zero transition: strategy, policy, and politics. Proc Natl Acad Sci 120(47):e2207727120
22. John K, Rivera TJ, Saleh F (2025) Proof-of-work versus proof-of-stake: a comparative economic analysis. Rev Financ Stud hhaf013
23. Munir MA, Habib MS, Hussain A, Shahbaz MA, Qamar A, Masood T, Sultan M, Mujtaba MA, Imran S, Hasan M, Akhtar MS (2022) Blockchain adoption for sustainable supply chain management: economic, environmental, and social perspectives. Front Energy Res 10:899632
24. Grabowska S, Saniuk S, Gajdzik B (2022) Industry 5.0: improving humanization and sustainability of Industry 4.0. Scientometrics 127(6):3117–3144

FSC
www.fsc.org
MIX
Papier aus verantwortungsvollen Quellen
Paper from responsible sources
FSC® C105338

If you have any concerns about our products,
you can contact us on
ProductSafety@springernature.com

In case Publisher is established outside the EU,
the EU authorized representative is:
Springer Nature Customer Service Center GmbH
Europaplatz 3, 69115 Heidelberg, Germany

Printed by Libri Plureos GmbH
in Hamburg, Germany